AUDEL®

W9-AJK-818

Mechanical Trades Pocket Manual

by Carl A. Nelson

This book is
the property of:

____

____

____

Please return if found

_______ - _____ - _______

Additional copies of this book are
always available from

MAINTENANCE TROUBLESHOOTING
273 Polly Drummond Road
Newark, Delaware 19711
(302) 738-0532 • FAX: (302) 738-3028

DEDICATED TO ELIMINATING PROBLEMS

THIRD EDITION

Copyright © 1986, 1990 by Macmillan Publishing Company, a division of Macmillan, Inc.
Copyright © 1983 by The Bobbs-Merrill Co., Inc.
Copyright © 1974 by Howard W. Sams & Co., Inc.

All rights reserved. No part of this book may be reproduced or transmitted in any form or by any means, electronic or mechanical, including photocopying, recording, or by any information storage and retrieval system, without permission in writing from the Publisher.

While every precaution has been taken in the preparation of this book, the Publisher assumes no responsibility for errors or omissions. Neither is any liability assumed for damages resulting from the use of the information contained herein.

Macmillan Publishing Company
866 Third Avenue, New York, NY 10022
Collier Macmillan Canada, Inc.

Production services by the Walsh Group, Yarmouth, ME

Library of Congress Cataloging-in-Publication Data

Nelson, Carl A.
Mechanical trades pocket manual / by Carl A. Nelson.—3rd ed.
p. cm.
"An Audel book."
ISBN 0-02-588665-7
1. Mechanical engineering—Handbooks, manuals, etc. I. Title.
TJ151.N418 1990
621.8—dc20 89-13830
CIP

Macmillan books are available at special discounts for bulk purchases for sales promotions, premiums, fund-raising, or educational use. For details, contact:

Special Sales Director
Macmillan Publishing Company
866 Third Avenue
New York, NY 10022

10 9 8 7

Printed in the United States of America

Foreword

The intention of this manual is to provide reference material for mechanical tradesmen. While it is primarily concerned with installation, maintenance, and repair of machinery and equipment, other fields of activity involved in this overall operation are included. This broad range of information on methods, procedures, equipment, tools, etc., is presented in convenient form and plain language to aid the mechanic in performance of day-to-day tasks.

The constant aim has been to present the subject as clearly, concisely, and simply as possible. To accomplish this, numerous sketches and practical examples are used, and the explanations are given as briefly and simply as possible. Discussions of background principles and theory have been limited to the essentials required for understanding of the subject matter.

Information is given on a wide range of subjects, including many new materials and methods. While intended primarily as an industrial mechanic's handbook, the data, information, directions, etc., apply to operations performed by many mechanical trades.

Carl A. Nelson

Contents

For easy reference the contents are arranged in subject matter categories. Specific subjects are listed under category headings. For more detailed listing of subjects consult the index.

MECHANICAL DRAWING

Mechanical drawing is a graphic language used to convey information, size, location, accuracy, etc., of mechanisms ranging from the simple to the most complex. Because there is a general standardization of techniques throughout the world, this language is understood in spite of native tongue and measurement differences.

To understand this graphic mechanical language requires an understanding of basic principles and concepts as well as knowledge of the conventions commonly followed. Lines define the shape, size, and details of an object, and through the correct use of lines it is possible to graphically describe an object so that it can be accurately visualized. A listing of these lines is called an "alphabet" of lines (Fig. 1).

Mechanical drawings are not a true representation of what the human eye actually sees. What is seen by the eye is most nearly represented by what is called a *Perspective Drawing*. The portion of the object that is closest to the observer appears the largest, while the parts farthest away are the smallest. Lines and surfaces become smaller and closer together as distance from the eye increases, seemingly to disappear at a point on an imaginary horizon called a "vanishing point." Perspective drawing is pictorial in nature, often influenced by artistic talent and not readily lending itself to mechanical methods and drawing practices. Fig. 2 shows an approximation of what the eye sees when viewing a simple object.

Perspective drawing is only one of many drawing systems in general use. The word "projection" is used when describing the various systems of mechanical drawing. This refers to the extension (projection) to a plane or sheet of paper, of what is seen by the eye as it views the object. The most widely used system of mechanical drawing is called *Orthographic Projection.* The word orthographic may seem a little overpowering, but it is an apt description of the system. For purposes of making it more understandable, the word

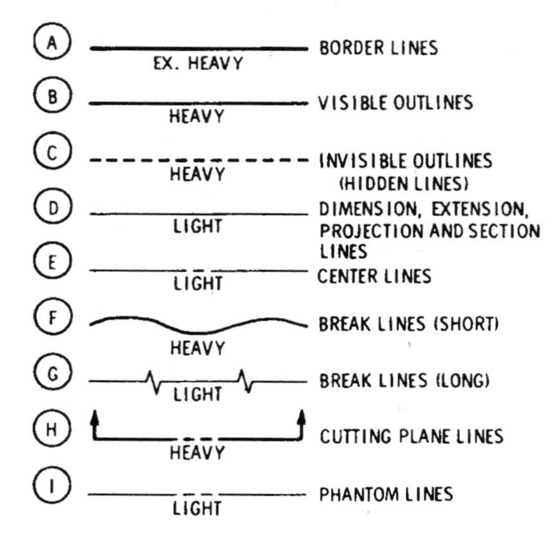

Fig. 1.

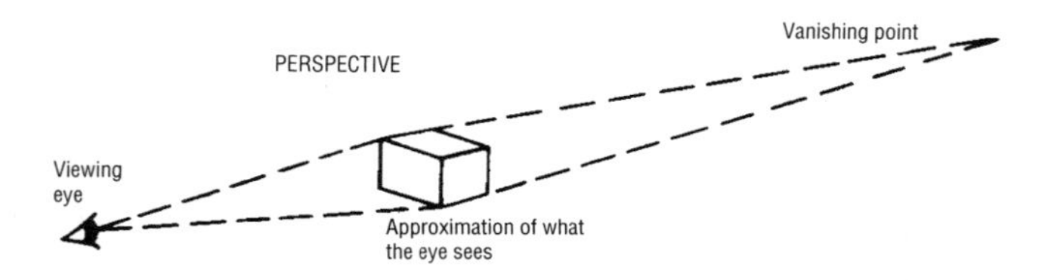

Fig. 2. Perspective drawing.

may be divided into two parts: *ortho* and *graphic*. The first part, *ortho*, is derived from the Greek (orthos) and means straight. The second part, *graphic*, is also from the Greek (graphein) meaning to write, diagram, or picture. The word *graphic* is usually defined as meaning lifelike, vivid, or pictorial. *Orthographic* then simply means that a picture is made by projecting the lines of sight in a straight line, or parallel, and at a right angle to the surface being viewed. Of course the eye does *not* do this when it views an object from a fixed point. As previously stated, mechanical drawings are not true representations of what the human eye actually sees. To project the lines representing the edges of an object parallel, it must be assumed that the eye is seeing the edge from a point where the line is projected. This concept of orthographic sight projection is quite difficult for many persons to grasp immediately, but becomes easy with practice. To visualize orthographic projection, the eye should be considered as moving in a pattern that exactly matches the shape of the object. In Fig. 3, if the eye is considered to move in a pattern matching the rectangular object, an outline of the object will be projected to the paper.

If the eye were to move in this pattern and the edges of the object were to be projected to the surface of the plane or paper, a true representation of the outline of the object could be drawn on the paper. Such a drawing represents only the outline of the surface of the object. Additional drawings from other viewing points would of course be necessary to show the various features of the object, such as thickness variations, other surface shapes, etc. These additional drawings, as well as the original, are called "views" (each taken from a specific viewing position). Several such drawings or views are required to clearly show various features of an object. Therefore, what is commonly called a "drawing" may be made up of several related drawings or views. Standardized principles and practices govern the positioning of these views on the drawing in relation to one another.

To show size, shape, and location, a number of "views" are necessary. Each view represents the true shape and size of the *surface* of the object being viewed, as it is seen by looking directly at it. They are arranged so that each view represents the surface adjacent to it. The top view is drawn above and vertically in line with

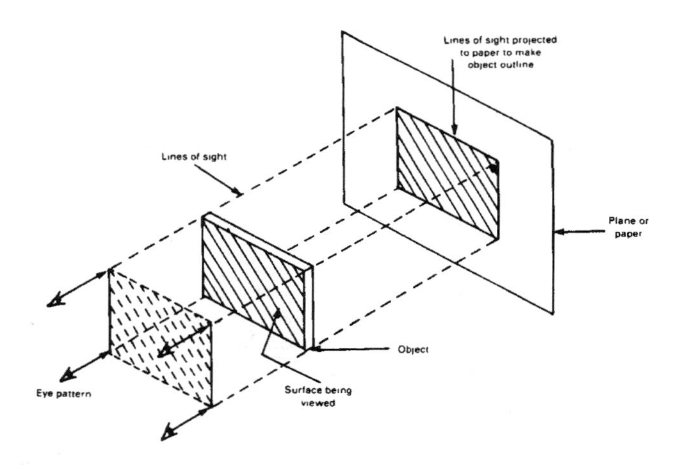

Fig. 3. Orthographic projection.

the front view. The side views, at the side of the front view and horizontally in line, are drawn to the right or left as the case may be. The three-view drawing (usually showing the top, front, and right side) is the most widely used combination of views. The six principal views used in orthographic projection are shown in Fig. 4.

When viewing the surface of an object, many of the edges and intersections behind the surface of the object are not visible. To be complete, a drawing must include lines which represent these edges and intersections. Lines made up of a series of small dashes, called *invisible outlines* or *hidden lines*, are used to represent these behind-the-surface edges and intersections. Fig. 5 shows the use of a dashed line to show a hidden or invisible edge.

When surfaces of an object are at right angles to one another, regular views are adequate for their representation. However, when one or more are inclined and slant away from either the horizontal

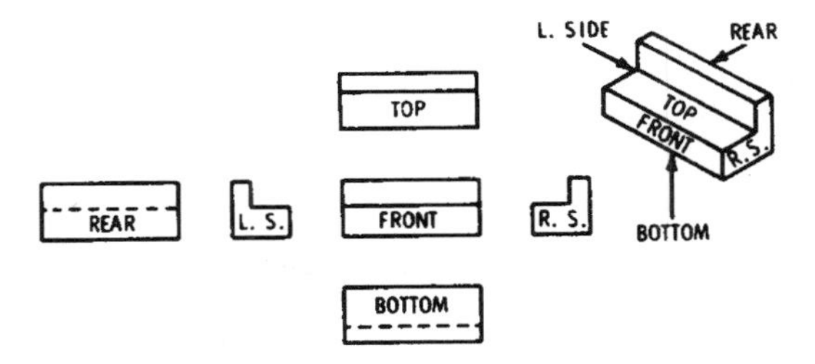

Fig. 4.

or vertical plane, regular views will not show the true shape of the inclined surface. To show the true shape, an *auxiliary view* is used. In an auxiliary view, the slanted surface is projected to a plane which is parallel to it. Shapes that would appear in distorted form in the regular view appear in their true shape and size in the auxiliary view.

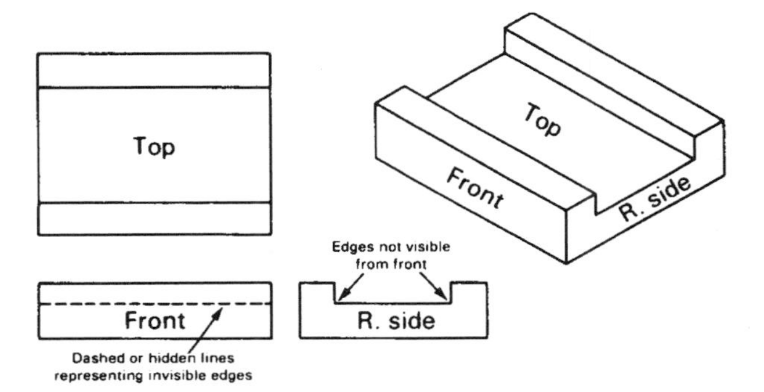

Fig. 5. Hidden lines.

Usually only the inclined surface is shown in the auxiliary view, presented in its true shape and size, as shown in Fig. 6.

As the internal details of a part become more complex and the hidden lines become more numerous, a point is reached where the drawing is difficult to interpret. A technique to simplify such a drawing is to cut away a portion and expose the inside surfaces. A view with this imaginary cut to reveal inside portions is called a *sectional view*. Such a view may be a *full section* where the imaginary cutting plane passes completely through the object, or a *part section* in which the plane extends only part way through the object.

On cutaway section views the invisible edges become visible and may be represented by solid-object lines. The exposed surfaces through which the imaginary cut is made are identified by slant lines called *section* or *cross-hatch* lines, as shown in Fig. 8. To indicate the

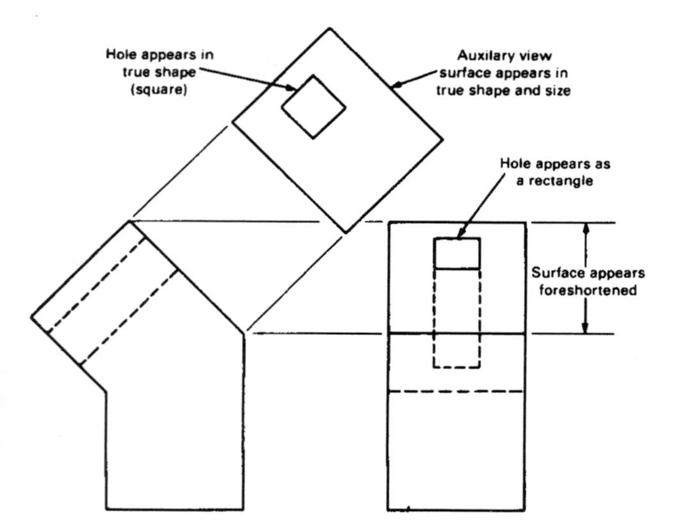

Fig. 6. Auxiliary view.

edge of the cutting plane and the directions in which the section is viewed, a *cutting-plane line* is used. Letters are usually placed near the direction arrowheads to identify the section (Fig. 7).

Screw threads in one form or another are used on practically all kinds of mechanical objects. They are most widely used on fasteners, adjusting devices, and to transmit power. For these purposes a number of thread forms are used, the most common of which are illustrated in Fig. 9.

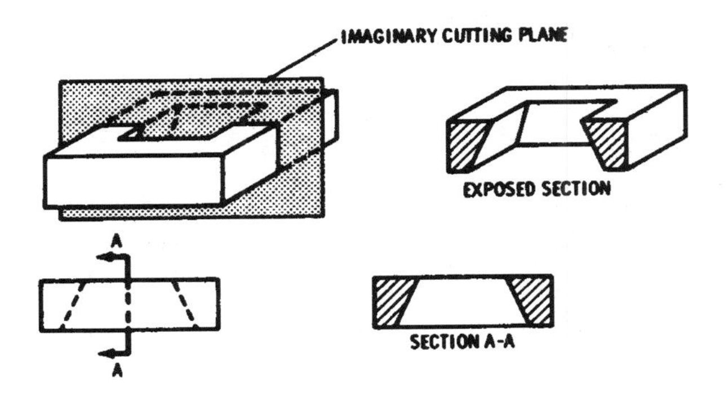

Fig. 7.

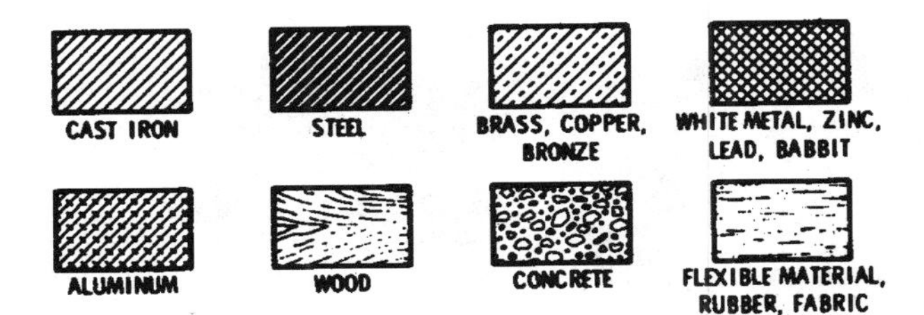

Fig. 8.

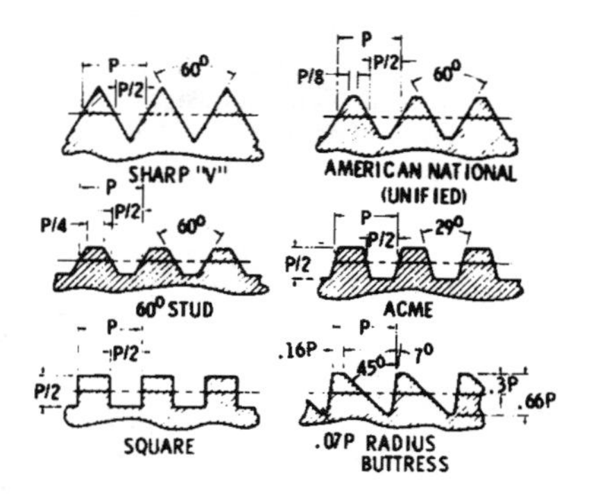

Fig. 9.

Because screw threads are so widely used there is a repeated necessity to show and specify them on mechanical drawings. As a true representation of a screw thread is an extremely laborious operation, it is almost never done. Instead, threads are given a symbolic representation suitable for general understanding. Three methods are used: *detailed representation*, which approximates the true appearance, *schematic representation*, which is nearly as pictorial and much easier to draw, and *simplified representation*, which is the easiest to draw and therefore the most commonly used (Fig. 10).

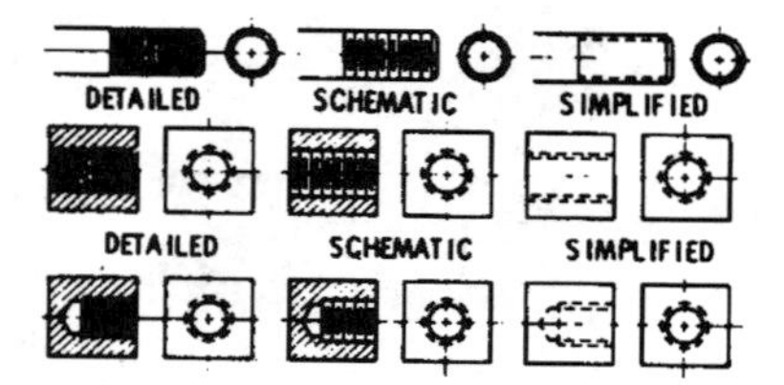

Fig. 10.

Drawings made up of lines to describe shape and contour must also have dimensions and notes to supply sizes and location. Dimensions are placed between a combination of points and lines. The dimension line indicates the direction in which the dimension applies, and the extension lines refer the dimension to the view. Leaders are used with notes to indicate the feature on the drawing to which the note applies.

Three systems of writing dimension values are in general use. Most widely used is the "common-fraction" system, with all dimension values written as units and common fractions. The second system uses decimal fractions when distances require precision greater than plus or minus 1/64 of an inch. The third system is the "complete decimal system" and uses decimal fractions for all dimensional values. Two-place decimals are used where common fractions are used in the other two systems. When greater precision is required, the value is written in three, four, or more places. The complete decimal system is increasing in use, particularly for machine parts, tools, and other precise mechanical type drawings.

When large values must be shown with the common fraction system, feet and inch units may be used. The foot(') and inch (") marks may be used to identify the units. Feet and inch dimensions should only be used for distances exceeding 72 inches. When dimensions are all in inches, the inch marks are preferably omitted from all dimensions and notes.

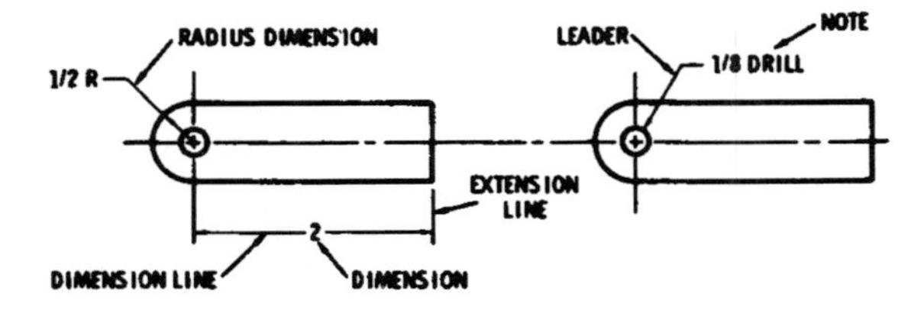

Fig. 11.

The *Unified Standard* thread, which incorporates the earlier National, American, SAE, and ASME standards, is the most widely used thread form. It is designated on drawings by a note specifying in sequence the nominal size, number of threads per inch, thread series symbol, and the thread class number. The symbols used are (UNC) denoting a coarse thread series, (UNF) denoting a fine thread series, and (UNEF) denoting an extra fine thread series. Thread class number 1 indicates a loose free fit, number 2 indicates a free fit with very little looseness, number 3 very close fit. The number 2 fit is used for most general applications. The letter A indicates an external thread and the letter B an internal thread (Fig. 12).

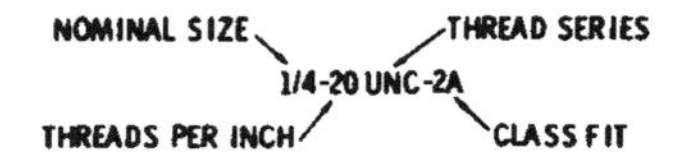

Fig. 12.

ISOMETRIC DRAWING

The *isometric* system of drawing is three-dimensional and more pictorial than the orthographic system. The word isometric means equal measure and has reference to the isometric position which is the basis of isometric drawing.

The isometric position may be developed by rotating a cube around its vertical axis and tilting it forward until all faces are foreshortened equally as shown in Fig. 13A. The overall outline is then a regular hexagon as shown in Fig. 13B. The three lines of the front corner of the cube in isometric position make equal angles with each other and are called the *isometric axes*, as shown in Fig. 13C.

On isometric drawings, horizontal object lines are drawn parallel to the 30-degree isometric axis, and vertical lines are drawn parallel to the vertical axis. Actual lengths are shown to scale, frequently resulting in a distorted appearance due to the foreshortening effect. Lines that are not parallel to the isometric axes do not appear in their true length. To draw such lines, their ends are located on isometric lines and the points connected as shown in Fig. 13D. A circle shows in isometric as an ellipse, and may be constructed by first making an isometric square, as shown in Fig. 13E. The midpoint of each side is located and the arcs of circles drawn to be tangent at the midpoints, as shown in Fig. 13E.

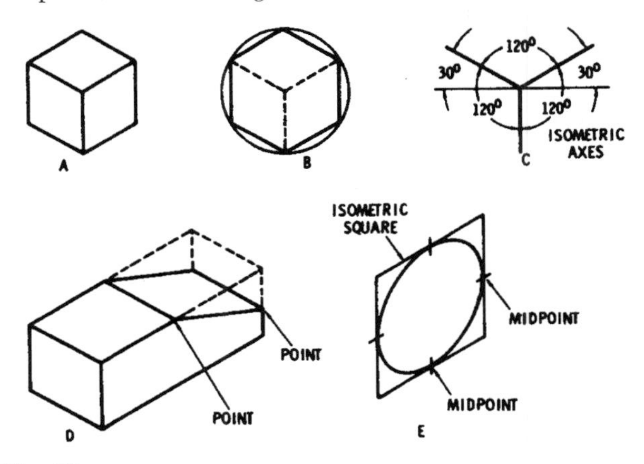

Fig. 13.

When making sketches of objects having nonisometric surfaces it may be helpful to imagine the object as contained in rectangular boxes. By sketching the rectangular boxes in isometric position, points may be accurately located and constructed simplified (Fig. 14).

SINGLE LINE ISOMETRIC PIPE DRAWING

The rectangular box system is also employed in making isometric pipe drawings. The piping systems are considered as being on the surface or contained inside the box. To simplfy the system, pipe is

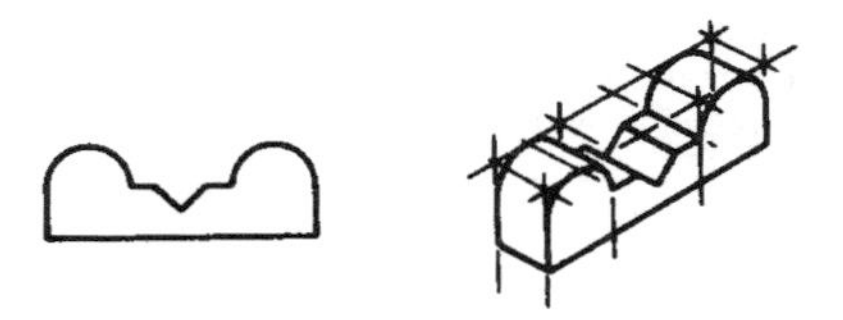

Fig. 14.

represented by a single line and fittings are shown by symbols (Fig. 15).

To aid in clarifying the drawing and to avoid confusion of direction when making it, an orientation diagram is used (Fig. 16). This diagram is a representation of the isometric axes, each labeled to indicate relative position and direction. The vertical axis is always labeled "UP" and "DOWN." The two horizontal axes are given appropriate direction labels such as "FRONT," "BACK," "NORTH," "SOUTH," etc.

As was the case in developing the isometric position, the point of observation for an isometric pipe drawing is always directly in front of the object. In respect to the orientation diagram, the observation point is directly in front of the vertical "up" and "down" axis. When the point has been selected from which the pipe system will be viewed, the diagram is labeled to conform to this point. The lines on the drawing are then made to indicate the actual direction of the pipe.

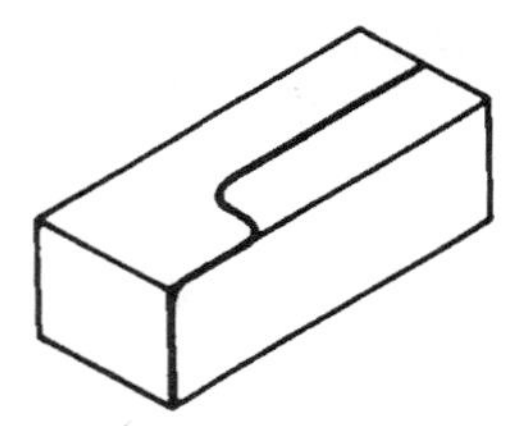

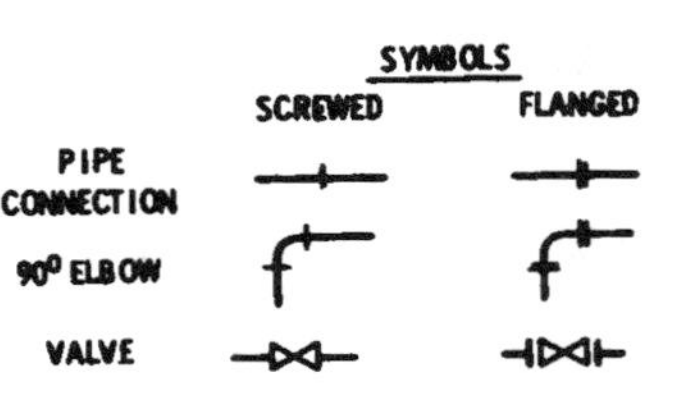

Fig. 15.

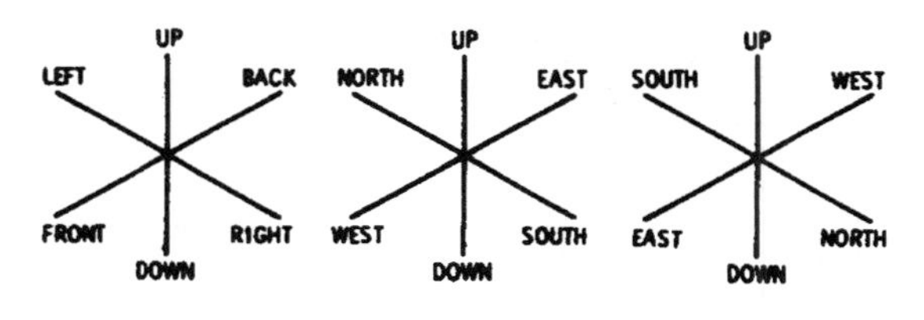

Fig. 16.

It is important when selecting a viewing position to consider which position will result in drawing with the most clarity. Two drawings of the same system when viewed from different points will result in one being much easier to read than the other, as illustrated in Fig. 17.

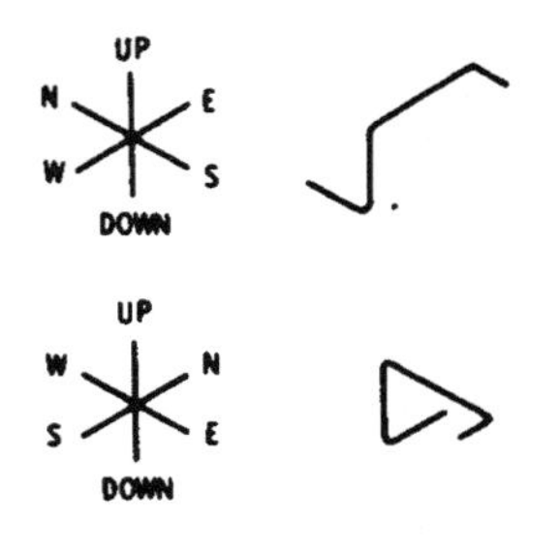

Fig. 17.

Chart 1. Pipe Fitting Symbols

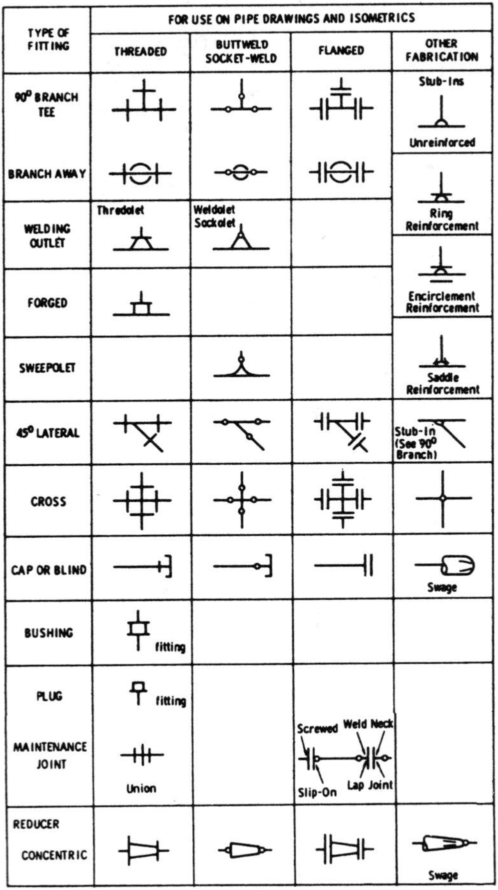

TYPE OF FITTING	FOR USE ON PIPE DRAWINGS AND ISOMETRICS			
	THREADED	BUTTWELD SOCKET-WELD	FLANGED	OTHER FABRICATION
90° BRANCH TEE				Stub-Ins Unreinforced
BRANCH AWAY				Ring Reinforcement
WELDING OUTLET	Thredolet	Weldolet Sockolet		Encirclement Reinforcement
FORGED				
SWEEPOLET				Saddle Reinforcement
45° LATERAL				Stub-In (See 90° Branch)
CROSS				
CAP OR BLIND				Swage
BUSHING	fitting			
PLUG	fitting			
MAINTENANCE JOINT	Union		Screwed Weld Neck Slip-On Lap Joint	
REDUCER CONCENTRIC				Swage

Chart 1. Pipe Fitting Symbols (Contined)

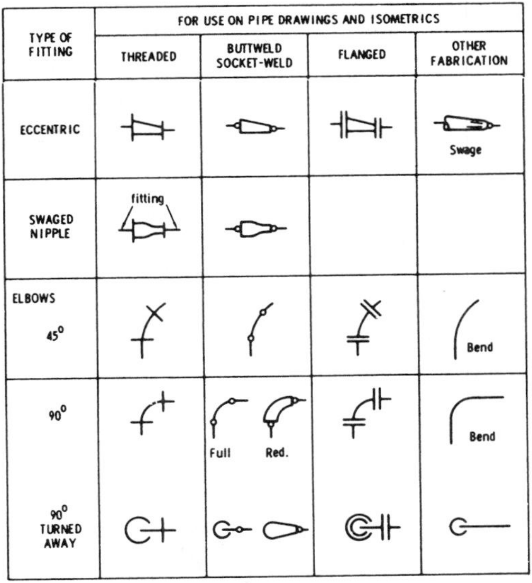

TYPE OF FITTING	FOR USE ON PIPE DRAWINGS AND ISOMETRICS			
	THREADED	BUTTWELD SOCKET-WELD	FLANGED	OTHER FABRICATION
ECCENTRIC				Swage
SWAGED NIPPLE	fitting			
ELBOWS 45°				Bend
90°		Full Red.		Bend
90° TURNED AWAY				

Chart 2. Valve Symbols

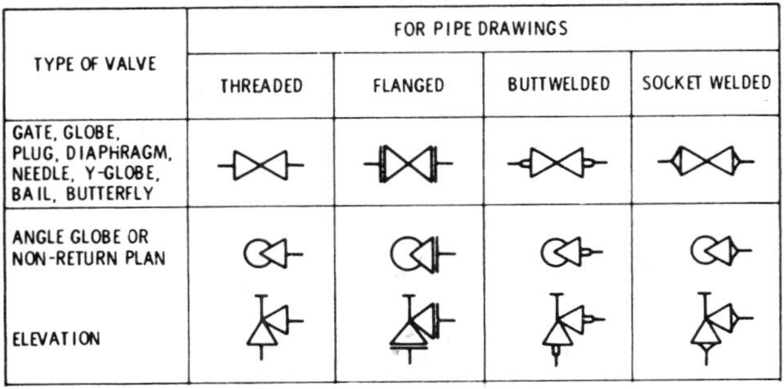

TYPE OF VALVE	FOR PIPE DRAWINGS			
	THREADED	FLANGED	BUTTWELDED	SOCKET WELDED
GATE, GLOBE, PLUG, DIAPHRAGM, NEEDLE, Y-GLOBE, BAIL, BUTTERFLY				
ANGLE GLOBE OR NON-RETURN PLAN				
ELEVATION				

Chart 2. Valve Symbols (Continued)

TYPE OF VALVE	FOR PIPE DRAWINGS			
	THREADED	FLANGED	BUTTWELDED	SOCKET WELDED
SWING, LIFT, TILT OR WAFER TYPE CHECK	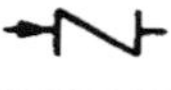			
THREE-WAY PLUG VALVE				

The direction of the crossing lines or symbol marks depends on the direction of the pipe. Flange faces on horizontal pipe runs are vertical, therefore fitting marks on horizontal runs should be "Up" and "Down." Vertical pipe run flange faces are horizontal and may be drawn on either horizontal axis.

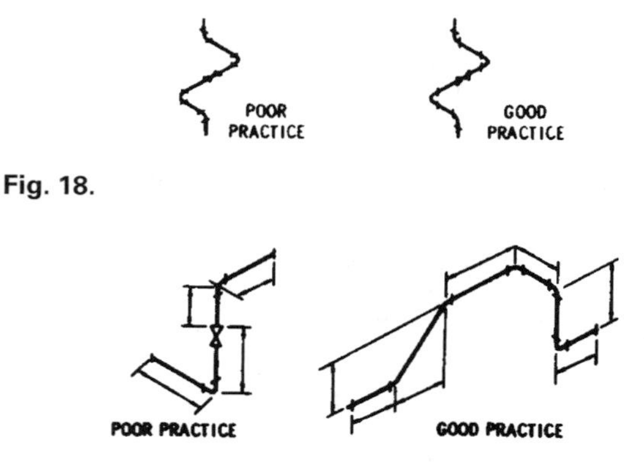

Fig. 18.

Fig. 19.

Isometric extension and dimension lines are drawn parallel to the isometric axis. Dimensions may extend to object lines, but preferred practice is to have arrowheads end on extension lines. If possible, all extension and dimension lines should be placed outside the object.

FIELD LAYOUT

The installation of machinery and equipment usually requires its location in respect to given points, objects, or surfaces. These may be building columns, walls, machinery, equipment, etc. To accomplish this in field layout work, *base lines* are located in respect to these references and the layout developed from the base lines. For plane surface layout, two base lines at right angles are usually sufficient. Specific lines and points are then located by laying out *center lines* parallel to these base lines. A common field layout problem therefore is layout of right angle base lines.

In Fig. 20, the base line (a) is laid out by measurement parallel to the building columns, and base line (b) is laid out at right angles to base line (a). The center lines for machinery installation are then laid out parallel to the base lines.

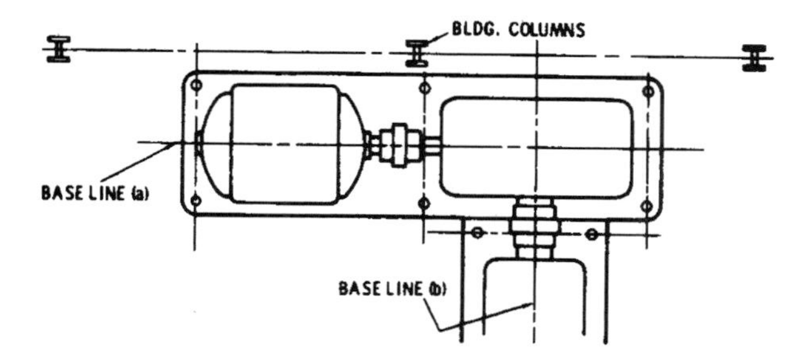

Fig. 20.

The large steel square, or carpenter framing square, while suitable for relatively small layout work, should not be used for large-scale work. A slight error at the square may be magnified to unacceptable proportions when lines are extended. A more dependable and accurate method is to develop the layout to suit the size of the job.

There are several right-angle layout methods that will give accurate results when used for large-scale field layout. The two most suitable methods are development by swinging arcs and the 3-4-5 triangle layout. In either case only simple measurements made with care are required. However, the larger the scale of the layout the greater the accuracy, since the effect of small errors diminishes as dimensions are increased.

Arc Layout Method

Use a steel tape, wire, or similar device that cannot stretch. Do not use a string line. Reasonable care in executing the following steps are necessary to obtain accurate results (Fig. 21).

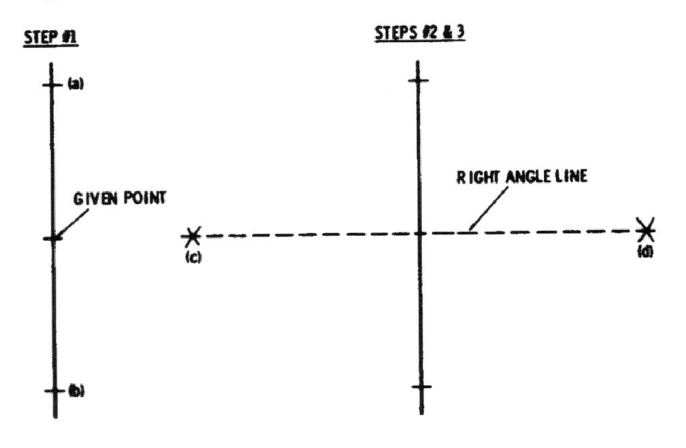

Fig. 21.

Step #1—Locate two points, (a) and (b), on the straight line at equal distances from the given point. This may be done by measurement, using a marked stick, or swinging a steel tape.

Step #2—Swing arcs from points (a) and (b) using a radius length about 1½ times the measurement used to locate these points. Exactly the same length radius must be used for both arcs. Locate point (c) where the arcs intersect.

Step #3—Construct a line from point (c) through the given point on the straight line. It will be at right angles to the original straight line.

Note: If conditions permit when swinging arcs to locate point (c), arcs locating point (d) may also be swung. This will provide a double check on the accuracy of the layout, as the three points, (c), given point, and (d) should form a straight line.

3-4-5 Layout Method

The 3-4-5 layout method is based on the fact that any triangle having sides with a 3-4-5 length ratio is a right triangle (Fig. 22).

Step #1—Select a suitable measuring unit. Use the largest practical, as the larger the layout the less effect minor errors will have. Measure three (3) units from the given point at approximately a right angle from the original line and swing arc (a).

Step #2—Measure four (4) units from the given point along the original line and locate point (b).

Step #3—Measure five (5) units from point (b) and locate point (c) on the arc line (a).

Step #4—Construct a line from point (c) through the given point. It will be at right angles to the original line.

Note: If conditions permit, a similar triangle may be constructed on the opposite side of the original line. This will provide a double

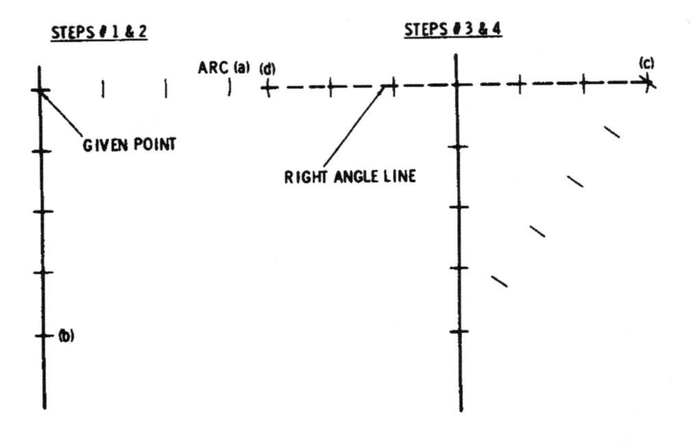

Fig. 22.

check on the accuracy of the layout, as the three points (d), given point, and (c) should form a straight line.

In many cases, establishment of lines in respect to reference points, and location of center lines and intersecting lines may be much more easily accomplished by instrument. The Engineer's Transit, in common use by surveyors, is an excellent instrument to use for this purpose. While the transit is designed for general surveying work, of which simple right-angle layout is an elementary operation, it can accomplish this operation with great accuracy.

To lay out a right-angle line with an engineer's transit, a point along the base line is selected as the reference location from which sightings are to be made. The transit is placed approximately over the point and the tripod legs moved as required to bring the plumb bob in approximate position. The instrument is then leveled approximately and shifted laterally on top of the tripod until the plumb bob is exactly over the reference point. The instrument is then leveled by means of the leveling screws and the level tubes, each level tube being first brought approximately to center, and then each bubble centered carefully. A sighting is then made through the telescope at a second point on the base line, to align the instrument with the base line. The horizontal circle graduations on the instrument are then set to exactly zero by the aid of the vernier adjustment. The establishment of a right-angle line now merely requires the rotation of the instrument 90 degrees and a sighting to locate a point on the desired right-angle line.

While the transit is normally handled by surveyors and/or engineers, this elementary operation may be accurately performed by any person experienced with precision measuring tools, if he has received proper instruction. A much more sophisticated procedure called industrial surveying makes use of this instrument and other similar related instruments for locating, aligning, and leveling in the course of manufacturing operations. The extremely close tolerances of measurement for location, alignment, and elevation required in the manufacture and assembly of large machines, aircraft, space vehicles, etc. had necessitated the development of this skill. The ordinary tools, jigs, and measuring devices are not adequate, and methods and apparatus similar to those of surveying have been substituted.

In cases where layout lines are also to be used as reference or alignment lines for machinery and equipment installation, it is recommended that piano wire lines be used. High tensile strength piano wire line is superior to fiber line because it will not stretch, loosen, or sag. When properly tightened it will stay taut and relatively stationary in space, retaining its setting and providing a high degree of accuracy.

Other lines and/or points may be precisely located from such a line by measurement, use of plumb lines, etc. A fixed-point location may be easily marked on a wire line by crimping a small particle of lead or other soft material to the wire.

To make a wire line taut it must be drawn up very tightly and placed under high tensile load. To accomplish this, and to hold the line in this condition, requires rigid fastenings at the line ends and a means of tightening and securing the line. Provisions must also be made for adjustment if the accuracy of setting, which is the principal advantage of a wire line, is to be accomplished.

Illustrated in Fig. 23 is a device which incorporates provisions for holding, tightening, adjusting, and securing a piano wire line. The "anchor plate" must be rigidly mounted to maintain the high tensile load in the line. Precise adjustment is accomplished by lightly tapping on the "adjusting plate" to move it in the direction required. When making adjustments, the "clamping screws" should be only lightly tightened, as the adjusting plate must move on the surface of the anchor plate. When the line is accurately positioned, the clamping screws should be securely tightened.

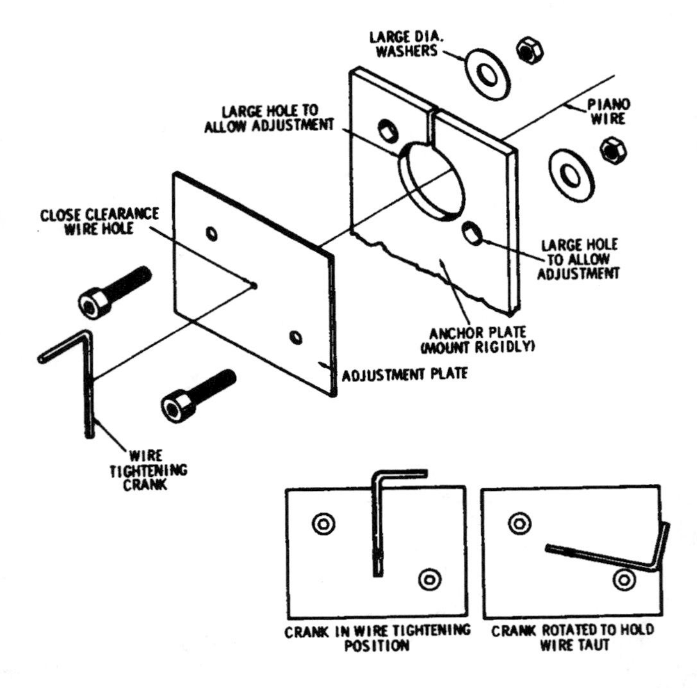

Fig. 23.

MACHINERY INSTALLATION

The first step in machinery installation is to provide a suitable base or support, termed a "foundation." It must be capable of carrying the applied load without settlement or crushing. Heavy machinery foundations are usually concrete type structures, structural steel being used for lighter applications and where space and economy are determining factors.

Anchor bolts are used to secure machinery rigidly to concrete foundations. The anchor bolts are usually equipped with a hook or some other form of fastening device to insure unity with the foundation concrete. The most common method of locating anchor bolts is making a template with holes corresponding to those in the machine to be fastened, as shown in Fig. 24. A simplified anchor bolt assembly made from readily available parts is shown in Fig. 24.

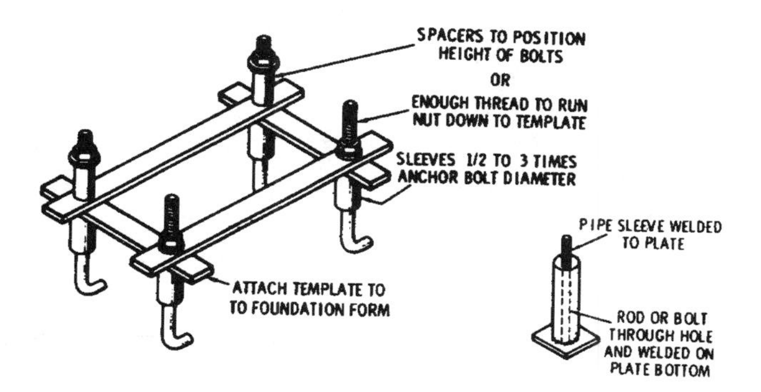

Fig. 24.

Machine units should not be mounted directly on a concrete foundation and secured with anchor bolts. Such a practice presents many difficult problems with units requiring alignment. Good design incorporates a bed plate, secured to the foundation with anchor bolts, upon which machine units are mounted and fastened. The bed plate provides a solid level supporting surface, allowing ease and accuracy of shimming and alignment. As the mounting bolts for the machinery are separate from the bed-plate anchor bolts, the bed plate is not disturbed as machine mounting bolts are loosened and tightened during alignment.

Because it is impractical to mount a bed plate directly on a concrete foundation, shims and grout are used. The shims provide a means of leveling the bed plate at the required elevation; the grout provides support and securely holds all parts in position. This method of foundation mounting is shown in the cross-section illustration of Fig. 25.

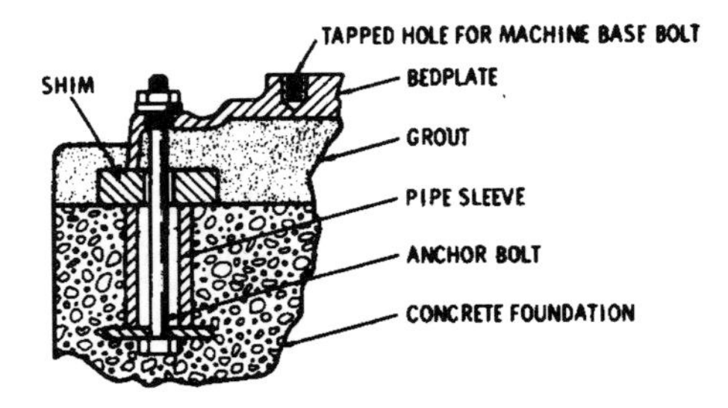

Fig. 25.

Grout is a fluid mixture of mortar-like concrete that is poured between the foundation and bed plate. It secures and holds the leveling shims and provides an intimate support surface for the bed plate. The foundation should be constructed with an elevation allowance of ¾" to 1½" for grout. Less than ¾" grout may crack and break up in service, while 1½" is a practical shim thickness limit.

The number and location of shims will be determined by the design of the bed plate. Firm support should be provided at points where weight will be concentrated and at anchor bolt locations. Flat shims of the heaviest possible thickness are recommended. The use of wedge-type shims simplifies the leveling of the bed plate; however, this must be done properly to give adequate support. Wedge shims should be double and placed in line with the bottom surface of the bed plate edge as shown in Fig. 26A. The improper practice of using single wedges at right angles to the bed plate edge is shown in Fig. 26B. As only a relatively small area of the single shim is supporting the bed plate, unit stress may be excessive and crushing may occur.

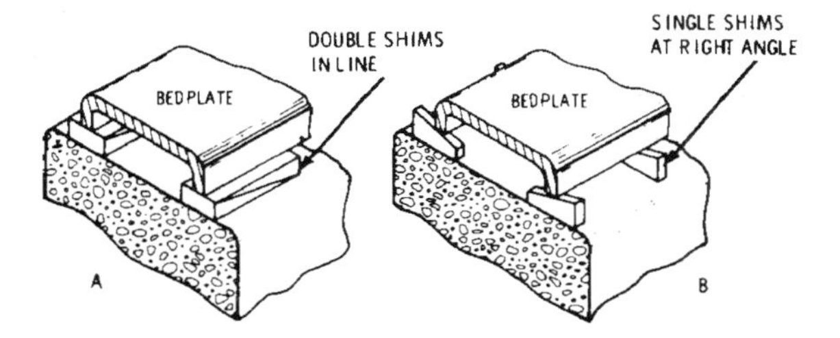

Fig. 26.

Leveling

The term "leveling," in respect to machinery installation, is the operation of placing machinery or equipment on a true horizontal plane. The tool used, called a "level," is an instrument incorporating a glass tube containing spirits. A bubble is formed in the tube when it is slightly less than completely filled with the spirits. The tube is then mounted in the body of the level and calibrated to indicate true horizontal position when the bubble is centered. The accuracy of a level may be checked by comparison of the bubble readings when placed on a horizontal plane, and then reversed end-for-end. The bubble readings should be identical in both positions.

When leveling machinery, the level should be used as a measuring instrument, not as a checking device for trial and error adjustments. The level should be used in conjunction with a feeler or thickness gauge to measure the amount of error or off level. The leveling shim thickness can then be calculated from this measurement.

Determining Shim Thickness

1 Insert thickness gauge as needed at low end of level to get zero bubble reading.
2. The shim thickness required will be as many times thicker than the gauge thickness as the distance between bearing points is greater than the level length (Fig. 27).

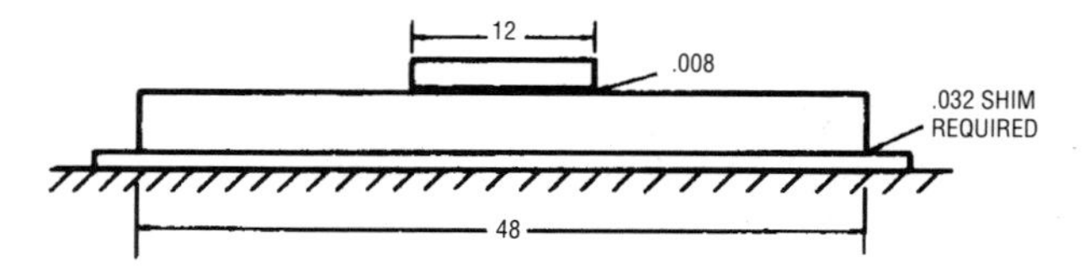

Fig. 27.

The shim thickness is 4 times the gauge thickness because the distance between bearing points is 4 times greater than the level length. The shim thickness is 4 times .008, or .032.

Elevation

In addition to the need for machinery to be installed level, in most cases it must also be placed at a given height in respect to other objects or surfaces. The usual procedure followed is to construct the foundation with an allowance for shimming. To attain a specific elevation and a level position the machine should first be roughly shimmed to approximate level, at slightly below the desired elevation. One bearing point can then be shimmed to the given elevation. As this point will be high, all other bearing points can then be raised to correct elevation by shimming them level.

Machinery Mounts

In many manufacturing plants the machinery is no longer permanently fastened in place with anchor bolts. This is especially true in metal manufacturing industries where frequent model changes or design improvements of the product are made. Flexibility and mobility are a necessity, since production lines are constantly being rearranged, with new or improved machines replacing obsolete units.

In some cases where a line of many machines may be shut down if a single unit fails, the mobility capability may be carried even further. Disabled units may be removed from the line and replaced by new or rebuilt replacement units to minimize production down time. To accomplish this the machine must be equipped with flexible quick-disconnecting coupling service lines for power, air, water, etc.

A prime requirement for this kind of mobility is devices that provide support for the machine and the capability of quick adjustment for leveling and alignment purposes. These devices are called "Machinery Mounts," since they eliminate the need for anchor bolts and make machines free standing. This type of machine mounting allows units to be picked up and moved to another location with a minimum of installation work. Using this system of mounting eliminates the need to remove anchor bolts or other fasteners, or to break floors. The leveling features built into such mounts allow leveling of the machine in a minimum amount of time to precise limits. Because the weight of the machine is the primary anchor for this system of mounting, it is very important that the weight be solidly transferred to the floor. Because a floor surface is seldom smooth and flat, mounts with a leveling feature are usually used.

Because of the ever-increasing variety of industrial machinery, the problem of vibration, shock, and noise, in addition to that of mobility, has become an important consideration. To overcome this problem, various types of special machine mounts have been developed. One widely used style of machine mount, with vibration and shock damping capability as well as adjustment and leveling features, is shown in Fig. 28. This style of mount may be used on machinery ranging from office and laboratory installations to heavy metal working machines and similar equipment subjected to high-impact shock loading and severe vibrations. This style mount eliminates the need for anchor bolts or floor lag bolts, yet keeps machinery firmly in place.

Fig. 28. Machinery mount incorporating vibration control and leveling features.

Use of this style mount makes possible quick and easy relocation of machinery while eliminating the problem of shock and vibration from one machine disturbing another. Sensitive high-precision machinery can be located for best work flow without danger of impact or vibrations from other machinery nearby. Noise, as well as vibrations transmitted from the base of the machine through the floor and building, is also reduced. Another very important benefit resulting from the use of vibration-damping mounts is the reduction of internal stresses in the machine. When vibrating equipment is rigidly bolted to the floor, an amplification of internal stresses occurs. The results often are misalignment of machine frames and undue wear on bearings and related parts.

MACHINE ASSEMBLY

Allowance for Fits

The American Standards Association (A.S.A.) classifies machine fits into eight groups, specified as Class #1 through Class #8. The standard specifies the limits for internal and external members for different sizes in each class. The groups below, listed by common shop terms, compare approximately to the A.S.A. classes as follows:

Running Fit—A.S.A. Class #2—This fit is for assemblies where one part will run in another under load with lubrication.

Table 1. Recommended Allowances

	RUNNING FIT		PUSH FIT	
Diameter Inches	Ordinary Loads	Severe Loads	Light Service	No Play
	(Clear)	(Clear)	(Clear)	(Inter)
Up to ½	.0005	.001	.00025	.0000
½ to 1	.001	.0015	.0003	.00025
1 to 2	.002	.0025	.0003	.00025
2 to 3½	.0025	.0035	.0003	.0003
3w to 6	.0035	.0045	.0005	.0005
	DRIVE FIT		FORCE or SHRINK FIT	
Diameter Inches	Field Assembly (Inter)	Shop Assembly (Inter)	Force (Inter)	Shrink (Inter)
Up to ½	.0002	.0005	.00075	.001
½ to 1	.0002	.0005	.001	.002
1 to 2	.0005	.0008	.002	.003
2 to 3½	.0005	.001	.003	.004
3½ to 6	.0005	.001	.004	.005

(Clear) Indicates clearance between members
(Inter) Indicates interference between members

Push Fit—*A.S.A. Class #4* & *#5*—This fit ranges from the closest fit that can be assembled by hand, through zero clearance, to very slight interference. Assembly is selective and not interchangeable.

Drive Fit—*A.S.A. Class #6*—This fit is used where parts are to be tightly assembled and not normally disassembled. It is an interference fit and requires light pressure. It is also used as a shrink fit on light sections.

Force or Shrink Fit—*A.S.A. Class #8*—This fit requires heavy force for cold assembly or heat to assemble parts as a shrink fit. It is used where the metal can be highly stressed without exceeding its elastic limit.

Keys, Key Seats, and Keyways

A *key* is a piece of metal placed so that part of it lies in a groove, called a *key seat*, cut in a shaft. The key then extends somewhat above the shaft and fits into a *keyway* cut in a hub. See Fig. 29.

The simplest key is the square key, placed half in the shaft and half in the hub. A flat key is rectangular in cross section and is used in the same manner as the square key for members with light sections. The gib head key is tapered on its upper surface and is driven in to form a very secure fastening.

A variation on the square key is the *Woodruff key*. It is a flat disc made in the shape of a segment of a circle. It is flat on top with a round bottom to match a semicylindrical keyseat.

Scraping

In metal working, slight errors in plane or curved surfaces are often corrected by hand scraping. Most machine surfaces that slide on one another, as in machine tools, are finished in this manner. Also plane bearing boxes are scraped to fit their shafts after having been bored, or in the case of babbitt bearings, after having been poured.

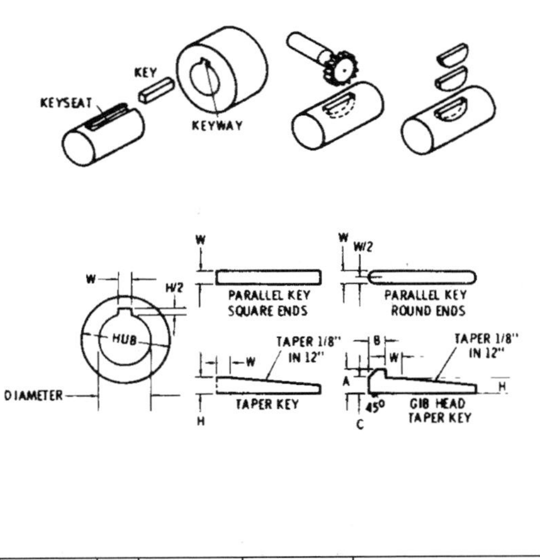

KEY NO.	NOMINAL SIZE AxB
204	1/16 x 1/2
304	3/32 x 1/2
305	3/32 x 5/8
404	1/8 x 1/2
405	1/8 x 5/8
406	1/8 x 3/4
505	5/32 x 5/8
506	5/32 x 3/4
507	5/32 x 7/8
606	3/16 x 3/4
607	3/16 x 7/8
608	3/16 x 1
609	3/16 x 1 1/8
807	1/4 x 7/8
808	1/4 x 1
809	1/4 x 1 1/8
810	1/4 x 1 1/4
811	1/4 x 1 3/8
812	1/4 x 1 1/2
1008	5/16 x 1
1009	5/16 x 1 1/8
1010	5/16 x 1 1/4
1011	5/16 x 1 3/8
1012	5/16 x 1 1/2
1210	3/8 x 1 1/4
1211	3/8 x 1 3/8
1212	3/8 x 1 1/2

SHAFT DIAMETER INCHES	W	H		H/2		GIB HEAD					
						SQUARE			FLAT		
		SQUARE	FLAT	SQUARE	FLAT	A	B	C	A	B	C
	INCHES										
1/2-9/16	1/8	1/8	3/32	1/16	3/64	1/4	7/32	5/32	3/16	1/8	1/8
5/8-7/8	3/16	3/16	1/8	3/32	1/16	5/16	9/32	7/32	1/4	3/16	5/32
3/16-1 1/4	1/4	1/4	3/16	1/8	3/32	7/16	11/32	11/32	5/16	1/4	3/16
1 5/16-1 3/8	5/16	5/16	1/4	5/32	1/8	9/16	13/32	13/32	3/8	5/16	1/4
1 7/16-1 3/4	3/8	3/8	1/4	3/16	1/8	11/16	15/32	15/32	7/16	3/8	5/16
1 13/16-2 1/4	1/2	1/2	3/8	1/4	3/16	7/8	19/32	5/8	5/8	1/2	7/16
2 5/16-2 3/4	5/8	5/8	7/16	5/16	7/32	1 1/16	23/32	3/4	3/4	5/8	1/2
2 7/8-3 1/4	3/4	3/4	1/2	3/8	1/4	1 1/4	7/8	7/8	7/8	3/4	5/8
3 5/8-3 3/4	7/8	7/8	5/8	7/16	5/16	1 1/2	1	1	1 1/16	7/8	3/4
3 7/8-4 1/2	1	1	3/4	1/2	3/8	1 3/4	1 3/16	1 3/16	1 1/4	1	13/16
4 3/4-5 1/2	1 1/4	1 1/4	7/8	5/8	7/16	2	1 7/16	1 7/16	1 1/2	1 1/4	1
5 3/4-6 1/2	1 1/2	1 1/2	1	3/4	1/2	2 1/2	1 3/4	1 3/4	1 3/4	1 1/2	1 1/4

Fig. 29.

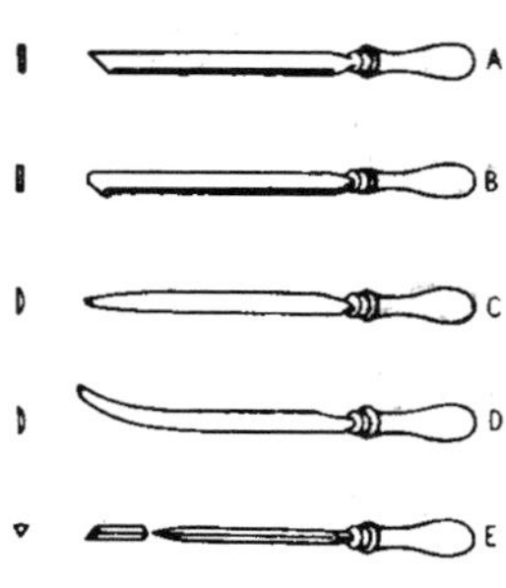

Fig. 30.

Various styles of hand scrapers are shown in Fig. 30. The flat scraper (A) is used for flat scraping. The hook scraper (B), also used on flat surfaces, is preferred by some. Flat and curved scrapers with a half-round cross section (C) and (D) are used for scraping bearings. The three-cornered scraper (E) is used to some extent on curved surfaces and to remove burrs and round the corners of holes.

Flat Surface Scraping

Coat the entire surface of a true surface plate with a suitable scraping dye such as Prussian blue.

Place the surface plate on the surface to be scraped, or if the work piece is small, place it on the surface plate.

Move the plate or the piece back and forth a few times to color the high points on the work piece.

Scrape the high spots on the work piece which are colored with the dye where they contacted the surface plate (Fig. 31).

Bearing Scraping

Coat the journal of the shaft with a thin layer of Prussian blue, spreading it with the forefinger.

Place the shaft in the bearing, or vice versa, tighten it, and turn one or the other several times through a small angle.

Scrape the high spots in the bearing which are colored where they contacted the shaft journal (Fig. 31).

TORQUE WRENCH

The words "torque wrench" are commonly used to describe a tool which is a combination wrench and measuring tool. It is used to apply a twisting force, as do conventional wrenches, and to simulta-

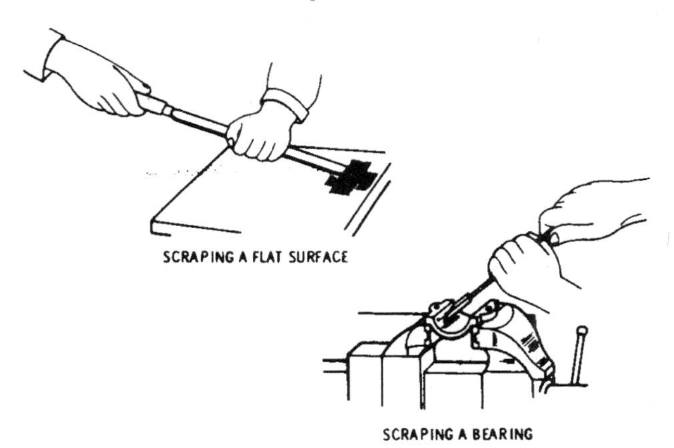

Fig. 31.

neously measure the magnitude of the force. This twisting force, which tends to turn a body about an axis of rotation, is called *torque*.

There are numerous types of torque wrenches, some that are direct reading, others with signaling mechanisms to warn when the predetermined torque is reached. All are based on the fundamental law of the level: *force times distance equals torque.*

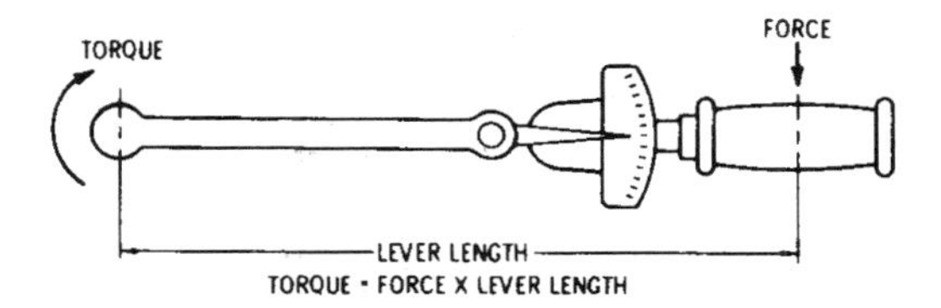

Fig. 32.

Units of Measure

Torque units of measure (inch pound and foot pound) are the product of a force measured in pound units and a lever length measured in either inch units or foot units.

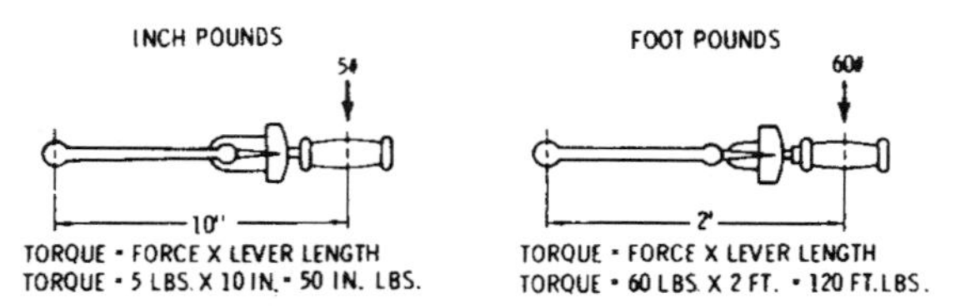

Fig. 33.

To convert inch pounds to foot pounds, divide by 12.
Example: 60 in.lbs. equals 5 ft.lbs. (60 divided by 12)
To convert foot pounds to inch pounds, multiply by 12.
Example: 12 ft.lbs. equals 144 in.lbs. (12 multiplied by 12)

TORQUE WRENCH TERMS

Push or Pull

Force should be applied to a torque wrench by pulling whenever possible. This is primarily because there is greater hazard to fingers or knuckles when pushing, should some part fail unexpectedly. While pulling is the preferred method, either way can produce accurate results.

Break-Away Torque

The torque required to loosen a fastener is generally some value lower than that to which it has been tightened. For a given size and type of fastener there is a direct relationship between tightening torque and breakaway torque. When this relationship has been determined by actual test, tightening torque may be checked by loosening and checking breakaway torque.

Set or Seizure

In the last stages of rotation in reaching a final torque reading, seizing or set of the fastener may occur. When this occurs there is a noticeable popping effect. To break the set, back off and then again apply the tightening torque. Accurate torque setting can not be made if the fastener is seized.

Run-Down Resistance

The torque required to rotate a fastener before makeup occurs is a measure of run-down resistance. To obtain the proper torque value where tight threads on locknuts produce a run-down resistance, add the resistance to the required torque value. Run-down resistance must be measured on the last rotation or as close to the makeup point as possible.

Wrench Size

The correct size wrench for a job is one that will read between 25% to 75% of the scale when the required torque is applied. This will allow adequate capacity and provide satisfactory accuracy. Avoid using an oversize torque wrench; obtaining correct readings as the pointer starts up the scale is difficult. Too small a wrench will not allow for extra capacity in the event of seizure or run-down resistance.

Torque and Tension

Torque and tension are distinctively different and must not be confused. Torque is twist, the standard unit of measure being foot-pounds; tension is straight pull, the unit of measure being pounds. Wrenches designed for measuring the tightness of a threaded fastener are distinctively torque wrenches and not tension wrenches.

Attachments

Many styles of attachments are available to fit various fasteners and to reach applications that may otherwise be impossible to torque. Most of these increase the wrench capacity as they lengthen the lever arm. Therefore, when using such attachments, scale

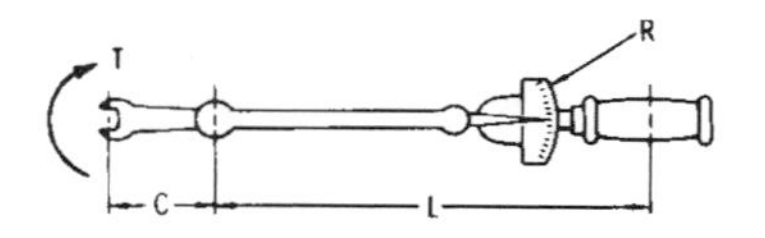

Fig. 34.

reading corrections must be made. The scale correction will be in reverse ratio to the increase in lever arm length. If the arm is doubled by adding an attachment to the wrench, its capacity is doubled and the scale shows only one-half of the actual torque applied. The following formula can be used to determine correct scale readings when using an attachment:

$$\text{Scale reading} = \frac{\text{torque required} \times \text{wrench length}}{(\text{wrench length} + \text{attachment length})}$$

$$R = \frac{T \times L}{(L + C)}$$

TORQUE SPECIFICATIONS

The suggested maximum torque values in Table 2 for fasteners of various materials should be used as a guide only. Manufacturers' specifications should be followed on specific torque applications.

Table 2. Torque in Foot Pounds

Fastener Diameter	Threads Per Inch	Mild Steel	Stainless Steel 18-8	Alloy Steel
1/4	20	4	6	8
5/16	18	8	11	16
3/8	16	12	18	24
7/16	14	20	32	40
1/2	13	30	43	60
5/8	11	60	92	120
3/4	10	100	128	200
7/8	9	160	180	320
1	8	245	285	490

Torque Values for Steel Fasteners

The strength of a bolted connection depends on the clamping force developed by the bolts. The tighter the bolt, the stronger the connection. The two principal factors that limit the clamping force the bolt may develop are the bolt size and its tensile strength.

The tensile strength of a bolt depends principally on the material from which it is made. Bolt manufacturers identify bolt materials by head markings which conform to SAE and ASTM specifications. These marks are known as grade markings. The most commonly specified grades are listed in Table 3.

The values in Table 3 do not apply if special lubricants such as colloidal copper or molybdenum disulphite are used. Use of special lubricants can reduce the amount of friction in the fastener assembly so the torque applied may produce far greater tension than desired.

Table 3. Suggested Torque Values for Graded Steel Bolts

Grade		SAE 1&2	SAE 5	SAE 6	SAE 8
Tensile Strength		64000 PSI	105000 PSI	130000 PSI	150000 PSI
Grade Mark					
Bolt Dia.	Threads. Per In.	Foot Pounds Torque			
1/4	20	5	7	10	10
5/16	18	9	14	19	22
3/8	16	15	25	34	37
7/16	14	24	40	55	60
1/2	13	37	60	85	92
9/16	12	53	88	120	132
5/8	11	74	120	169	180
3/4	10	120	200	280	296
7/8	9	190	302	440	473
1	8	282	466	660	714

The above values are based on approximately 75% of yield strength.
Fasteners must be lubricated (Petroleum Lubricant)

MECHANICAL POWER TRANSMISSION

The three principal systems used for the transmission of rotary mechanical power between adjacent shafts are *belts, chains,* and *gears.* Understanding of certain terms and concepts that are basic to all three systems is a prerequisite to understanding of the systems.

Pitch

A word commonly used in connection with machinery and mechanical operations meaning: *The distance from a point to a corresponding point.*

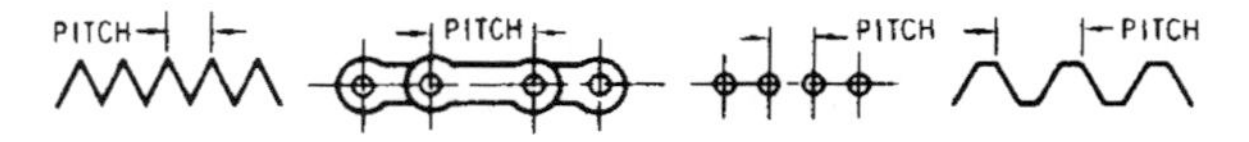

Pitch Diameter and Pitch Circle

The *pitch diameter* specifies the distance across the center of the *pitch circle.* Pitch diameter dimensions are specific values even though the *pitch circle* is imaginary. Rotary power transmission calculations are based on the concept of circles or cylinders in contact. These circles are called *pitch circles.*

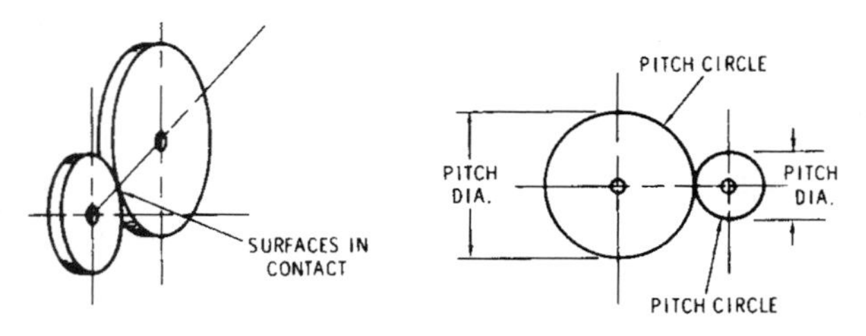

As the shafts rotate, the surfaces of the pitch circles travel equal distances at equal speeds (assuming no slippage). Shafts then rotate at speeds proportional to the circumference of the pitch circles and therefore proportional to the pitch diameters.

This concept of pitch circles in contact applies to belt and chain drives, although the pitch circles are actually separated. This is true because the belt or chain is in effect an extension of the pitch circle surface.

A. As rotation occurs the pitch circle surfaces will travel the same distance at the same surface speed.
B. Because the circumference of a 4" pitch circle is double that of a 2" pitch circle, its rotation will be one-half as much.
C. The rotation speed of the 4" pitch circle will be one-half that of the 2" pitch circle.

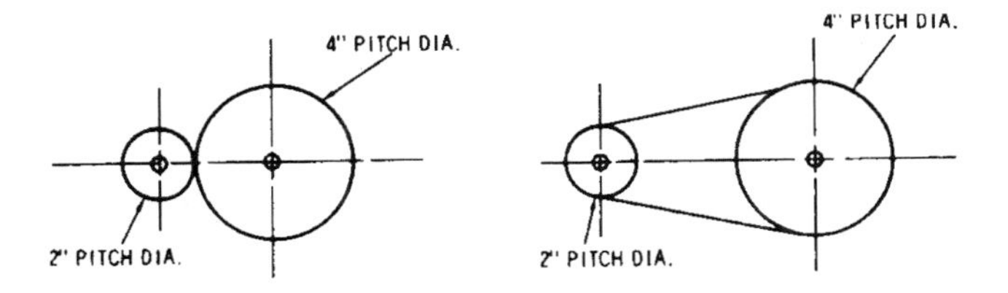

Calculations

Rotational speed and pitch diameter calculations for belts, chains, and gears are based on the concept of pitch circles in contact. The relationship that results from this concept may be stated as follows:

Shaft speeds are inversely proportional to pitch diameters.

In terms of rotational speeds and pitch diameters, this relationship may be expressed in equation form as follows:

$$\frac{\textit{Driver Rotational Speed}}{\textit{Driven Rotational Speed}} = \frac{\textit{Driven Pitch Diameter}}{\textit{Driver Pitch Diameter}}$$

To simplify the use of the equation, letters and numbers instead of words are used to represent the terms.

S1 for *DRIVER ROTATIONAL SPEED*
S2 for *DRIVEN ROTATIONAL SPEED*
P1 for *DRIVER PITCH DIAMETER*
P2 for *DRIVEN PITCH DIAMETER*

The basic equation then becomes $\frac{S1}{S2} = \frac{P2}{P1}$

The equation may be arranged into the following forms, one for each of the four values. To find an unknown value the known values are substituted in the appropriate equation.

$$S1 = \frac{P2 \times S2}{P1} \quad S2 = \frac{P1 \times S1}{P2} \quad P1 = \frac{S2 \times P2}{S1} \quad P2 = \frac{S1 \times P1}{S2}$$

The above equations may also be stated as rules. Following these rules is another convenient way to calculate unknown shaft speeds and pitch diameters.

To Find — *Driving Shaft Speed*
Multiply *driving pitch diameter* by speed of *driven shaft* and divide by *driving pitch diameter.*

To Find — *Driven Shaft Speed*
Multiply *driving pitch diameter* by speed of *driving shaft* and divide by speed of *driven pitch diameter.*

To Find — *Driven Pitch Diameter*
Multiply *driven pitch diameter* by speed of *driven shaft* and divide by speed of *driving shaft.*

To Find — *Driven Pitch Diameter*
Multiply *driving pitch diameter* by speed of *driving shaft* and divide by speed of *driven shaft.*

In gear and sprocket calculations the number of teeth is used rather than the pitch diameters. The equations and rules hold true because the number of teeth in a gear or sprocket is directly proportional to its pitch diameter.

Calculations (Examples)

To find an unknown value, substitute the known values in the appropriate equation or follow the appropriate rule.

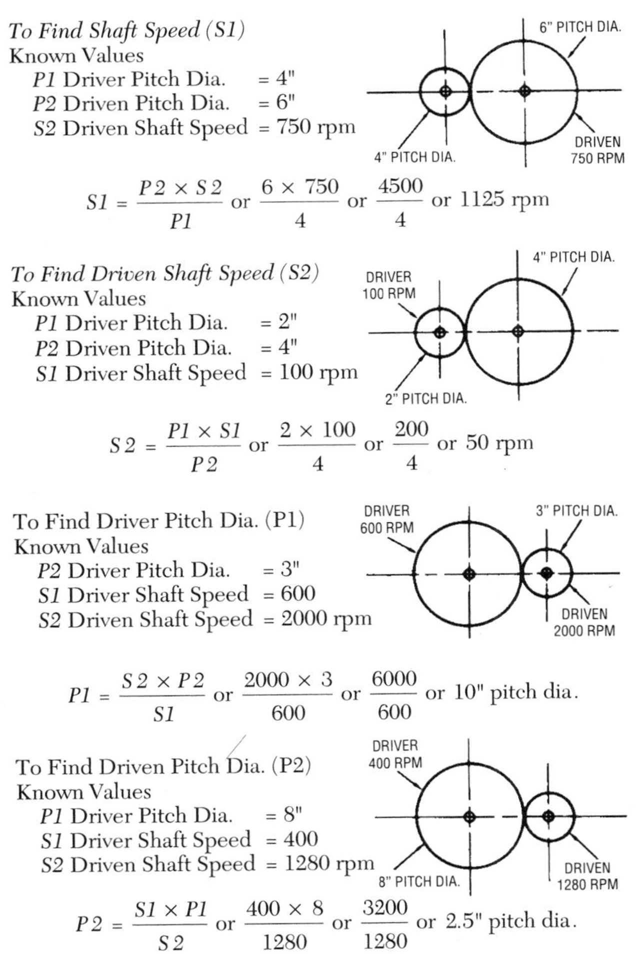

To Find Shaft Speed (S1)

Known Values

- *P1* Driver Pitch Dia. = 4"
- *P2* Driven Pitch Dia. = 6"
- *S2* Driven Shaft Speed = 750 rpm

$$S1 = \frac{P2 \times S2}{P1} \text{ or } \frac{6 \times 750}{4} \text{ or } \frac{4500}{4} \text{ or } 1125 \text{ rpm}$$

To Find Driven Shaft Speed (S2)

Known Values

- *P1* Driver Pitch Dia. = 2"
- *P2* Driven Pitch Dia. = 4"
- *S1* Driver Shaft Speed = 100 rpm

$$S2 = \frac{P1 \times S1}{P2} \text{ or } \frac{2 \times 100}{4} \text{ or } \frac{200}{4} \text{ or } 50 \text{ rpm}$$

To Find Driver Pitch Dia. (P1)

Known Values

- *P2* Driver Pitch Dia. = 3"
- *S1* Driver Shaft Speed = 600
- *S2* Driven Shaft Speed = 2000 rpm

$$P1 = \frac{S2 \times P2}{S1} \text{ or } \frac{2000 \times 3}{600} \text{ or } \frac{6000}{600} \text{ or } 10\text{" pitch dia.}$$

To Find Driven Pitch Dia. (P2)

Known Values

- *P1* Driver Pitch Dia. = 8"
- *S1* Driver Shaft Speed = 400
- *S2* Driven Shaft Speed = 1280 rpm

$$P2 = \frac{S1 \times P1}{S2} \text{ or } \frac{400 \times 8}{1280} \text{ or } \frac{3200}{1280} \text{ or } 2.5\text{" pitch dia.}$$

"V" BELTS

The "V" belt has a tapered cross-sectional shape which causes it to wedge firmly into the sheave groove under load. Its driving action takes place through frictional contact between the sides of the belt and the sheave groove surfaces. While the cross-sectional shape varies slightly with make, type, and size, the included angle of most "V" belts is about 42 degrees. There are three general classifications of "V" belts: Fractional Horsepower, Standard Multiple, and Wedge.

Fractional Horsepower

Used principally as single belts on fractional horsepower drives, they are designed for intermittent and relatively light loads. These belts are manufactured in four standard cross-sectional sizes as shown in Fig. 35A.

Standard belt lengths vary by one-inch increments between a minimum length of 10 inches and a maximum length of 100 inches. In addition, some fractional horsepower belts are made to fractional inch lengths.

The numbering system used indicates the cross-sectional size and the nominal outside length. The last digit of the belt number indicates tenths of an inch. Because the belt number indicates length along the outside surface, belts are slightly shorter along the pitch line than the nominal size number indicates.

Examples

Belt Number	*Size*	*Outside Length*
3L 470	3L	47"
4L 425	4L	42½"

Standard Multiple

Standard multiple belts are designed for the continuous service usually encountered in industrial applications. As the name indicates, more than one belt provides the required power transmission capacity. Most manufacturers furnish two grades, a standard and a premium quality. The standard belt is suitable for the majority of industrial drives that have normal loads, speeds, center distances, sheave diameters, and operating conditions. The premium quality is made for drives subjected to severe loads, shock, vibration, temperatures, etc.

The standard multiple "V" belt is manufactured in five standard cross-sectional sizes designated: "A," "B," "C," "D," "E," as shown in Fig. 35B.

The actual pitch length of standard multiple belts may be from one to several inches greater than the nominal length indicated by the belt number. This is because the belt numbers indicate the length of the belt along its inside surface. As belt length calculations

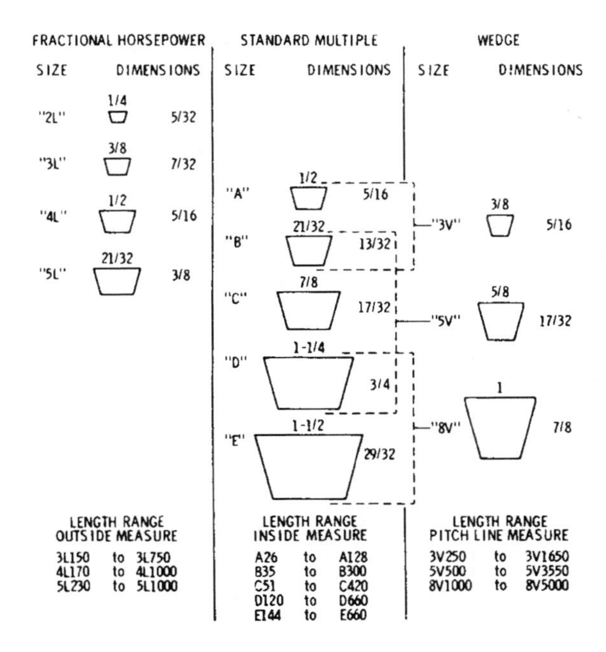

Fig. 35. A B C

are in terms of belt length on the pitch line, a table of pitch line belt lengths is recommended when selecting belts.

Wedge

The wedge belt is an improved design "V" belt which makes possible a reduction in size, weight, and cost of "V" belt drives. Utilizing improved materials, these multiple belts have a smaller cross-section per horsepower and use smaller diameter sheaves at shorter center distances than is possible with standard multiple belts. Because of the premium-quality heavy-duty construction, only three cross-sectional belt sizes are used to cover the duty range of the five sizes of standard multiple belts. The dimensions of the three standard "wedge" belt cross-sectional sizes "3V," "5V," "8V" are shown in Fig. 35C.

The wedge belt number indicates the number of ⅛ inches of top width of the belt. As shown in Fig. 35C, the "3V" belt has a top width of ⅜ inch, the "5V" a width of ⅝ inch, and the "8V" a full 1 inch of top width.

The belt length indicated by the wedge belt number is the effective pitch line length of the belt. As belt calculations are in terms of pitch line lengths, nominal belt numbers can be used directly when choosing wedge belts.

"V" Belt Matching

Satisfactory operation of multiple belt drive requires that each belt carry its share of the load. To accomplish this, all belts in a drive must be essentially of equal length. Because it is not economically practical to manufacture belts to exact length, most manufacturers follow a practice of code marking to indicate exact length (Table 4).

Each belt is measured under specific tension and marked with a code number to indicate its variation from nominal length. The number 50 is commonly used as the code number to indicate a belt within tolerance of its nominal length. For each ⅒ of an inch over nominal length, the number 50 is increased by 1. For each ⅒ of an inch under nominal length, 1 is subtracted from the number 50. Most manufacturers code mark as shown in Fig. 36.

Table 4. Standard "V" Belt Lengths

A BELTS

BELT NUMBER Standard	Pitch Length	Outside Length
A26	27.3	28.0
A31	32.3	33.0
A35	36.3	37.0
A38	39.3	40.0
A42	43.3	44.0
A46	47.3	48.0
A51	52.3	53.0
A55	56.3	57.0
A60	61.3	62.0
A68	69.3	70.0
A75	76.3	77.0
A80	81.3	82.0
A85	86.3	87.0
A90	91.3	92.0
A96	97.3	98.0
A105	106.3	107.0
A112	113.3	114.0
A120	121.3	122.0
A128	129.3	130.0

B BELTS

BELT NUMBER Standard	Pitch Length	Outside Length
B35	36.8	38.0
B38	39.8	41.0
B42	43.8	45.0
B46	47.8	49.0
B51	52.8	54.0
B55	56.8	58.0
B60	61.8	63.0
B68	69.8	71.0
B75	76.8	78.0
B81	82.8	84.0
B85	86.8	88.0
B90	91.8	93.0
B97	98.8	100.0
B105	106.8	108.0
B112	113.8	115.0
B120	121.8	123.0
B128	129.8	131.0
B136	137.8	139.0
B144	145.8	147.0
B158	159.8	161.0
B173	174.8	176.0
B180	181.8	183.0
B195	196.8	198.0
B210	211.8	213.0
B240	240.3	241.5
B270	270.3	271.5
B300	300.3	301.5

C BELTS

BELT NUMBER Standard	Pitch Length	Outside Length
C51	53.9	55.0
C60	62.9	64.0
C68	70.9	72.0
C75	77.9	79.0
C81	83.9	85.0
C85	87.9	89.0
C90	92.9	94.0
C96	98.9	100.0
C105	107.9	109.0
C112	114.9	116.0
C120	122.9	124.0
C128	130.9	132.0
C136	138.9	140.0
C144	146.9	148.0
C158	160.9	162.0
C162	164.9	166.0
C173	175.9	177.0
C180	182.9	184.0
C195	197.9	199.0
C210	212.9	214.0
C240	240.9	242.0
C270	270.9	272.0
C300	300.9	302.0
C360	360.9	362.0
C390	390.9	392.0
C420	420.9	422.0

D BELTS

BELT NUMBER Standard	Pitch Length	Outside Length
D120	123.3	125.0
D128	131.3	133.0
D144	147.3	149.0
D158	161.3	163.0
D162	165.3	167.0
D173	176.3	178.0
D180	183.3	185.0
D195	198.3	200.0

3V BELTS		5V BELTS		8V BELTS	
Belt No.	Belt Length	Belt No.	Belt Length	Belt No.	Belt Length
3V250	25.0	5V500	50.0	8V1000	100.0
3V265	26.5	5V530	53.0	8V1060	106.0
3V280	28.0	5V560	56.0	8V1120	112.0
3V300	30.0	5V600	60.0	8V1180	118.0
3V315	31.5	5V630	63.0	8V1250	125.0
3V335	33.5	5V670	67.0	8V1320	132.0
3V355	35.5	5V710	71.0	8V1400	140.0
3V375	37.5	5V750	75.0	8V1500	150.0
3V400	40.0	5V800	80.0	8V1600	160.0

Table 4. Standard "V" Belt Lengths (Cont'd)

D BELTS

D210	213.3	215.0
D240	240.8	242.5
D270	270.8	272.5
D300	300.8	302.5
D330	330.8	332.5
D360	360.8	362.5
D390	390.8	392.5
D420	420.8	422.5
D480	480.8	482.5
D540	540.8	542.5
D600	600.8	602.5

E BELTS

BELT NUMBER Standard	Pitch Length	Outside Length
E180	184.5	187.5
E195	199.5	202.5
E210	214.5	217.5
E240	241.0	244.0
E270	271.0	274.0
E300	301.0	304.0
E330	331.0	334.0
E360	361.0	364.0
E390	391.0	394.0
E420	421.0	424.0
E480	481.0	484.0
E540	541.0	544.0
E600	601.0	604.0

3V BELTS		5V BELTS		8V BELTS	
3V425	42.5	5V850	85.0	8V1700	170.0
3V450	45.0	5V900	90.0	8V1800	180.0
3V475	47.5	5V950	95.0	8V1900	190.0
3V500	50.0	5V1000	100.0	8V2000	200.0
3V530	53.0	5V1060	106.0	8V2120	212.0
3V560	56.0	5V1120	112.0	8V2240	224.0
3V600	60.0	5V1180	118.0	8V2360	236.0
3V630	63.0	5V1250	125.0	8V2500	250.0
3V670	67.0	5V1320	132.0	8V2650	265.0
3V710	71.0	5V1400	140.0	8V2800	280.0
3V750	75.0	5V1500	150.0	8V3000	300.0
3V800	80.0	5V1600	160.0	8V3150	315.0
3V850	85.0	5V1700	170.0	8V3350	335.0
3V900	90.0	5V1800	180.0	8V3550	355.0
3V950	95.0	5V1900	190.0	8V3750	375.0
3V1000	100.0	5V2000	200.0	8V4000	400.0
3V1060	106.0	5V2120	212.0	8V4250	425.0
3V1120	112.0	5V2240	224.0	8V4500	450.0
3V1180	118.0	5V2360	236.0	*8V5000	500.0
3V1250	125.0	5V2500	250.0		
3V1320	132.0	5V2650	265.0		
3V1400	140.0	5V2800	280.0		
		5V3000	300.0		
		5V3150	315.0		
		5V3350	335.0		
		5V3550	355.0		

For example, if the 60-inch "B" section belt shown is manufactured 3/10 of an inch longer, it will be code marked 53 rather than the 50 shown. If made 3/10 of an inch shorter, it will be code marked 47. While both of these belts have the belt number B60 they cannot be used satisfactorily in a set because of the difference in their actual length.

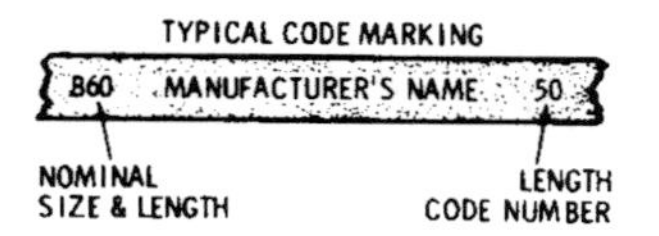

Fig. 36

It is possible for the length of belts to change slightly during storage. Under satisfactory conditions, however, changes will not exceed measuring tolerances. Therefore, belts may be combined by matching code numbers. Ideally, sets should be made up of belts having the same code numbers; however, the resiliency of the belts allows some length variation. Table 5 lists the maximum recommended variations for standard multiple belts when making up matched belt sets.

Table 5.

Matching Number Range	Belt Lengths				
	A	B	C	D	E
2	26-180	35-180	51-180		
3		195-315	195-255	120-255	144-240
4			270-360	270-360	270-360
6			390-420	390-660	390-660

"V" Belt Drive Alignment

The life of a "V" belt is dependent on first, the quality of materials and manufacture, and second, on installation and maintenance. One of the most important installation factors influencing operating life is belt alignment. In fact, excessive misalignment is probably the most frequent cause of shortened belt life.

While "V" belts, because of their inherent flexibility, can accommodate themselves to a degree of misalignment not tolerated by other types of power transmission, they still must be held within reasonable limits. Maximum life can be attained only with true alignment, and as misalignment increases belt life is proportionally reduced. If misalignment is greater than 1/16 inch for each 12 inches of center distance, very rapid wear will result.

Misalignment of belt drives results from shafts being out of angular or parallel alignment, or from the sheave grooves being out of axial alignment. These three types of misalignment are illustrated in Fig. 37.

Because the shafts of most "V" belt drives are in a horizontal plane, angular shaft alignment is easily obtained by leveling the shafts. In those cases where shafts are not horizontal, a careful check must be made to ensure the angle of inclination of both shafts is the same.

Before any check is made for parallel-shaft and axial-groove alignment of most drives can be done simultaneously if the shafts and sheaves are true.

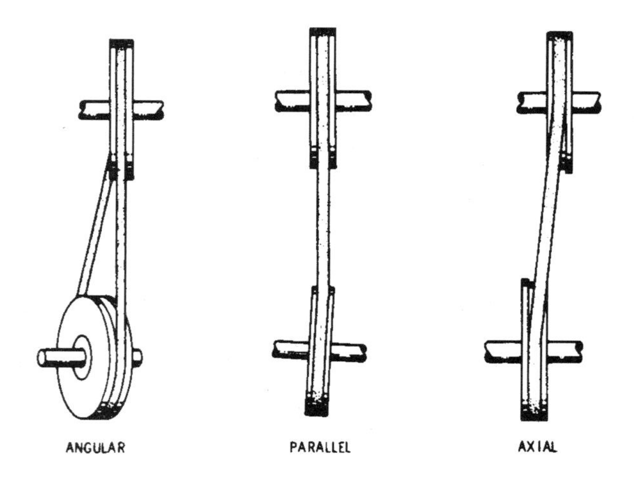

Fig. 37.

"V" Belt Alignment

The most satisfactory method of checking parallel-shaft and axial-groove alignment is with a straightedge. It may also be done with a taut line; however, when using this method care must be exercised, as the line is easily distorted. The straightedge checking method is illustrated in Fig. 38, with arrows indicating the four check points. When sheaves are properly aligned no light should be visible at these four points.

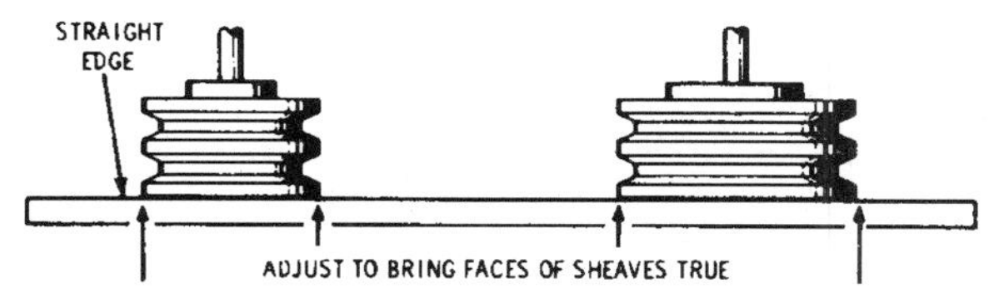

Fig. 38.

"V" Belt Installation

"V" belts should never be "run-on" to sheaves. To do so places excessive stress on the cords, usually straining or breaking some of them. A belt damaged in this manner will flop under load and turn over in the sheave groove. The proper installation method is to loosen the adjustable mount, reduce the center distance, and slip the belts loosely into the sheave grooves.

The following six general rules should be followed when installing "V" belts.

1. Reduce centers so belts can be slipped on sheaves.
2. Have all belts slack on the same side (top of drive).
3. Tighten belts to approximately correct tension.
4. Start unit and allow belts to seat in grooves.
5. Stop—retighten to correct tension.
6. Recheck belt tension after 24 to 48 hours of operation.

Checking Belt Tension

Belt tension is a vital factor in operating efficiency and service life. Too low a tension results in slippage and rapid wear of both belts and sheave grooves. Too high a tension stresses the belts excessively and unnecessarily increases bearing loads.

The tensioning of fractional horsepower and standard multiple belts may be done satisfactorily by tightening until the proper "feel" is attained. The proper "feel" is when the belt has a live springy action when struck with the hand. If there is insufficient tension the belt will feel loose or dead when struck. Too much tension will cause the belts to feel taut, as there will be no give to them.

Wedge Belt Tension

The 3V, 5V, and 8V wedge belts operate under very high tension, since there are fewer belts and/or smaller belts per horsepower. When properly tightened they are taut and have little give; therefore, tightening by "feel" is not dependable. A better method is to use the belt-tension measuring tool shown in Fig. 39.

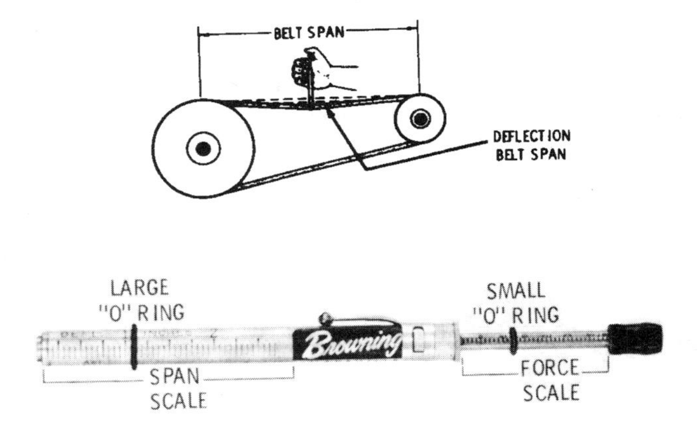

Fig. 39.

Browning Belt Tension Checker

Instructions

To determine the lbs. force required to tension a drive with the Browning Belt Tensioner you simply do the following:

1. Measure the Belt Span as shown.
2. Divide belt span by 64 to get belt deflection needed to check tension.
3. Set large "O" ring on span scale at required belt deflection. This scale is in 1/16" increments.
4. Set small "O" ring at zero on the "Force Scale" (plunger).
5. Place the larger end of the tension checker squarely on one belt at the center of the belt span. Apply force on the plunger until the bottom of the large "O" ring is even with the top of the next belt or with the bottom of a straightedge laid across the sheaves.
6. Read the force scale under the small "O" ring to determine the force required to give the needed deflection.
7. Compare the force scale reading with the correct value for the belt style and cross section used, as given in table below. The force should be between the minimum and maximum values shown.
8. If there is too little deflection force, the belts should be tightened. If there is too much deflection force, the belts should be loosened.

Belt Cross Section	Small P.D. Range	Deflection Force Lbs.	
		Min.	Max.
3V	2.65 - 3.65	3	4
	4.12 - 6.90	4	6
5V	7.1 - 10.9	8	12
	11.8 - 16.0	10	15
8V	12.5 - 17.0	18	27
	18.0 - 22.4	20	30

"V" Belt Replacement

When replacing "V" belts, care must be exercised that the correct type is selected. Errors in choice might be made since the top width of some of the sizes in the three types is essentially the same. Also, belts from different manufacturers should not be mixed on the same drive because of variations from nominal dimensions.

When determining the length belt required for most drives it is not necessary to be exact: first, because of the adjustment built into most drives, and second, because belt selection is limited to the standard lengths available. Since the standard lengths vary in steps of several inches, an approximate length calculation is usually adequate. For these reasons the following easy method of belt calculation can be used for most "V" belt drives:

1. Add the pitch diameters of the sheaves and multiply by 1½
2. To this add twice the distance between centers.
3. Select the nearest *longer* standard belt.

This method should not be used when centers are fixed or if there are extreme pitch-diameter differences on short centers.

Sheave Groove Wear

All "V" belts and sheaves will wear to some degree with use. As wear occurs the belts will ride lower in the grooves. Generally a new belt should not seat more than 1⁄16 inch below the top of the groove. While belt wear is usually noticed, sheave-groove wear is often overlooked.

As wear occurs at the contact surfaces on the sides of the grooves, a dished condition develops. This results in reduced wedging action, loss of gripping power, and accelerated wear as slippage occurs. Installing new belts in worn grooves will give temporary improvement in operation, but belt wear will be rapid. When changing belts, therefore, sheave-groove wear should be checked with gauges or templates.

Care must be taken when checking grooves that the correct gauge or template in respect to type, size, and pitch diameter is used. As sheave grooves are designed to conform to the belt cross-section change as it bends, small diameter sheaves have less angle than larger diameter sheaves. The variation in sheave-groove includes angles ranging from 34 degrees for small diameter sheaves up to 42 degrees for the largest diameter sheaves.

Sheave Grooves

When sheave groove wear becomes excessive, shoulders will develop on the groove side walls. If the sheave is not repaired or replaced, these shoulders will quickly chew the bottom corners off new belts and ruin them.

The more heavily loaded a drive, the greater the effect of groove wear on its operation. Light to moderately loaded drives may

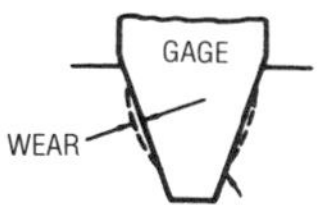

Maximum Allowable Wear
.025" Moderately Loaded Drives
.015" Heavily Loaded Drives

Table 6. Standard Groove Dimension Table

V-Belt	Minimum Recommended Pitch Diam.	Pitch Diam.	Groove °	Angle W	D	X	S	E
A	3.0	2.6 to 5.4	34°	.494	.490	.125	⅝	⅜
		Over 5.4	38°	.504				
B	5.4	4.6 to 7.0	34°	.637	.580	.175	¾	½
		Over 7.0	38°	.650				
C	9.0	7.0 to 7.99	34°	.879	.780	.200	1	11/16
		8.0 to 12.0	36°	.887				
		Over 12.0	38°	.895				
D	13.0	12.0 to12.99	34°	1.259	1.050	.300	17/32	⅞
		13.0 to 17.0	36°	1.271				
		Over 17.0	38°	1.283				
E	21.0	18.0 to 24.0	36°	1.527	1.300	.400	1¾	1⅛
		Over 24.0	38°	1.542				

FILEBREAK ALL SHARP CORNERS
GROOVE ANGLE
W D OD PD S E H

FACE WIDTH OF STANDARD V-BELT PULLEYS

FACE WIDTH = S(N+1) + 2E
WHERE:
N = NUMBER OF GROOVES

Belt	GROOVE DIAMETER IN INCHES										
	A	B	C	D	E	W	T	U	V	Angle of Groove	Use on O.D.
3V	11/32	13/32	.025	.325	.350	.350	.056	.123	.334	36°	under 3.5
								.109	.333	38°	3.5 to 6.0
								.096	.332	40°	6.01 to 12.0
								.081	.331	42°	12.01, over
								.187	.566	38°	under 10.0
5V	½	11/16	.05	.550	.600	.600	.0875	.163	.564	40°	10.1 to 16.0
								.139	.562	42°	16.01, over
								.312	.931	38°	under 16.0
8V	¾	1⅛	.10	.900	1.000	1.000	.125	.272	.927	40°	6.0 to 22.4
								.232	.923	42°	22.41, over

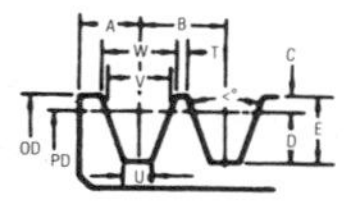

Fig. 40.

tolerate as much as 1/32" wear, whereas heavily loaded drives will be adversely affected by .010" to .015" of wear. Wear should be checked with the appropriate gauge at the point illustrated in Fig. 40.

POSITIVE-DRIVE BELTS

Positive-drive belts, also called timing belts and gear belts, combine the flexibility of belt drives with the advantages of chain and gear drives. Power is transmitted by positive engagement of belt teeth with pulley grooves, as in chain drives, rather than by friction as in belt drives. This positive engagement of belt teeth with pulley grooves eliminates slippage and speed variations. There is no metal-to-metal contact and no lubrication required.

The positive-drive belt is constructed with gear-like teeth which engage mating grooves in the pulley. Unlike most other belts, they do not derive their tensile strength from their thickness. Instead, these belts are built thin, their transmission capacity resulting from the steel-cable tension members and tough molded teeth as shown in Fig. 41.

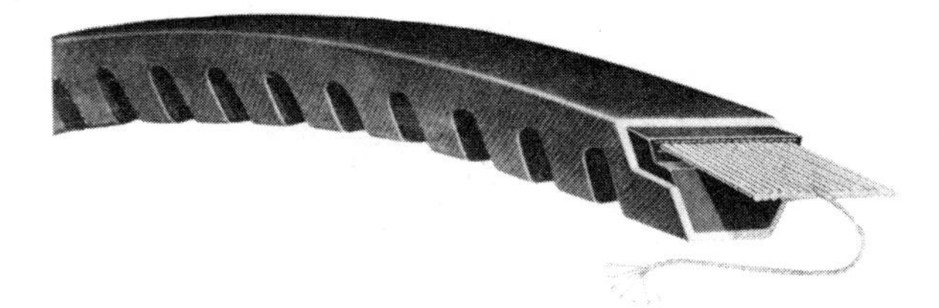

Fig. 41. Positive-drive belt internal construction.

High-strength synthetic materials are used in positive-drive construction. The four basic components are shown in Fig. 42: neoprene backing, steel tension members, neoprene teeth with nylon facing. The nylon covering on the teeth is highly wear resistant, and after a short "run-in" period becomes highly polished with resultant low friction. Belts are so constructed that tooth

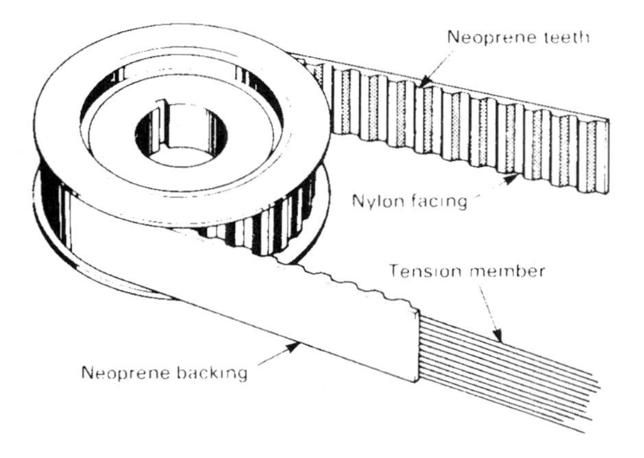

Fig. 42. Basic components of positive-drive belt.

strength exceeds the tensile strength of the belt when six or more teeth are meshed with a mating pulley.

With positive-drive belts, as with gear and chain drives, pitch is a fundamental consideration. In this case *circular pitch* is the distance between tooth centers (measured on the pitch line of the belt) or the distance between groove centers (measured on the pitch circle of the pulley) as indicated in Fig. 43.

The *pitch line* of a positive-drive belt is located within the cable tension member. The *pitch circle* of a positive-drive pulley coincides with the *pitch line* of the belt mating with it. The pulley *pitch diameter* is *always greater* than its face diameter. All positive-drive belts must run with pulleys of the same pitch. A belt of one pitch *cannot* be used successfully with pulleys of a different pitch.

Positive drive belts are made in five (5) stock pitches. The following code system is used to indicate the pitch of positive-drive systems:

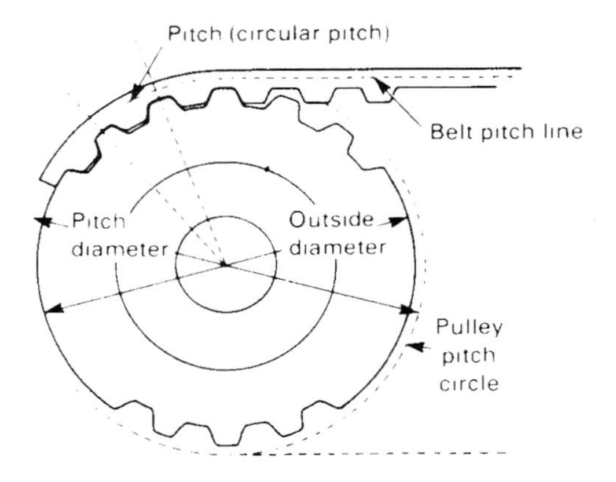

Fig. 43. Positive-drive belt-measurements.

Code	*Code Meaning*	*Pitch*
XL	Extra Light	1/5 inch
L	Light	3/8 inch
H	Heavy	1/2 inch
XH	Extra Heavy	7/8 inch
XXH	Double Extra Heavy	1¼ inch

The standard positive-drive belt numbering system is made up of three parts: first, the pitch length of the belt, which is the actual pitch length multiplied by ten (10); second, the code for the pitch of the drive; third, the belt width multiplied by one hundred (100).

Examples

390L100—39.0 Pitch Length—Light Size 3/8" pitch—1" wide
480H075—48.0 Pitch Length—Heavy Size 1/2" pitch—3/4" wide
560XH200—56.0 Pitch Length—Extra Heavy Size 7/8" pitch—2" wide

The following list is the stock sizes of positive-drive belts carried by most manufacturers. In addition, much wider and longer drives can be furnished on special order.

Code	*Pitch*	*Stock Widths*	*Length Range*
XL	⅕"	¼"-5⁄16-⅜"	6" to 26"
L	⅜"	½"-¾"-1"	12" to 60"
H	½"	¾"-1"-1½-2"-3"	24" to 170"
XH	⅞"	2"-3"-4"	50" to 175"
XXH	1¼"	2"-3"-4"-5"	70" to 180"

Because of a slight side thrust of positive-drive belts in motion, at least one pulley in a drive must be flanged. When center distance between shafts is eight or more times the diameter of the small pulley, or when drive is operating on vertical shafts, both pulleys should be flanged.

When installing positive-drive belts, the center distance should be reduced and the belt placed loosely on the pulleys. Belts should not be forced in any way over the pulley or flange, as damage to the belt will result. The belt should be tightened to a snug fit. Since the positive-drive belt does not rely on friction, there is no need for high initial stress. However, if torque is unusually high, a loose belt may "jump grooves." In such cases, the tension should be increased gradually until satisfactory operation is attained. Care must be exercised that shaft parallelism is not disturbed while doing this. On heavily loaded drives, with wide belts, it may be necessary to use a tension measuring tool to accurately tighten the belt. Belt manufacturers should be consulted for their recommendations of equipment and procedures to follow for special situations of this type.

FLAT BELTS

While there are very few new installations of flat-belt drives in industry, there are still many installations in operation. Flat belting may be made from either leather, rubber, or canvas. Leather belts are by far the most commonly used. Rubber belting is usually used where there is exposure to weather conditions or moisture, since they do not stretch under these conditions. Canvas or similar fabrics, usually impregnated with rubber, is used when the materials (such as liquids) in contact with the belt would have an adverse effect on leather or rubber.

Leather belting is specified by thickness and width. The two general thickness classifications are single and double. These are further divided into light, medium, and heavy. The thickness specifications for first-quality leather belting are as follows:

Medium	Single	5⁄32 to 3⁄16 inch
Heavy	Single	3⁄16 to 7⁄32 inch
Light	Double	15⁄64 to 17⁄64 inch
Medium	Double	9⁄32 to 5⁄16 inch
Heavy	Double	21⁄64 to 23⁄64 inch

In the installation of leather belts, precautions should be taken to put the flesh side on the outside and the grain, or smooth, side toward the pulley. It has been found by experience that when the belt is installed in this manner it will wear much longer and deliver more power than it put on in the reverse way.

The terms commonly used in conjunction with flat belts are *slip* and *creep*. Slip occurs when the load on the belt is increased above a point where the friction between pulley and belt are sufficient to drive the load. Slippage occurs on either or both pulleys and may, if continued, cause the belt to be thrown off.

Creep is a physical characteristic of the belt itself, inherent in power transmission when flat belting is employed. When a flat belt is transmitting power, the pulling part of the belt is under a much greater tension than the slack part and is therefore slightly stretched. As a result of this stretching, a slightly elongated belt is delivered to the driving pulley, which, when it reaches the slack side of the pulley, returns to normal length. The belt is therefore creeping ahead on the pulley. This creeping ahead on the driving pulley causes the slack side and the driven pulley to run at slightly slower surface speed. Since creeping is caused by the elasticity of the belt, it increases with the load; however, in a well-designed drive it is usually in the area of one percent at full load.

Flat belts are made endless by means of a belt joint. There are several methods by which this is done: Cementing, Lacing, and Patented Hooks.

Cemented joints are made by carefully tapering the ends of the belt and joining them together with cement and pressure. This is a specialized operation, requiring special tools, clamps, cements, etc., and is probably so seldom required of most mechanics that it is best left to a specialist to perform.

Laced joints are used very frequently when leather belting is being employed. When properly done, this type of joint is very strong and can be rapidly made. It has no metallic parts, as with metal hook lacing, and therefore does not present a hand injury hazard. Although there are no specific rules in making a laced joint for various size belts, two methods are shown in Fig. 44 and Fig. 45. A general practice is to punch two rows of holes in each end of the belt, the first row one inch from the end of the belt, and the second row two inches from the end. The number of holes across the belt should equal the number of inches in the width, with the holes on the side ⅜ inch from the edge.

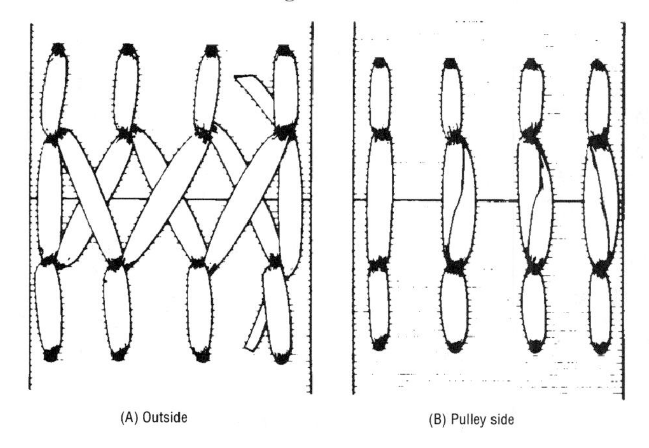

Fig. 44. Typical 4-inch belt lacing.

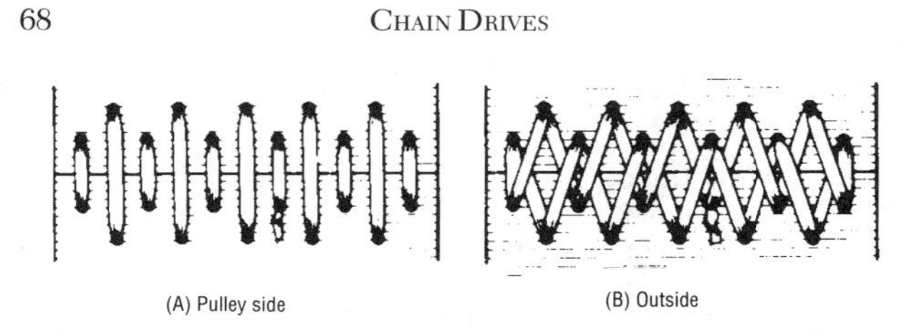

Fig. 45. Lacing on a 11-inch rubber conveyor belt.

The lacing is started from one edge of the belt, the ends of the lacing being passed from the outside to the inside. Each end is then passed through the hole in the opposite end of the belt, making two thicknesses of lacing on the inner side, as shown in Fig. 44. Each lacing end is then passed through the hole in the back row and again through the same hole as before in the first row, thus filling the four holes at the side and leaving the lacing straight with the length of the belt.

Both ends of the lacing now being on the outside of the belt, they must be put through the second hole in the first row on the opposite end, thus causing the lacing to cross on the back or outside of the belt. The ends are now put through these same holes again, opposite end through opposite hole, making two strands of lacing between this set of holes on the inside. The ends again being on the back of the belt are now put through the second row of holes and again through the corresponding holes in the first row, again bringing the ends to the back side. The lacings are again crossed and the same procedure is followed, until all four rows of holes have been filled in the same manner.

To draw the lacing through the holes and pull it tight, it will be necessary to use pliers. After the pull through the inside hole of the last row to the back side, the ends of the lacing can be cut off as shown in Fig. 44.

CHAIN DRIVES

A chain drive consists of an endless chain whose links mesh with toothed wheels called sprockets. Chain drives maintain a positive ratio between the driving and driven shafts, as they transmit power without slip or creep. The *roller chain* is the most widely used of the various styles of power transmission chain.

Roller Chain

Roller chain is composed of an alternating series of *roller links* and *pin links*. The roller links consist of two pin-link plates, two bushings and two rollers. The rollers turn freely on the bushings which are press-fitted into the link plates. The pin links consist of two link plates into which two pins are press-fitted.

In operation the pins move freely inside the bushings while the rollers turn on the outside of the bushings. The relationship of the pins, bushings, rollers, and link plates is illustrated in Fig. 46.

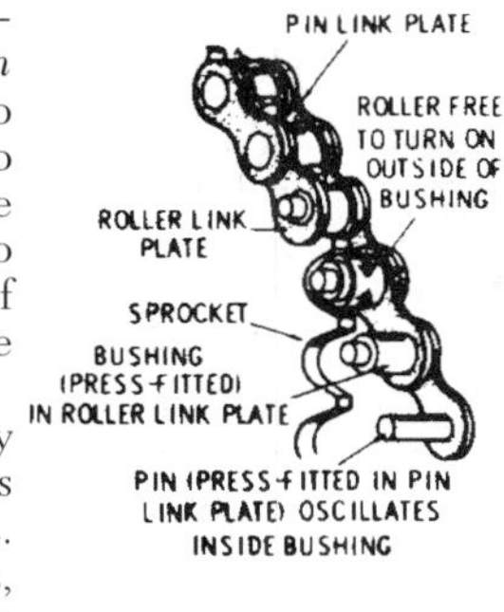

Fig. 46.

Roller Chain Dimensions

The principal roller chain dimensions are *pitch*, *chain width*, and *roller diameter*. These dimensions are standardized, and although there are slight differences between manufacturers' products, because of this standardization chains and sprockets of different manufacturers are interchangeable.

Pitch is the distance from a point on one link to a corresponding point on an adjacent link. *Chain width* is the minimum distance between link plates of a roller link. *Roller diameter* is the outside diameter of a roller, and is approximately ⅝ of the pitch.

Standard series chains range in pitch from ¼ inch to 3 inches. There is also a heavy series chain ranging from ¾-inch to 3-inch pitch. The heavy series dimensions are the same as the standard, except that the heavy series has thicker link plates.

Standard Roller Chain Numbers

The standard roller chain numbering system provides complete identification of a chain by number. The right-hand digit in the chain number is 0 for chain of the usual proportions, 1 for lightweight chain, and 5 for a rollerless bushing chain. The number to the left of the right-hand figure denotes the number of ⅛ inches in the pitch. The letter H following the chain number denotes the heavy series. The hyphenated 2 suffixed to the chain number denotes a double strand chain, 3 a triple strand, etc.

For example, the number 60 indicates a chain with six ⅛th's, or ¾-inch, pitch. The number 41 indicates a narrow lightweight ½-inch pitch chain, the number 25 indicates a ¼-inch pitch rollerless

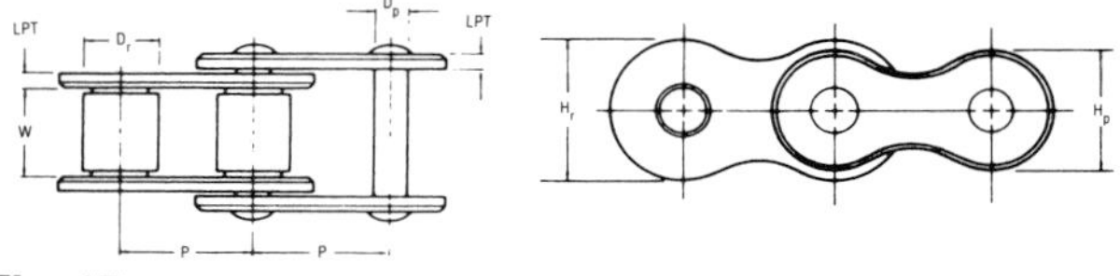

Fig. 47.

Table 7. American Standard Roller Chain Dimensions

USA STANDARD CHAIN NO. STD.	HEAVY	PITCH (P)	MAX ROLLER DIAMETER (D_R)	(WIDTH) (W)	PIN DIAMETER (D_P)	THICKNESS LINK PLATE (LPT) STD.	HEAVY
25 *	—	¼	0.130*	⅛	0.0905	0.030	—
35 *	—	⅜	0.200*	3/16	0.141	0.050	—
41 †	—	½	0.306	¼	0.141	0.050	—
40	—	½	5/16	5/16	0.156	0.060	—
50	—	⅝	0.400	⅜	0.200	0.080	—
60	60H	¾	15/32	½	0.234	0.094	.125
80	80H	1	⅝	⅝	0.312	0.125	.156
100	100H	1 ¼	¾	3/3	0.375	0.156	.187
120	120H	1 ½	⅞	1	0.437	0.187	.219
140	140H	1 ¾	1	1	0.500	0.219	.250
160	160H	2	1 ⅛	1 ¼	0.562	0.250	.281
180	180H	2 ¼	1 13/32	1 13/32	0.687	0.281	.312
200	200H	2 ½	1 9/16	1 ½	0.781	0.312	.375
240	240H	3	1 ⅞	1 ⅞	0.937	0.375	.500

*Without rollers.

†Lightweight Chain.

chain, and the number 120 a chain has twelve xth's, or 1½ inches, pitch. In multiple strand chains 50-2 designates two strands of 50 chain, 50-3 triple strand, etc. General chain dimensions for standard roller chain from q-inch pitch to 3-inch pitch are tabulated in Table 7.

Roller Chain Connections

A length of roller chain before it is made endless will normally be made up of an even number of pitches. At either end will be an unconnected roller link with an open bushing. A special type of pin link called a *connecting link* is used to connect the two ends. The partially assembled connecting link consists of two pins press-fitted and riveted in one link plate. The pin holes in the free link plate are sized for either a slip fit or a light press fit on the exposed pins. The plate is secured in place either by a cotter pin as shown in Fig. 48A, or by a spring clip as shown in Fig. 48B.

Fig. 48.

If an odd number of pitches is required, an *offset link* may be substituted for an end roller link. A more stable method of providing an odd number of pitches is by use of the *offset section*. A standard connecting link is used with the offset section (Fig. 49).

Roller Chain Sprockets

Roller chain sprockets are made to standard dimensions, tolerances, and tooth form. The standard includes four types (Fig. 50): type A has a plain sprocket without hubs; type B has a hub on one side; type C has a hub on both sides; type D has a detachable hub.

Fig. 49.

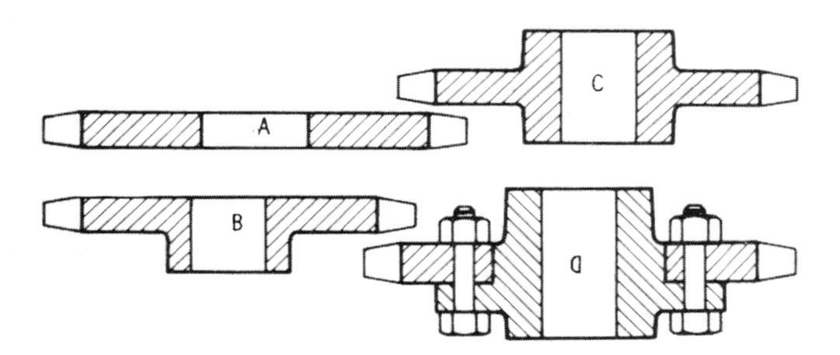

Fig. 50.

Roller Chain Installation

Correct installation of a roller chain drive requires that the shafts and the sprockets be accurately aligned. Shafts must be set level, or if inclined from a level position, both shafts must be at exactly the same angle. The shafts must also be positioned parallel within very close limits. The sprockets must be in true axial alignment for correct sprocket tooth and chain alignment.

Horizontal shafts may be aligned with the aid of a spirit level. The bubble in the level will tell when they are both in exact horizontal position. Shafts may be adjusted for parallel alignment as shown in Fig. 51. Any suitable measuring device such as calipers, feeler bars, etc., may be used. The distance between shafts on both sides of the sprockets should be equal. For an adjustable shaft drive, make the distance less than final operating distance for easier chain installation. For drives with fixed shafts, the center distance must be set at the exact dimension specified.

To set axial alignment of the sprockets, apply a straightedge to the machined side surfaces as shown in Fig. 52. Tighten the set screws in the hubs to hold the sprockets and keys in position. If one of the sprockets is subject to end float, locate the sprocket so that it will be aligned when the shaft is in its normal running position. If the center distance is too great for the available straightedge, a taut piano wire may be used.

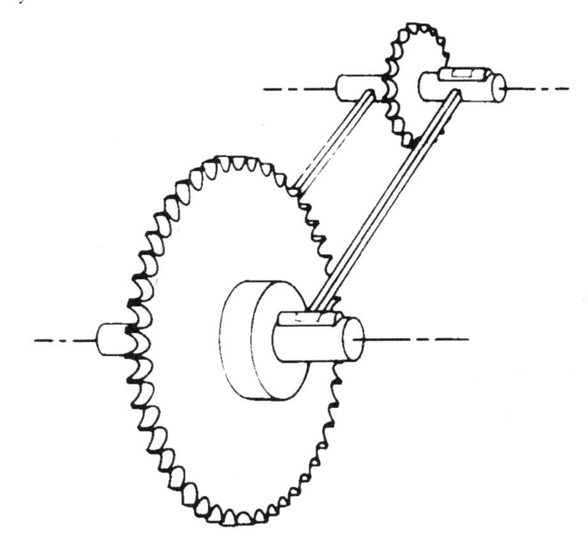

Fig. 51.

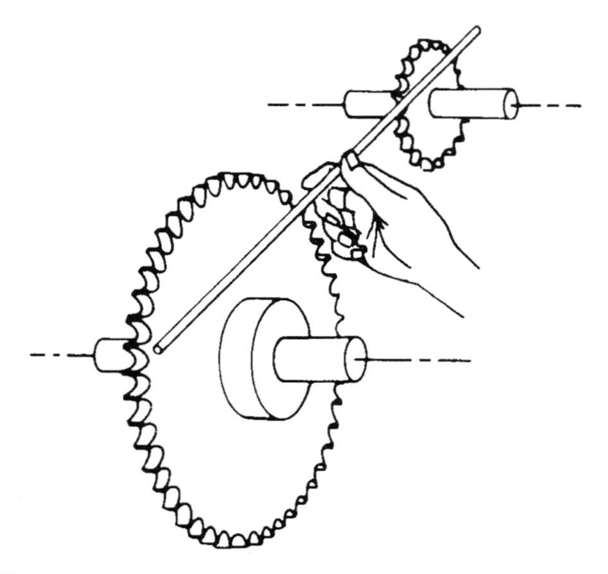

Fig. 52.

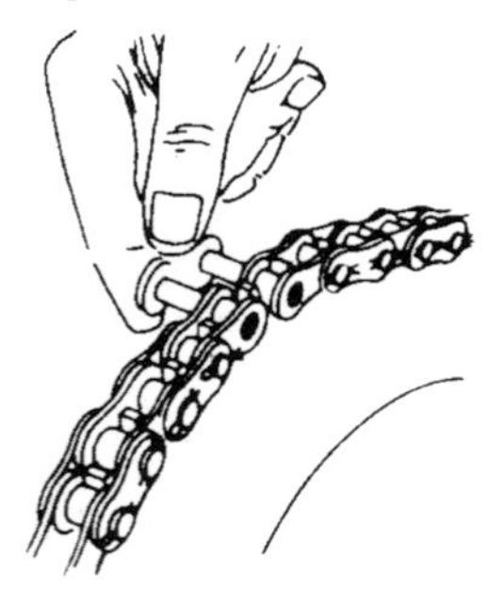

Fig. 53.

To install roller chain, fit it on both sprockets, bringing the free chain ends together on one sprocket. Insert the pins of the connecting link in the two end links of the chain as shown in Fig. 53; then install the free plate of the connecting link. Fasten the plate with the cotters or spring clip depending on type used. When fastened, tap back the ends of the connecting link pins so the outside of the free plate comes snugly against the fastener. This will prevent the connecting link from squeezing the sprocket teeth, which might interfere with free flexing of the joint and proper lubrication.

Adjustable drives must be positioned to provide proper chain tension. Horizontal and inclined drives should have an initial sag equal to 2% of the shaft centers. Measurements are made as shown in Fig. 54. The table shows measurements for various center distances to obtain approximately the recommended 2% sag.

Shaft Centers	Sag
18"	3/8"
24"	1/2"
30"	5/8"
36"	3/4"
42"	7/8"
48"	1 "
54"	1 1/8"
60"	1 1/4"
70"	1 1/2"
80"	1 5/8"
90"	1 7/8"
100"	2 "
125"	2 1/2"

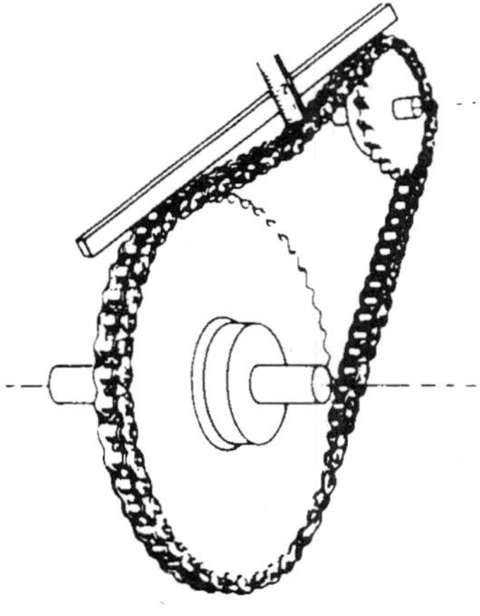

Fig. 54.

To measure the amount of sag, pull the bottom side of the chain taut so that all of the excess chain will be in the top span. Pull the top side of the chain down at its center and measure the sag as shown in Fig. 54, then adjust the centers until the proper amount is obtained. Make sure the shafts are rigidly supported and securely anchored to prevent deflection or movement which would destroy alignment.

Silent Chain

Silent chain, also called *inverted-tooth* chain, is constructed of leaf links having inverted teeth so designed that they engage cut tooth wheels in a manner similar to the way a rack engages a gear. The chain links are alternately assembled, either with pins or a combination of joint components.

Silent chain and sprockets are manufactured to a standard that is intended primarily to provide for interchangeability between chains and sprockets of different manufacturers. It does not provide for a standardization of joint components and link plate contours, which differ in each manufacturer's design. However, all manufacturers' links are contoured to engage the standard sprocket tooth, so joint centers lie on pitch diameter of the sprocket. The general proportions and designations of a typical silent chain link are shown in Fig. 55.

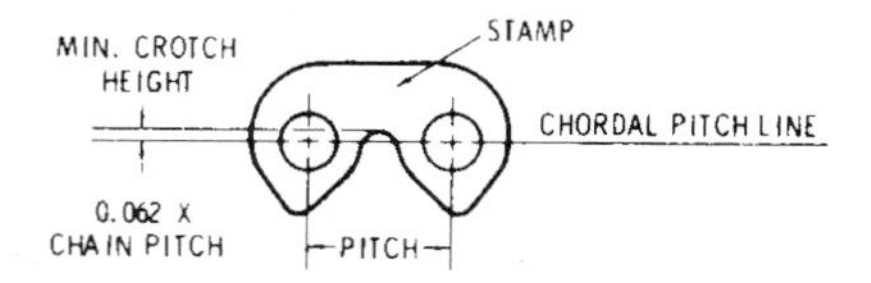

Fig. 55.

Silent chain is manufactured in a wide range of pitches and widths in various styles. Chain under ¾-inch pitch has outside guide links which engage the sides of the sprocket. The most widely used style is the *middle-guide* design with one or more rows of guide links that fit guide grooves in the sprockets. Some manufacturers use the term "wheel" rather than sprocket.

Silent chains are designated by a combined letter and number symbol as follows:

1. A two-letter symbol: SC.
2. One or two numerical digits indicating the pitch in eights of inches (usually stamped on each chain link).
3. Two or three numerical digits indicating the chain width in quarter inches.

For example, the number "SC302" designates a silent chain of ⅜-inch pitch and ½-inch width. Or the number "SC1012" designates a silent chain of 1¼-inch pitch and 3 inches width.

Chain Replacement

During operation chain pins and bushings slide against each other as the chain engages, wraps, and disengages from the sprockets. Even when parts are well lubricated, some metal-to-metal contact does occur, and these parts eventually wear. This progressive joint wear elongates chain pitch, causing the chain to lengthen and ride higher on the sprocket teeth. The number of teeth in the large sprocket determines the amount of joint wear that can be tolerated before the chain jumps or rides over the ends of the sprocket teeth. When this critical degree of elongation is reached, the chain must be replaced.

Chain manufacturers have established tables of maximum elongation to aid in the determination of when wear has reached a critical point and replacement should be made. By placing a certain number of pitches under tension, elongation can be measured. When elongation reaches the limits recommended in the table, the chain should be replaced.

The recommended measuring procedure is to remove the chain and suspend it vertically with a weight attached to the bottom. When the chain must be measured while on sprockets, remove all slack and apply sufficient tension to keep the chain section that is being measured taut.

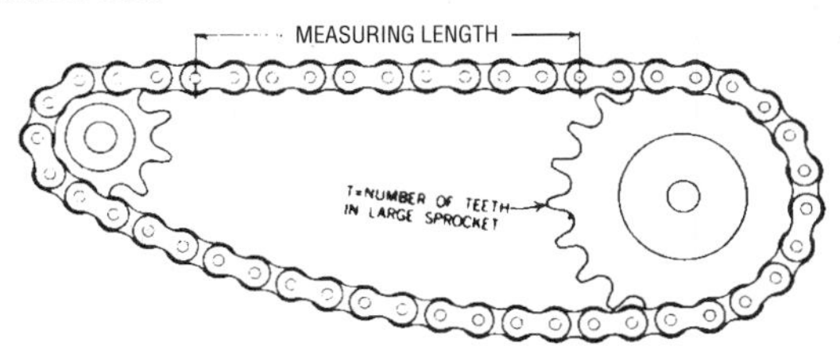

Fig. 56.

Table 8. Chain Elongation Limits

Chain Number	Pitch Inches	Measuring Length: Number of pitches	Measuring Length: Nominal length inches	Length of chain when replacement is required*: Number of teeth in largest sprocket (T)							
				Up to 67	68-73	74-81	82-90	91-103	104-118	119-140	141-173
35	3/8	32	12	12 3/8	12 11/32	12 5/16	12 9/32	12 1/4	12 7/32	12 3/16	12 5/32
40	1/2	24	12	12 3/8	12 11/32	12 5/16	12 9/32	12 1/4	12 7/32	12 3/16	12 5/32
50	5/8	20	12 1/2	12 7/8	12 19/32	12 13/16	12 25/32	12 3/4	12 25/32	12 11/16	12 21/32
60	3/4	16	12	12 3/8	12 11/32	12 5/16	12 9/32	12 1/4	12 7/32	12 3/16	12 5/32
80	1	24	24	24 3/4	24 11/16	24 5/8	24 9/16	24 1/2	24 7/16	24 3/8	24 5/16
100	1 1/4	20	25	25 3/4	25 11/16	25 5/8	25 9/16	25 1/2	25 7/16	25 3/8	25 5/16
120	1 1/2	16	24	24 3/4	24 11/16	24 5/8	24 9/16	24 1/2	24 7/16	24 3/8	24 5/16
140	1 3/4	14	24 1/2	25 1/4	25 3/16	25 1/8	25 1/16	25	24 15/16	24 7/8	24 13/16
160	2	12	24	24 3/4	24 11/16	24 5/8	24 9/16	24 1/2	24 7/16	24 3/8	24 5/16
180	2 1/4	11	24 3/4	25 1/2	25 7/16	25 3/8	25 5/16	25 1/4	25 3/16	25 1/8	25 1/16
200	2 1/2	10	25	25 3/4	25 11/16	25 3/8	25 9/16	25 1/2	25 7/16	25 3/8	25 5/16
240	3	8	24	24 3/4	24 11/16	24 5/8	24 9/16	24 1/2	24 7/16	24 3/8	24 5/16

* Valid for drives with adjustable centers or drives employing adjustable idler sprockets.

SPUR GEARS

The spur gear might be called the basic gear, as all other types have been developed from it. Its teeth are straight and parallel to the bore center line. Spur gears may run together with other spur gears on parallel shafts; with internal gears on parallel shafts; and with a rack. A rack is in effect a straight line gear. The smaller of a pair of gears is often called a *pinion.* On large heavy-duty drives the larger of a pair of gears is often called a *bullgear.*

The involute profile or form is the one commonly used for gear teeth. It is a curve that is traced by a point on the end of a taut line unwinding from a circle. The larger the circle the straighter the curvature, and for a rack, which is essentially an infinitely large gear, the form is straight or flat.

The involute system of gearing is based on a rack having straight or flat-sided teeth. The sides of each tooth incline toward the center top at an angle called the *pressure angle.* The 14½-degree pressure angle was standard for many years; however, the use of the 20-degree pressure angle has been growing, until today 14½-degree gearing is generally limited to replacement work. The advantages of 20-degree gearing are greater strength and wear resistance, and in

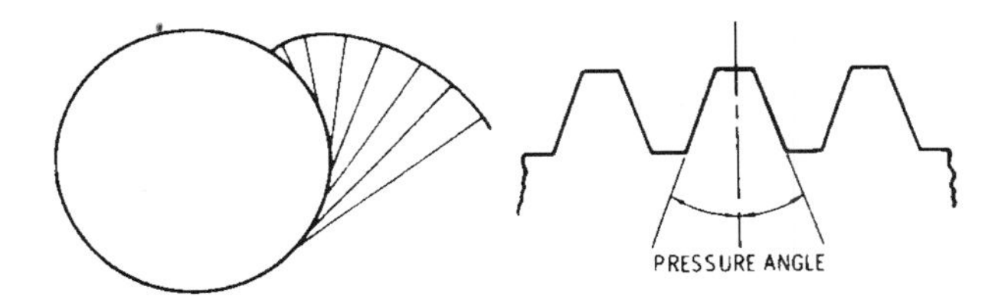

Fig. 57.

addition it permits the use of pinions with a few less teeth.

It is extremely important that the pressure angle be known when gears are mated, as all gears that run together *must have the same pressure angle.*

Many types and designs of gears have been developed from the spur gear. While they are commonly used in industry, many are complex in design and manufacture. Some of the types in wide use are: bevel gears, helical gears, herringbone gears, and worm gears. Each type in turn has many specialized design variations.

Pitch Diameters and Center Distance

Pitch circles are the imaginary circles that are in contact when two standard gears are in correct mesh. The diameters of these circles are the pitch diameters of the gears. The center distance of two gears in correct mesh is equal to one-half the sum of the two pitch diameters.

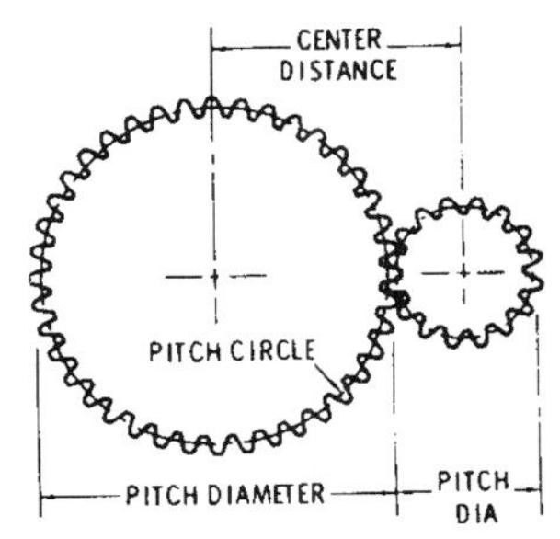

Fig. 58.

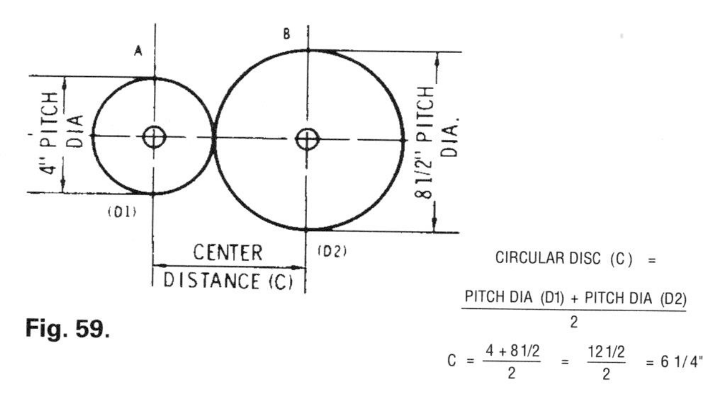

Fig. 59.

Circular Pitch

The size and proportions of gear teeth are designated by a specific type of pitch. In gearing terms there are specific types of pitch. They are *circular pitch* and *diametral pitch.* Circular pitch is simply the distance from a point on one tooth to a corresponding point on the next tooth, measured along the pitch line as illustrated in Fig. 60. Large gears are frequently made to circular pitch dimensions.

Diametral Pitch

The *diametral-pitch* system is the most widely used gearing system, with practically all common size gears being made to diametral-pitch dimensions. Diametral-pitch numbers designate the

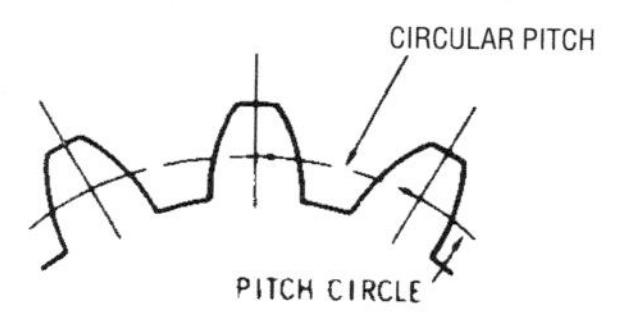

Fig. 60.

size and proportions of gear teeth by specifying the number of teeth per inch of pitch diameter. For instance, a 12 diametral-pitch number indicates there are 12 teeth in the gear for each inch of pitch diameter. Stated another way, diametral-pitch numbers specify the number of teeth in 3.1416 inches along the gear's pitch line.

Fig. 61 shows a gear with 4 inches of pitch diameter and its 3.1416 inches of pitch-circle circumference for each 1 inch of pitch diameter. In addition it illustrates that specifying the number of teeth for 1 inch of pitch diameter is in fact specifying the number of teeth in 3.1416 inches along the pitch line. The reason for this is that for each 1 inch of pitch diameter there are pi (π) inches or 3.1416 inches of pitch-circle circumference.

The fact that the diametral-pitch number specifies the number of teeth in 3.1416 inches along the pitch line may be more easily visualized when applied to the rack. As shown in Fig. 62, the pitch line of a rack is a straight line and a measurement may be easily made along it.

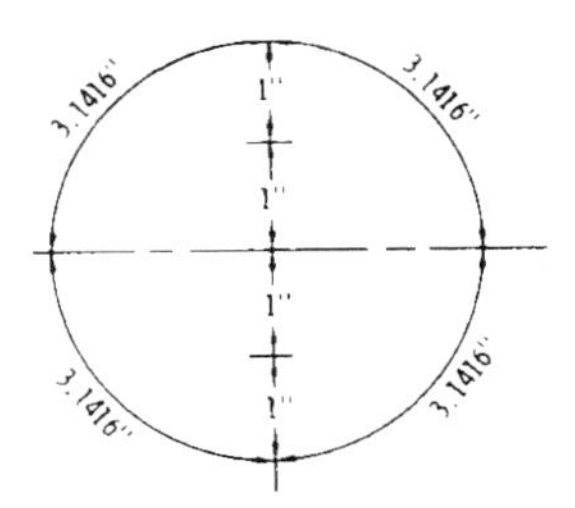

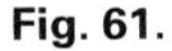
Fig. 61.

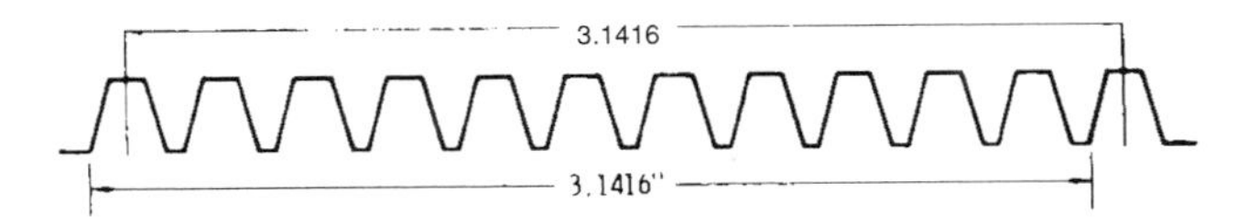

Fig. 62.

Diametral Pitch Approximation

In cases where it is necessary to determine the diametral pitch of a gear, this may be done easily without the use of precision measuring tools, templates, or gauges. Measurements need not be exact because diametral-pitch numbers are usually whole numbers. Therefore, if an approximate calculation results in a value close to a whole number, that whole number is the diametral-pitch number of the gear. There are three easy methods of determining the approximate diametral pitch of a gear. A common steel rule, preferably flexible, is adequate to make the required measurements.

Method #1—Count the number of teeth in the gear, add two (2) to this number and divide by the outside diameter of the gear. For example, the gear shown in Fig. 63 has 40 teeth and its outside diameter is about 4 7/32". Adding 2 to 40 gives 42, dividing 42 by 4 7/32" gives an answer of 9 31/32". As this is approximately 10, it can be safely stated that the gear is a 10 diametral-pitch gear.

Method #2—Count the number of teeth in the gear and divide this number by the measured pitch diameter. The pitch diameter of the gear is measured from the root or bottom of a tooth space to the top of a tooth on the opposite side of the gear.

Using the same 40-tooth gear shown in Fig. 63, the pitch measured from the bottom of a tooth space to the top of the opposite tooth is about 4". Dividing 40 by 4 gives an answer of 10. In this case

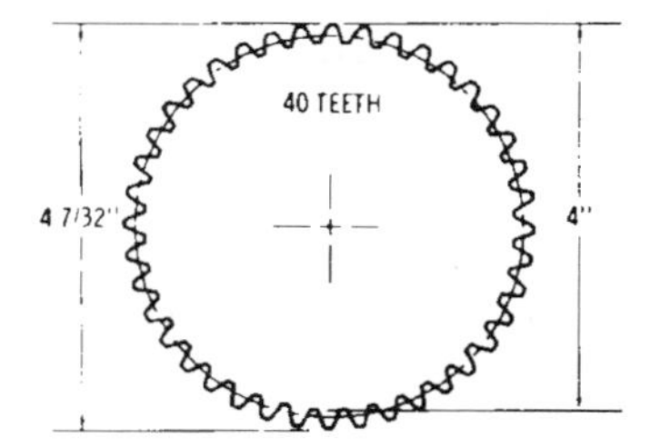

Fig. 63.

the approximate whole-number pitch-diameter measurement results in a whole-number answer. This method also indicates that the gear is a 10 diametral-pitch gear.

Method #3—Using a flexible scale, measure approximately 3⅛" along the pitch line of the gear (Fig. 64). To do this, bend the scale to match the curvature of the gear and hold it about midway between the base and the top of the teeth. This will place the scale approximately on the pitch line of the gear. If the gear can be rotated, draw a pencil line on the gear to indicate the pitch line; this will aid in positioning the scale. Count the teeth in 3⅛" to determine the diametral pitch number of the gear.

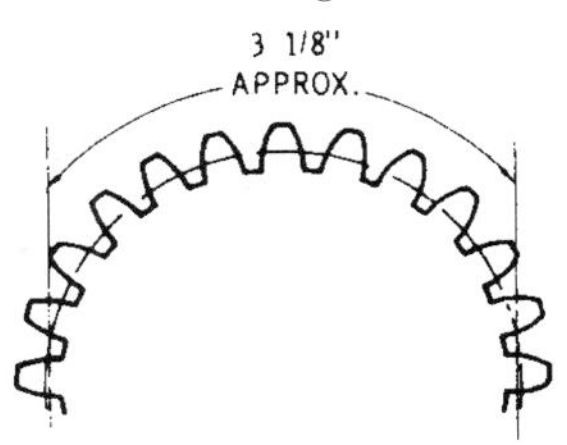

Fig. 64.

Gear Tooth Parts (Fig. 65)

Addendum — The distance a tooth projects above, or outside of, the pitch line or circle.

Dedendum — The depth of a tooth space below, or inside of, the pitch line or circle.

Clearance — The amount by which the dedendum of a gear tooth exceeds the addendum of a mating gear tooth.

Whole Depth—The total height of a tooth or the total depth of a tooth space.

Working Depth — The depth of tooth engagement of two mating gears. It is the sum of their addendums.

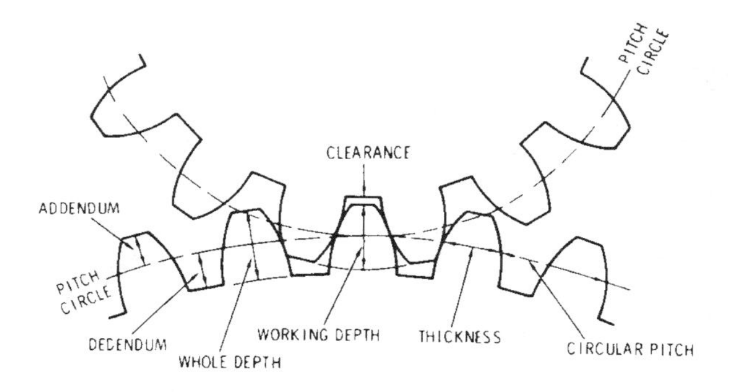

Fig. 65.

Tooth Thickness — The distance along the pitch line or circle from one side of a gear tooth to the other.

Gear Dimensions

The *outside diameter* of a spur gear is the pitch diameter plus two addendums.

The *bottom*, or *root diameter*, of a spur gear is the outside diameter minus two whole depths.

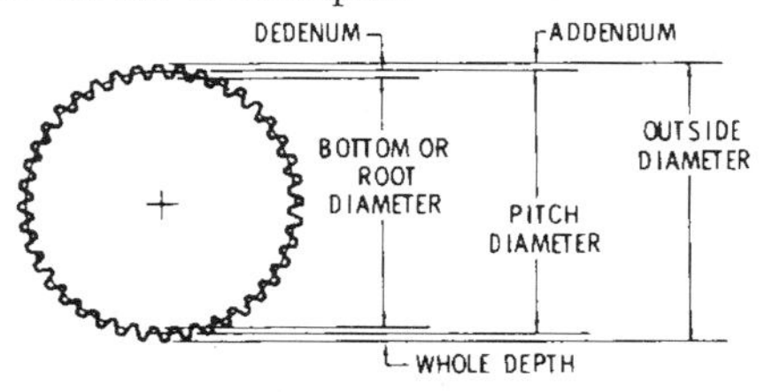

Fig. 66.

$$\text{Number of Teeth} = \text{N}$$

$$\text{Diametral Pitch} = P = \frac{N}{P}$$

$$\text{Pitch Diameter} = D = \frac{N}{P}$$

$$\text{Addendum} = A = \frac{1}{P}$$

$$\text{Whole Depth} = WD = \frac{2.2}{P} + .002''$$

$$\text{Outside Diameter} = \text{OD} = \text{D} + 2\text{A}$$

$$\text{Root Diameter} = \text{RD} = \text{OD} - 2\text{WD}$$

Backlash

Backlash in gears is the play between teeth that prevents binding. In terms of tooth dimensions, it is the amount by which the width of tooth spaces exceeds the thickness of the mating gear teeth. Backlash may also be described as the distance, measured along the pitch line, that a gear will move when engaged with another gear that is fixed or unmovable.

Normally there must be some backlash present in gear drives to provide running clearance. This is necessary as binding of mating gears can result in heat generation, noise, abnormal wear, possible overload, and/or failure of the drive. A small amount of backlash is also desirable because of the dimensional variations involved in manufacturing tolerances.

Backlash is built into standard gears during manufacture by cutting the gear teeth thinner than normal by an amount equal to one-half the required figure. When two gears made in this manner are run together, at standard center distance, their allowances combine to provide the full amount of backlash required.

On nonreversing drives, or drives with continuous load in one direction, the increase in backlash that results from tooth wear does not adversely affect operation. However, on reversing drives and drives where timing is critical, excessive backlash usually cannot be tolerated.

Table 9 lists the suggested backlash for a pair of gears operating at standard center distance.

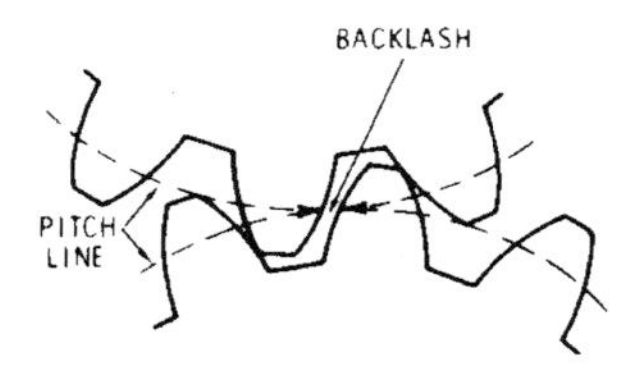

Fig. 67.

Table 9.

Pitch	Backlash
3 P	.013
4 P	.010
5 P	.008
6 P	.007
7 P	.006
8 - 9 P	.005
10 - 13 P	.004
14 - 32 P	.003
33 - 64 P	.0025

COUPLINGS

Power transmission couplings are the usual means of connecting coaxial shafts so that one can drive the other. For example, they are used to connect an electric motor to a pump shaft, or to the input shaft of a gear reducer, or to connect two pieces of shafting together to obtain a long length, as with line shafting. Power transmission couplings for such shaft connections are manufactured in a great variety of types, styles, and sizes. They may, however, be divided into two general groups or classifications, i.e., *rigid* couplings, also called *solid* couplings, and a second group called *flexible* couplings.

Solid Couplings

Rigid or *solid* couplings, as the names indicate, connect shaft ends together rigidly, making the shafts so connected into a single continuous unit. They provide a fixed union that is equivalent to a shaft extension. They should only be used when *true* alignment and a solid or rigid coupling are required, as with line shafting, or where provision must be made to allow parting of a rigid shaft. A feature of rigid couplings is that they are self-supporting and automatically align the shafts to which they are attached when the coupling halves on the shaft ends are connected. There are two basic rules that should be followed to obtain satisfactory service from rigid couplings.

First — A force fit must be used in assembly of the coupling halves to the shaft ends.

Second — After assembly, a runout check of all surfaces of the coupling must be made and any surface found to be running out must be machined true.

Checking of surfaces is especially necessary if the coupling halves are assembled by driving or bumping, rather than pressing. Rigid couplings should *NOT* be used to connect shafts of independent machine units that must be aligned at assembly. Fig. 68A shows a typical rigid-type coupling. Fig. 68B shows the mating surfaces that hold the coupling in rigid alignment when assembled.

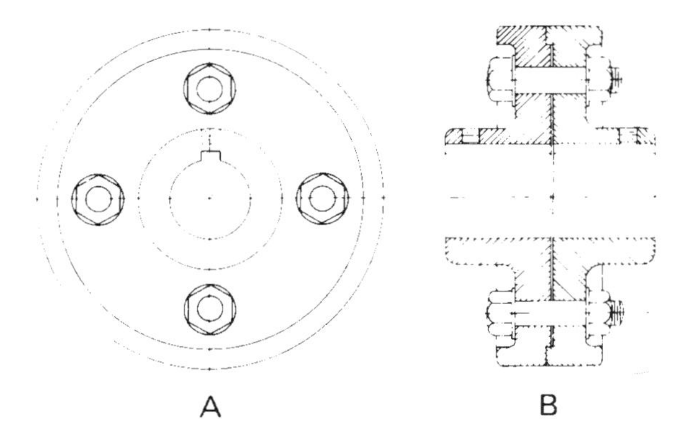

Fig. 68. Typical rigid coupling.

Flexible Couplings

The transmission of mechanical power often requires the connection of two independently supported coaxial shafts so one can drive the other. Prime movers, such as internal combustion engines and electric motors, connected to reducers, pumps, variable speed drives, etc., are typical examples. For these applications a flexible coupling is used because perfect alignment of independently supported coaxial shaft ends is practically impossible. In addition to the impracticality of perfect alignment, there is always wear and damage to the connected components and their shaft bearings, as well as the possibility of movement from temperature changes and external forces.

In addition to enabling coaxial shafts to operate satisfactorily with slight misalignment, flexible couplings allow some axial movement, or end float, and may, depending on the type, allow torsional movement as well. Another benefit which may result when flexible couplings have nonmetallic connecting elements is electrical insulation of connected shafts. In summary, there are four (4) conditions which may exist when coaxial shaft ends are connected. Flexible couplings are designed to compensate for some or all of these conditions:

1. Angular Misalignment
2. Parallel Misalignment
3. Axial Movement (End Float)
4. Torsional Movement

Illustrated in Fig. 69 are the four types of flexibility that may be provided by flexible couplings.

Flexible couplings are intended to compensate only for the slight unavoidable minor misalignment that is inherent in the design of machine components, and the practices followed when aligning coaxial shafts of connected units. If the application is one where misalignment must exist, universal joints, some style of flexible shafting, or special couplings designed for offset operation may be necessary. Flexible couplings should not be used in attempting to compensate for deliberate misalignment of connected shaft ends.

Most flexible couplings are made up of three basic parts, i.e. two hubs that attach to the shaft ends to be connected, and a flexing member or element that transmits power from one hub to the other. There are a variety of flexible coupling designs, all having characteristics and features to meet specific needs. Most flexible couplings fall into one or more of the following groupings.

Chain and Gear Couplings

Angular and parallel misalignment are allowed by couplings in this group, as well as end float and limited torsional flexibility. Lubrication is usually required. A chain coupling is essentially two sprockets with hubs for attachment to the shafts. The two sprockets are connected by assembling around them a length of double chain having a number of links that corresponds to the number of sprocket teeth. Gear couplings are made up of meshing internal and external gears or splines. Single-engagement types use one of each. Double-engagement types, which are the types most widely used, use two.

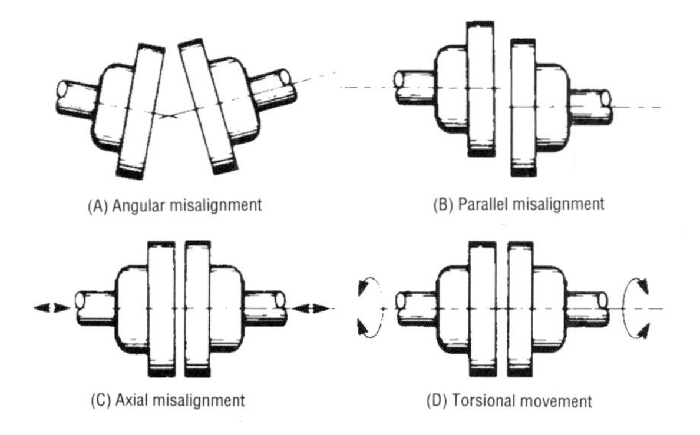

Fig. 69. Four types of flexibility.

Flexibility in these results from both the fit of the teeth and from special shaping of the teeth that permits them to pivot relatively. In general, these couplings permit more angular deflection, parallel misalignment, and end float than chain types. With few exceptions, both chain and gear couplings must be lubricated, and are designed with a complete housing, seals, and provisions for relubrication.

Jaw and Slider Couplings

Couplings in this category may be described as having tightly fitted parts that are constructed to allow sliding action. In general, these couplings have two hubs which attach to the shaft ends and are constructed with surfaces designed to receive a sliding member. This sliding element, which provides the coupling's flexibility, is referred to as the *slider.* The slider is driven by jaws, keys, or openings, and it in turn drives the other hub by the same method. Usually slider couplings have to be lubricated.

Resilient Element Couplings

This style of coupling is probably made by more manufacturers and in more design variations than the two other styles combined. The flexible elements used in these couplings may be metal, rubber, leather, or plastic. Most widely used are metal and rubber. Couplings in this category will allow angular and parallel misalignment and end float. Torsional flexibility covers a wide range. Some have virtually none, while certain of those using rubber flexible elements have more torsional flexibility than other types of couplings. Most of the couplings in this grouping do not need lubrication, excepting those using flexible metallic elements.

Coupling Alignment

The designation *coupling alignment* is the accepted term to describe the operation of bringing coaxial shaft ends into alignment. The most common situation is one where a coupling is used to connect and transmit mechanical power from shaft end to shaft end. The term implies that the prime function in performing this operation is to align the *surfaces* of the coupling. However, the principal objective is actually to bring the *center lines of the coaxial shafts* into alignment. While this may seem to be drawing a fine distinction, it is important that the difference be understood. This point plays an important part in understanding the procedure that must be followed to correctly align a coupling.

To illustrate this important point, consider the coupling halves shown in Fig. 70. It may be stated that when the coupling faces are aligned, the center lines of the shafts are also aligned. There is, however, this very important proviso: The proviso: The *surfaces* of the coupling halves must run *true* with the center lines of the shafts.

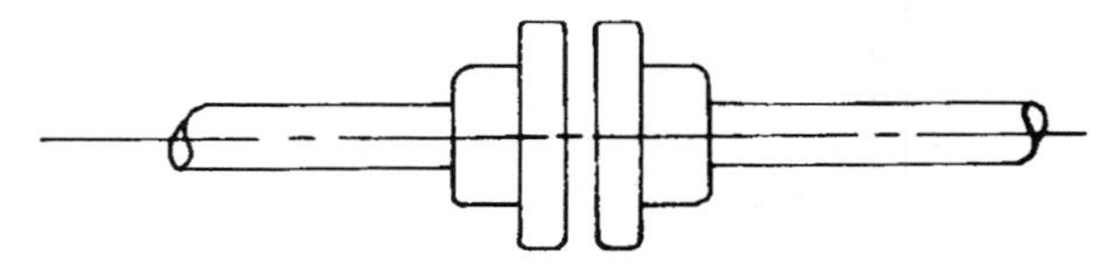

Fig. 70. Alignment of center lines.

If the surfaces do not run true with the shaft center lines, alignment of the untrue surfaces will result in *misalignment* of the shaft center lines. True running surfaces therefore are a basic requirement if the alignment procedures that follow are to be successful.

As the operation is one to accomplish alignment, it may also be described as an operation to correct or eliminate misalignment. We must then define and locate misalignment if we are to eliminate it. In respect to the shafts that are to be coupled, they may be misaligned in two ways. They may be at an angle, rather than in a straight line (Angular Misalignment), or they may be offset (Parallel Misalignment). These two kinds of misalignment are illustrated in Fig. 71.

Fig. 71.

The misalignment of the shaft center lines may be located in any plane within the full circle of 360 degrees. Angular misalignment may be a tilt up, down, or to the sides of one shaft in respect to the other. Parallel misalignment may be one shaft high, low, or to the side of the other. A practical method of correcting this misalignment is to align the center lines in two planes at right angles. The practice commonly followed, because it is convenient, is to check and adjust in the vertical and horizontal planes. The vertical and horizontal planes are illustrated in Fig. 72.

To align center lines of coaxial shafts in two planes, vertically and horizontally, in respect to both angularity and parallelism, requires *four* (4) separate operations: an angular and parallel alignment in the vertical plane, and the same in the horizontal plane. Too frequently these operations are performed in random order, and adjustments are made by trial and error. This results in a time-consuming series of operations, some of the adjustments often disturbing settings made during prior adjustments.

An organized procedure can eliminate the need to repeat operations. When a definite order is established, one operation is completed before another is started. In many cases, when the correct order is followed it is possible to satisfactorily align a coupling by going once through the four operations. When extreme accuracy is required the operations may be repeated, but always in the correct order. As adjustments are made by insertion of shims at low support points, an important preparatory step is to check the footings of the units. If there is any rocking motion, the open point must be eliminated by shimming, so all support points rest solidly on the base plate.

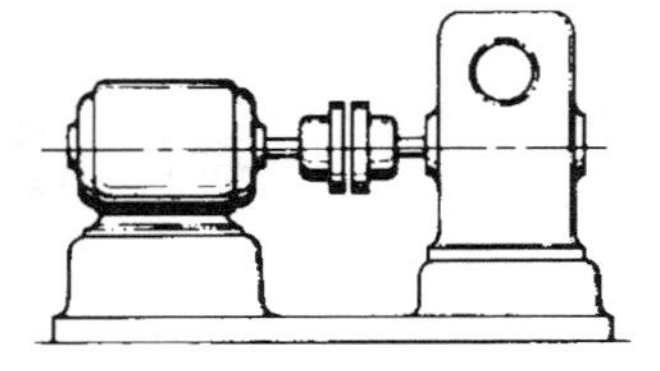

VERTICAL ALIGNMENT

MEASUREMENTS TAKEN IN A VERTICAL PLANE, AT TOP AND BOTTOM OF COUPLING

HORIZONTAL ALIGNMENT

MEASUREMENTS TAKEN IN A HORIZONTAL PLANE: PLANE AT SIDES OF COUPLING

Fig. 72.

The standard *thickness* gauge, also called a *feeler* gauge, is a compact assembly of high quality heat-treated steel leaves of various thickness. This is the measuring instrument commonly used to determine the precise dimension of small openings or gaps such as those that must be measured in the course of aligning a coupling. To determine the dimension of an opening or gap, the leaves are inserted, singly or in combination, until a leaf or combination is found that fits snugly. Then the dimension is ascertained by the figure marked on the leaf surface, or if several are used, by totaling the surface figures.

Another precision tool, not as widely known or used, but ideally suited for coupling alignment, is the *taper gauge* shown in Fig. 73.

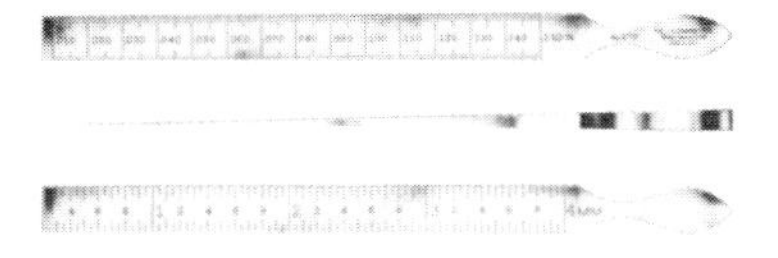

L. S. Starrett Co.

Fig. 73. Taper gauge.

The principal advantage of the *taper gauge*, sometimes called *gap gauge*, for coupling alignment is that it is direct reading, not requiring trial-and-error "feeling" to determine a measurement. The tip end is inserted into an opening or gap and the opening size is read on the graduated face. The tool shown is graduated in one thousands of an inch on one side and millimeters on the other.

Two methods of coupling alignment are widely used: *the straightedge-feeler gauge method* and *the indicator method*. In both methods, four (4) alignment operations are performed in a specific order. Only when performed in the correct order can adjustments be made at each step without disturbing prior settings. The four steps in the order of performance are:

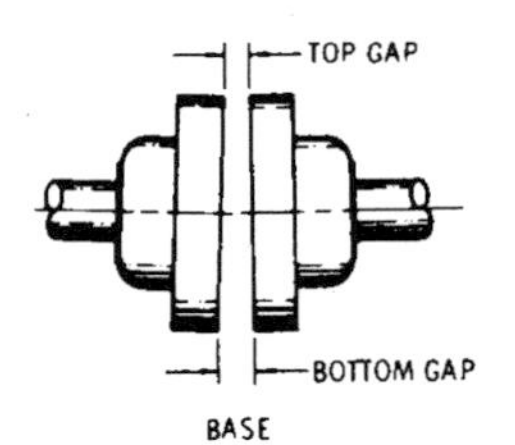

Step #1. Vertical Face Alignment — The first adjustment made to correct angular misalignment in the vertical plane. This is accomplished by tipping the unit as required. The gap at the top and bottom of the coupling is measured, and adjustment is made to bring these faces true.

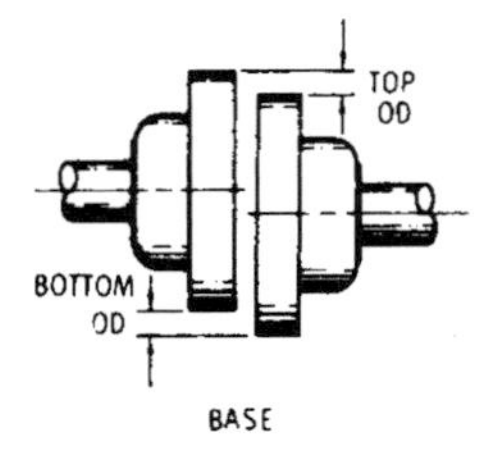

Step #2. Vertical Height Alignment — This adjustment corrects parallel misalignment in the vertical plane. The unit is raised without changing its angular position. Height difference from base to center line is determined by measuring on the OD of the coupling at top and/or bottom.

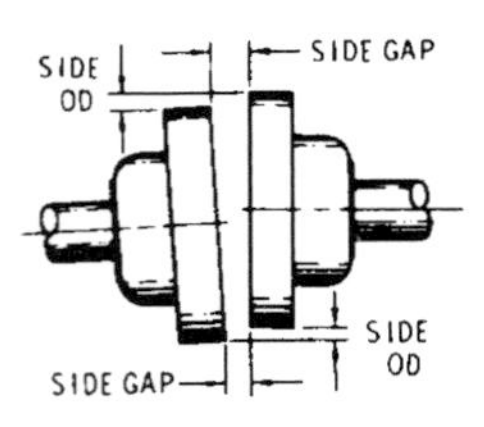

Step #3 and #4. Horizontal Face and OD Alignment — When the units are in alignment vertically, shimming is complete. The horizontal alignment operations may then be done simultaneously. The unit is moved as required to align the faces and OD's at the sides of the coupling.

Vertical Face Alignment Correction

The initial alignment step (vertical face alignment) can be a time-consuming operation when done in an unorganized manner. Correction of angular misalignment requires tilting one of the units into correct position using shims. Selection of shim thickness is commonly done by trial and error. Much time can be saved (and accuracy gained) by using simple proportion to determine the shim thickness.

The tilt required, or the angle of change at the base, is the same as the angle of misalignment at the coupling faces. Because of this angular relationship, the shim thickness is proportional to the misalignment. For example, in Fig. 74 the misalignment at the coupling faces is 0.006" in 5 inches; therefore each 5 inches of base must be tilted 0.006" to correct misalignment. As the base length is twice 5 inches, the shim thickness must be twice 0.006", or 0.012".

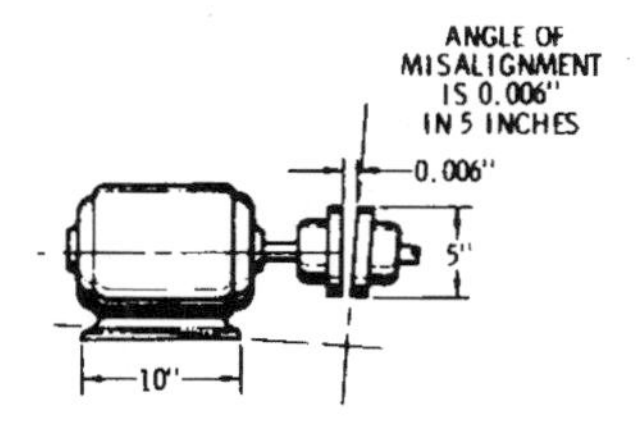

Fig. 74.

A simple rule for shim-thickness selection that gives accurate results and enables correction is to be made in one setting is: *Shim thickness is as many times greater than the misalignment as the base length is greater than the coupling diameter.*

Selection of Unit To Be Adjusted

In preparing to align a coupling it must be determined which unit is to be adjusted — the "driver" or the "driven." Common practice, and the one generally recommended, is to position, level, and secure the driven unit at the required elevation. Then, adjust the driver to align with it. Connections to the driven unit, such as pipe connections to a pump or output shaft connections to a reducer, should be completed prior to proceeding with coupling alignment. The driven unit should be set with its shaft center line slightly higher than the driver to allow for alignment shims.

The Straightedge-Feeler Gauge Method

Practically all flexible couplings on drives operating at average speeds will perform satisfactorily when misaligned as much as .005". Some will tolerate much greater misalignment. Alignment well within .005" is easily and quickly attainable using straightedge and feeler gauge when correct methods are followed.

Step #1. Vertical Face Alignment — Using the feeler gauge, measure the width of gap at top and bottom between the coupling faces. Using the difference between the two measurements, determine the shim thickness required to correct alignment. (It will be as many times greater than the misalignment as the driver base length is greater than the coupling diameter). Shim under low end of driver to tilt into alignment with the driven unit.

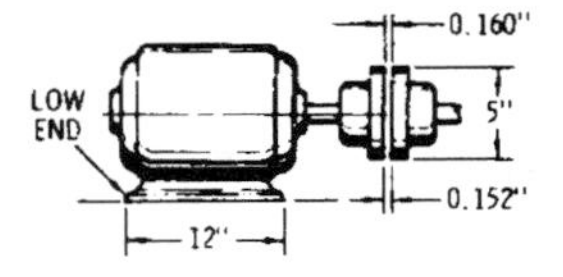

Fig. 75.

Example: Assume that measured misalignment is .160" minus .152" equaling .008" misalignment in 5 inches. Base length is about 2½ times coupling diameter; therefore, shim required is 2½ times .008", or .020". A shim .020" thick placed under the low end of the driver will tilt it into approximately angular alignment with the driven unit. The slight error involved is well within tolerance of flexible coupling alignment.

Step #2. Vertical Height Alignment — Using a straightedge and feeler gauge, measure the height difference between driver and driven units on the OD surfaces of the coupling. Place shims at all driver support points equal in thickness to the measured height difference.

Step #3 and #4. Horizontal Face and OD Alignment — Using a straightedge, check alignment of OD's at sides of coupling. Using a feeler gauge, check the gap between coupling faces at the sides of the coupling. Adjust driver as necessary to align the OD's and to set the gap equal at the sides. Do not disturb shims.

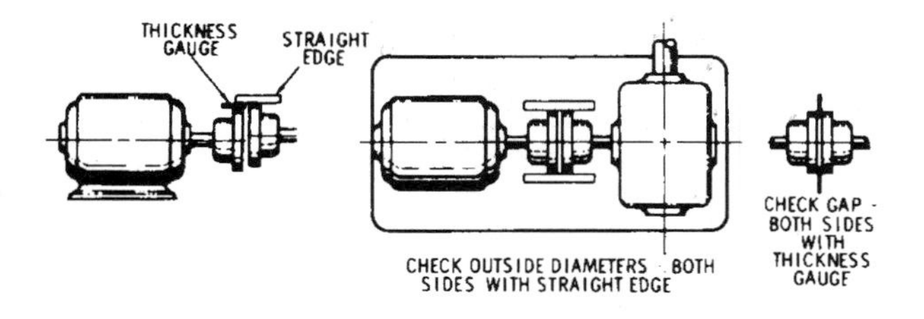

Fig. 76.

The Indicator Method (Fig. 77)

When an indicator is used as a measuring instrument to align coaxial shafts, it must be attached to the shaft or coupling of one of the units so it can be rotated. A stationary indicator will only show runout, when in contact with an object being rotated. When the indicator is rotated its tip will describe a circle which is concentric with the shaft bearings. The mating unit shaft should also be rotated, as the point contacted by the indicator tip will also describe a true circle. When these two circles coincide (zero runout on the indicator) the shafts will be in alignment in the plane being measured.

While the ideal practice is to rotate both units while indicating, this is sometimes not possible. In such cases the indicator is attached to the unit that can be rotated and the mating unit remains stationary. When this is done the surfaces of the coupling half remaining stationary must be true. Any surface errors or runout will result in a corresponding error in shaft alignment.

If both units are rotated while measuring, coupling surface runout does not affect accuracy of alignment. This is an advantage of the indicator alignment method, as it is possible to accurately align shafts in spite of coupling surfaces that are not running true.

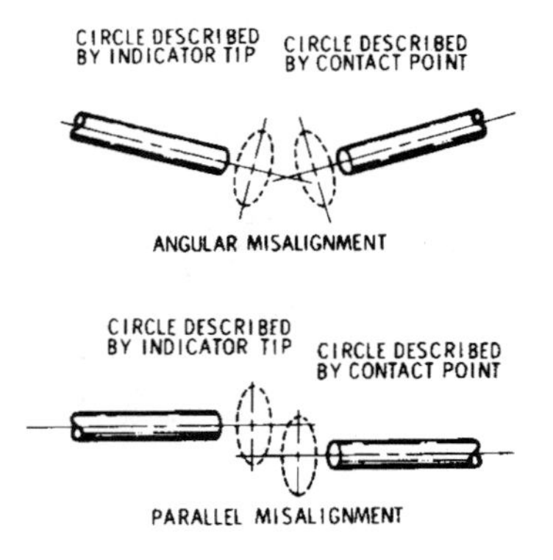

Fig. 77.

Step #1. Vertical Face Alignment — Attach indicator to shaft or coupling half of driver. Place indicator tip in contact with the face of the coupling half on the driven unit. Rotate shafts of both units together. Note indicator readings at top and bottom. Total indicator runout is a measure of the vertical angular misalignment. Place shims under driver at low end, tipping it into alignment with driven unit. The shim thickness will be as many times greater than the coupling face misalignment as the driver base length is greater than the coupling diameter.

Example: Assume total indicator runout is .007" (minus .005" to plus .002") as shown in Fig. 78. The driver base length is approximately 2½ times the coupling diameter (12" to 5"). The shim thickness should be 2½ times .007", or .017".

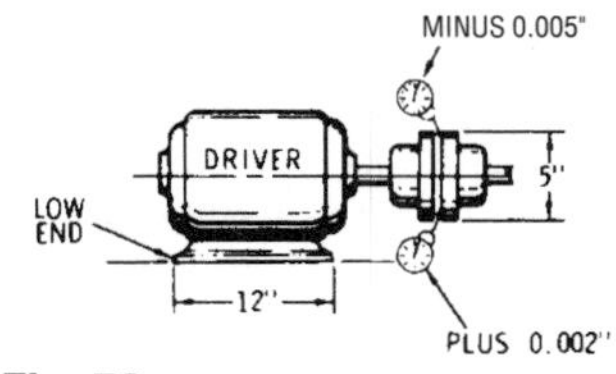

Fig. 78.

Step #2. Vertical Height Alignment — Place indicator tip in contact with outside surface of driven unit coupling half. Rotate shafts of both units together. Note indicator reading at top and bottom. Height difference is one-half total indicator runout. Place shims at all driver support points equal in thickness to one-half total indicator runout.

Example: Assume a total indicator runout of .018" (plus .012" to minus .006") as shown in Fig. 79. The required shim thickness is .009", one-half the .018" total indicator runout.

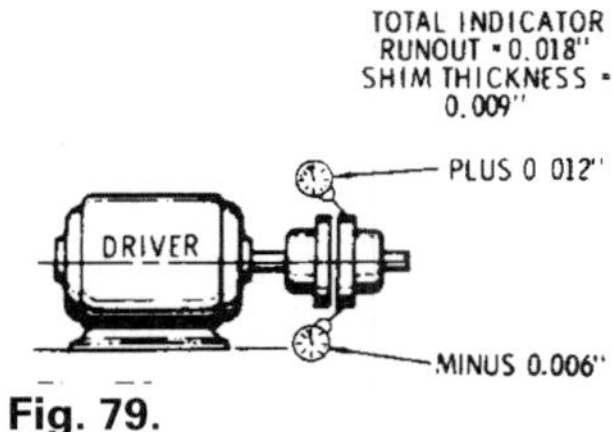

Fig. 79.

Steps #3 and #4. Horizontal Face and OD Alignment — Place indicator tip in contact with face of driven unit coupling half. Move driver as necessary to obtain zero reading on indicator. Place indicator in contact with OD surface of driven unit coupling half. Move driver as necessary to obtain zero reading on indicator. Repeat operations as necessary to obtain zero readings on both face and OD surfaces at sides of coupling. Do not disturb shims during horizontal alignment adjustments.

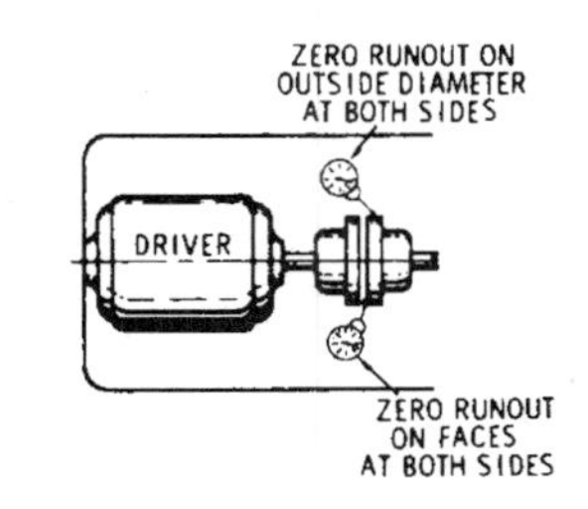

Fig. 80.

Temperature Change Compensation

To compensate for temperature difference between installation conditions and operating conditions, it may be necessary to set one unit high, or low, when aligning. For example, centrifugal pumps handling cold water, and directly connected to electric motors, require a low motor setting to compensate for expansion of the motor housing as its temperature rises. If the same units were handling liquids hotter than the motor operating temperature, it might be necessary to set the motor high. Manufacturers' recommendations should be followed for initial setting when compensation for temperature change is made at cold installation.

Final alignment of equipment with appreciable operating temperature difference should be made after it has been run under actual operating conditions long enough to bring both units to operating temperatures.

SCREW THREADS

The Unified Screw Thread Standards superseded the American Standards in 1948, when an accord was signed by the standardizing bodies of Canada, the United Kingdom, and the United States. The Unified Standards apply to the form, designation, dimensions, etc. of triangular threads in the inch system. It is substantially the same form as the American thread and is mechanically interchangeable. The design form and proportions of the Unified thread are illustrated in Fig. 81.

Unified Standards are established for various thread series, thread series being groups of diameter-pitch combinations distinguished by the number of threads per inch applied to a specific diameter.

Coarse-Thread Series — *UNC — Unified National Coarse*

Designated by the symbol *UNC*, it is generally used for bolts, screws, nuts, and other general classifications.

Fine-Thread Series — *UNF — Unified National Fine*

Designated by the symbol *UNF*, it is suitable for bolts, screws, nuts, etc. where a finer thread than that provided by the coarse series is required.

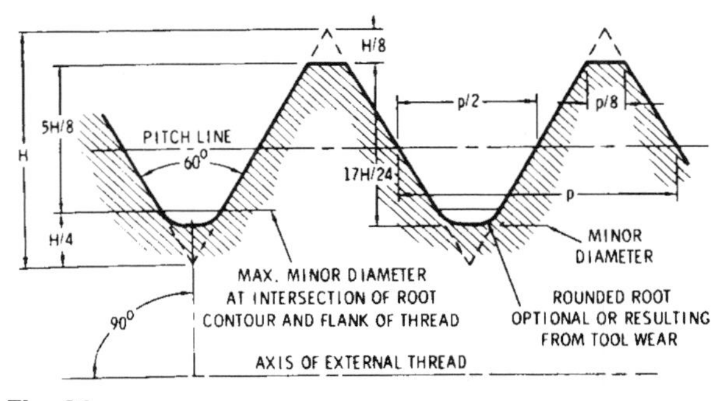

Fig. 81.

Extra-Fine Thread Series — *UNEF — Unified National Ex-Fine*

Designated by the symbol *UNEF*, it is used for short lengths of engagement, thin-walled tubes, ferrules, couplings, etc., where very fine pitches of threads are required.

Constant-Pitch Series — *UN — Unified National Form*

Designated by the symbol *UN*, various pitches are used on a variety of diameters. Preference is given whenever possible to the use of the 8-, 12-, and 16-thread series.

Thread Classes

The Unified Standards establish limits of tolerance called "classes." Classes 1A, 2A, and 3A are for external threads, and classes 1B, 2B, and 3B are for internal threads. Classes 1A and 1B provide the maximum fit allowances, classes 2A and 2B provide optimum fit allowances and classes 3A and 3B provide minimum allowances.

Classes 2A and 2B are the most commonly used standards for general applications and production items such as bolts, screws, nuts, etc. Classes 3A and 3B are used when close tolerances are desired. Classes 1A and 1B are used where a liberal allowance is required to permit ready assembly, even with slightly bruised or dirty threads.

Unified Thread Designation

The standard method of designation is to specify in sequence the nominal size, number of threads per inch, thread series symbol, and thread class symbol. For example, a ¾-inch Unified coarse series thread for a common fastener would be designated as follows:

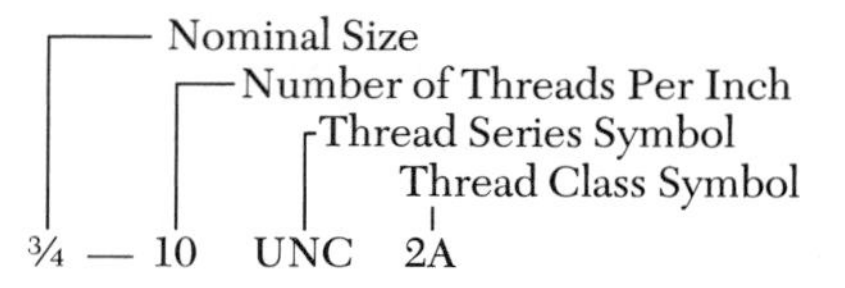

Unless otherwise specified, threads are righthand. A left-hand thread is designated by adding the letters LH after the thread class symbol.

Screw Thread Terms

Fig. 82 illustrates the meaning of the more important screw thread terms. While "lead" is not included in the illustration, it also is an important screw thread term.

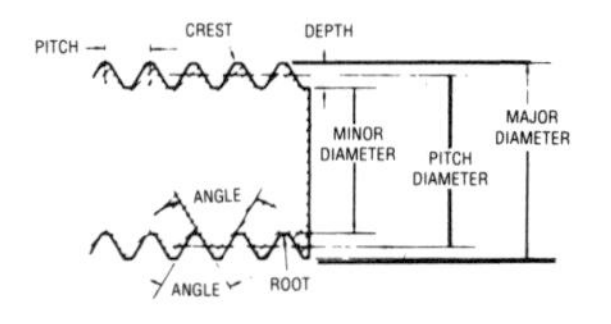

Fig. 82.

Lead — The distance a screw advances axially in one turn. On a single-thread screw, the lead and pitch are the same. On a double-thread screw, the lead is twice the pitch. On a triple-thread screw, the lead is three times the pitch, etc.

Double Depth of Screw Threads

The *double depth* of a screw thread is equal to twice the thread depth (2 x depth). Its principal use is to determine the hole size required to produce an internal thread of a given pitch (threads per inch). Table 10 lists the double depth for various pitch Unified Form threads. Double depth may also be stated as the difference between the major and minor thread diameters. Therefore, minor thread diameter may be determined by subtracting the double depth from the major thread diameter. For example, hole size for a 2.000" x 20 threads per inch internal thread may be determined by subtracting the double depth for 20 threads per inch from 2.000 (2.000 – .064 = 1.935). Thread depth as well as major and minor diameters are illustrated in Fig. 82.

A hole larger than the minor diameter is required to provide mating clearance. For precise machined threads this may be a very small amount. For tapped threads a greater amount is allowed. Practice for tapped threads is to provide 25% clearance, resulting in a 75% full thread. Fig. 88 illustrates the relationship between full and 75% full threads.

Table 10. Double Depth of Unified Form Threads

Threads per Inch	Double Depth	Threads per Inch	Double Depth
2	.6495	22	.0590
3	.433	24	.0541
4	.3247	26	.0500
5	.2598	28	.0464
6	.2165	30	.0433
7	.1856	32	.0406
8	.1624	36	.0361
9	.1443	40	.0325
10	.1230	44	.0295
12	.1082	48	.0271
14	.0928	56	.0232
16	.0812	64	.0203
18	.0722	72	.0180
20	.0649	80	.0162

Translation Threads

Screw threads used to move machine parts for adjustment, setting, transmission of power, etc., are classified as "translation threads." To perform these functions a stronger form than the triangular "V" is often required. The *square, acme,* and *buttress* thread forms have the required strength and are widely used for translation thread applications.

Square Thread

The square thread is a strong and efficient thread, but it is difficult to manufacture. The theoretical proportions of an external square thread are shown in Fig. 83. The mating nut must have a slightly larger thread space than the screw to allow a sliding fit. Similar clearance must also be provided on the major and minor diameters.

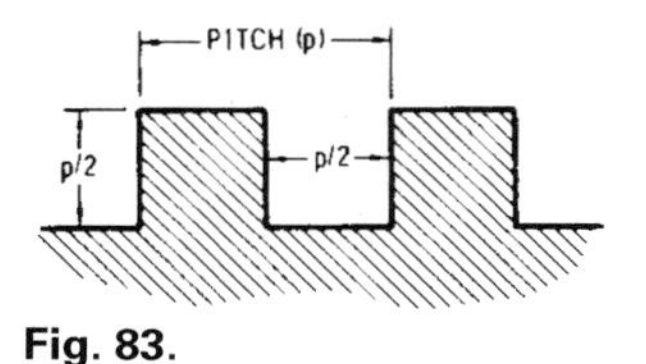

Fig. 83.

Acme Thread

The acme thread, while not quite as strong as the square thread, is preferred because it is fairly easy to machine. The angle of an acme thread, measured in an axial plane, is 29 degrees. The basic proportions of an acme thread are shown in Fig. 84. Standards for acme screw threads establish thread series, fit classes, allowances and tolerances, etc., similar to the standards for the unified thread form.

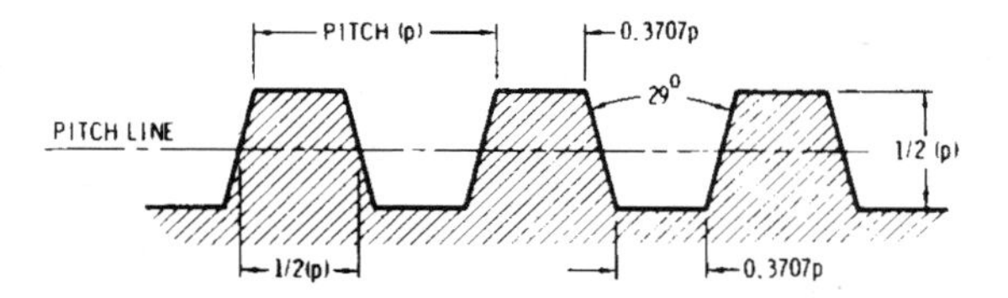

Fig. 84.

Buttress Thread

The buttress thread has one side cut approximately square and the other side slanting. It is used when a thread having great strength along the thread axis in one direction only is required.

Because one side is cut nearly perpendicular to the thread axis, there is practically no radial thrust when the thread is tightened. This feature makes the thread form particularly applicable where relatively thin tubular members are screwed together. The basic thread form of a simple design of buttress thread is shown in Fig. 85. Other buttress thread forms are complex, with the load side of the thread inclined from the perpendicular to facilitate machining. The angle of inclination of the American Standard buttress thread form is 7 degrees.

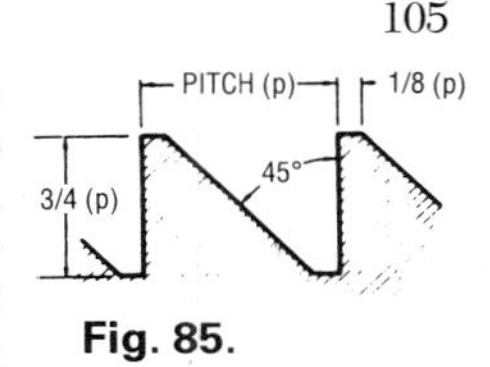

Fig. 85.

Metric Screw Threads

Much interest and a great deal of effort has been devoted to the subject of metric measurement in recent years. Some have urged rapid conversion to worldwide use of the metric system, while others have advocated a more cautious and gradual approach. In the area of mechanical fasteners using screw threads, there has been a dramatic increase in the use of metric system threads. This can be attributed to the import and export business, most particularly to machine tools and automobile imports. As the import volume of tools, automobiles, etc. with metric threads increased, there was a resultant increased usage of metric-related tools, machinery, and equipment of all types. Consequently, there developed a need for metric thread information and domestic standardization. Efforts to establish standards resulted in adoption by the American National Standards Institute (ANSI) of a 60-degree symmetrical screw thread with a basic International Organization for Standardization (ISO) profile. This is termed the ISO 68 profile. In the ANSI standard this 60-degree thread profile is designated as the M-Profile. The ANSI M-Profile Standard is in basic agreement with ISO screw thread standards, and features detailed information for diameter-pitch combinations selected as preferred standard sizes. The basic M-Profile is illustrated in Fig. 86.

The designations, terms, etc. used in the new metric standard differ in many respects from those used in the familiar Unified Standard. Because the Unified Standard incorporates terms and

Table 11. Unified Standard Screw Thread Series

Sizes		BASIC MAJOR DIAMETER	THREADS PER INCH							
			Series with graded pitches			Series with constant pitches				
Prim.	Sec.		Course UNC	Fine UNF	Extra fine UNEF	4UN	6UN	8UN	12UN	16UN
0		0.0600	—	80	—	—	—	—	—	—
	1	0.0730	64	72	—	—	—	—	—	—
2		0.0860	56	64	—	—	—	—	—	—
	3	0.0990	18	56	—	—	—	—	—	—
4		0.1120	40	48	—	—	—	—	—	—
5		0.1250	40	44	—	—	—	—	—	—
6		0.1380	32	40	—	—	—	—	—	—
8		0.1640	32	36	—	—	—	—	—	—
10		0.1900	24	32	—	—	—	—	—	—
	12	0.2160	24	28	32	—	—	—	—	—
1/4		0.2500	20	28	32	—	—	—	—	—
5/16		0.3125	18	24	32	—	—	—	—	—
3/8		0.3750	16	24	32	—	—	—	—	UNC
7/16		0.4375	14	20	28	—	—	—	—	16
1/2		0.5000	13	20	28	—	—	—	—	16
9/16		0.5625	12	18	24	—	—	—	UNC	16
5/8		0.6250	11	18	24	—	—	—	12	16
	11/16	0.6875	—	—	24	—	—	—	12	16
3/4		0.7500	10	16	20	—	—	—	12	UNF
	13/16	0.8125	—	—	20	—	—	—	12	16
7/8		0.8750	9	14	20	—	—	—	12	16
	15/16	0.9375	—	—	20	—	—	—	12	16
1		1.0000	8	12	20	—	—	UNC	UNF	16
	1 1/16	1.0625	—	—	18	—	—	8	12	16
1 1/8		1.1250	7	12	18	—	—	8	UNF	16
	1 3/16	1.1875	—	—	18	—	—	8	12	16
1 1/4		1.2500	7	12	18	—	—	8	UNF	16
	1 5/16	1.3125	—	—	18	—	—	8	12	16
1 3/8		1.3750	6	12	18	—	UNC	8	UNF	16
	1 7/16	1.4375	—	—	18	—	6	8	12	16
1 1/2		1.5000	6	12	18	—	UNC	8	UNF	16
	1 9/16	1.5625	—	—	18	—	6	8	12	16
1 5/8		1.6250	—	—	18	—	6	8	12	16
	1 11/16	1.6875	—	—	18	—	6	8	12	16
1 3/4		1.7500	5	—	—	—	6	8	12	16
	1 13/16	1.8125	—	—	—	—	6	8	12	16
1 7/8		1.8750	—	—	—	—	6	8	12	16
	1 15/16	1.9375	—	—	—	—	6	8	12	16
2		2.0000	4 1/2	—	—	—	6	8	12	16
	2 1/8	2.1250	—	—	—	—	6	8	12	16
2 1/4		2.2500	4 1/2	—	—	—	6	8	12	16
	2 3/8	2.3750	—	—	—	—	6	8	12	16

Table 11. Unified Standard Screw Thread Series (Cont'd)

Sizes		BASIC MAJOR DIAMETER	THREADS PER INCH							
			Series with graded pitches			Series with constant pitches				
Prim.	Sec.		Course UNC	Fine UNF	Extra fine UNEF	4UN	6UN	8UN	12UN	16UN
2 ½		2.5000	4	—	—	UNC	6	8	12	16
	2 ⅝	2.6250	—	—	—	4	6	8	12	16
2 ¾		2.7500	4	—	—	UNC	6	8	12	16
	2 ⅞	2.8750	—	—	—	4	6	8	12	16
3		3.0000	4	—	—	UNC	6	8	12	16
	3 ⅛	3.1250	—	—	—	4	6	8	12	16
3 ¼		3.2500	4	—	—	UNC	6	8	12	16
	3 ⅜	3.3750	—	—	—	4	6	8	12	16
3 ½		3.5000	4	—	—	UNC	6	8	12	16
	3 ⅝	3.6250	—	—	—	4	6	8	12	16
3 ¾		3.7500	4	—	—	UNC	6	8	12	16
	3 ⅞	3.8750	—	—	—	4	6	8	12	16
4		4.0000	4	—	—	UNC	6	8	12	16
	4 ⅛	4.1250	—	—	—	4	6	8	12	16
4 ¼		4.2500	—	—	—	4	6	8	12	16
	4 ⅜	4.3750	—	—	—	4	6	8	12	16
4 ½		4.5000	—	—	—	4	6	8	12	16
	4 ⅝	4.6250	—	—	—	4	6	8	12	16
4 ¾		4.7500	—	—	—	4	6	8	12	16
	4 ⅞	4.8750	—	—	—	4	6	8	12	16
5		5.0000	—	—	—	4	6	8	12	16
	5 ⅛	5.1250	—	—	—	4	6	8	12	16
5 ¼		5.2500	—	—	—	4	6	8	12	16
	5 ⅜	5.3750	—	—	—	4	6	8	12	16
5 ½		5.5000	—	—	—	4	6	8	12	16
	5 ⅝	5.6250	—	—	—	4	6	8	12	16
5 ¾		5.7500	—	—	—	4	6	8	12	16
	5 ⅞	5.8750	—	—	—	4	6	8	12	16
6		6.0000	—	—	—	4	6	8	12	16

practices of long standing with which mechanics have knowledge and experience, an explanation of the M-Profile Metric system may best be accomplished by a comparison of Metric M-Profile Standards to Unified Standards.

A major difference between the standards is in the designation used to indicate the thread form. The Unified standard uses a series of capital letters which not only indicate that the thread is of the Unified form, but also indicate if it is a coarse (UNC), a fine (UNF),

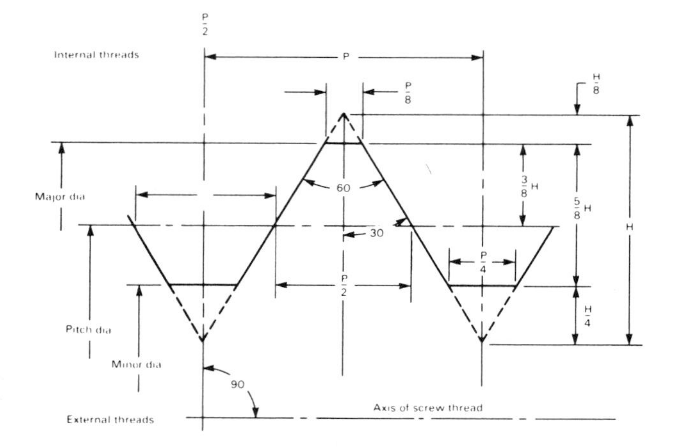

Fig. 86. Basic M thread profile (ISO basic profile).

an extra fine (UNEF), or a constant-pitch (UN), which is a series of pitches used on a variety of diameters. In the Metric M-Profile standard the capital letter M is used to indicate the thread is Metric M-Profile, with no reference to pitch classification.

A second major difference is that the Unified standard designates the thread pitch in terms of the number of threads in one inch of thread length. The Metric M-Profile standard states the specific pitch measurement, i.e., the dimension in millimeters from the center line of one thread to the center line of the next thread.

A third major difference is that the Unified system establishes limits of tolerances called classes. Classes 1A, 2A, and 3A supply to external threads only, and classes 1B, 2B, and 3B apply to internal threads only; classes 3A and 3B provide a minimum and classes 1A and 1B a maximum. In the Metric M-Profile standard the tolerance designation uses a number and a letter to indicate the pitch diameter tolerance, as well as a number and a letter to designate crest diameter tolerance. This results in four symbols for tolerance designation, rather than the two used in the Unified system. In the Metric M-Profile standard the lower-case letter g is used for external threads and the capital letter H is used for internal threads. The numbers 4, 5, 6,7, 8 are used to indicate degree of internal thread tolerance, and the numbers, 3, 4, 5, 6, 7, 8, 9 for the degree of external thread tolerance. Thus two nominally similar threads are designated as follows:

Unified	*Metric M*
⅜-16-UNC-2A	M10 × 1.5-6g6g

Common shop practice in designating Unified screw threads is to state only the nominal diameter and the threads-per-inch; for example, ⅜-16 for a coarse thread, and ⅜-24 for a fine thread. Most mechanics recognize the thread series by the threads-per-inch number. As to the thread class symbol, it is usually assumed that if no class number is stated, the thread is a class 2A or 2B. This is a safe assumption because class 2A and 2B are commonly used tolerances for general applications, including production of bolts, screw, nuts, and similar fasteners. When required to produce screw threads, most mechanics use commercial threading tools, i.e., taps and dies, or if chasing threads, commercial gauging tools for size checking. These commonly used tools are made to class 2A or 2B tolerances; therefore, the mechanic depends on tool manufacturers to maintain screw thread accuracy. Because manufacturers produce products to extremely close tolerances, this practice is highly satisfactory and relieves the mechanic of practically all concern for maintaining sizes and fits within stated tolerances.

To reach this same condition was one of the objectives during efforts to establish Metric M-Profile Standards. Because metric threads are in wide use throughout the world, a highly desirable condition would be one where the standard is compatible with those in world use. As the International Organization for Standardization (ISO) metric standard was for all practical purposes recognized as the worldwide metric standard, the Metric M-Profile Standard is in large measure patterned after it. It has a profile in basic agreement with the ISO profile, and features detailed information for diameter-pitch combinations selected as preferred standard sizes.

Metric M-Profile screw threads are identified by letter (M) for thread form, followed by the nominal diameter size and pitch expressed in millimeters, separated by the sign (X) and followed by the tolerance class separated by a dash (-) from the pitch. For example, a coarse pitch Metric M-Profile thread for a common fastener would be designated as follows:

External Thread M-Profile, Right Hand

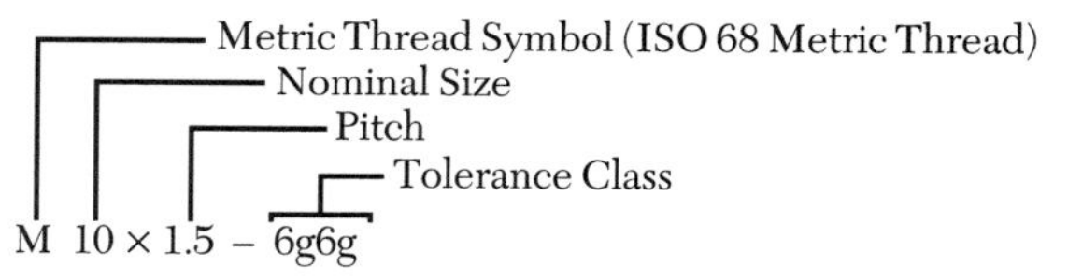

The simplified international practice for designating coarse pitch ISO screw threads is to leave off the pitch. Thus a M14 X 2 thread is designated just M14. In the ANSI standard, to prevent misunderstanding, it is mandatory to use the value for pitch in all designations. Thus a 10 mm coarse thread is designated M10 X 1.5. When no tolerance classification is stated, it is assumed to be classification 6g6g (usually stated as simply 6g), which is equivalent to the Unified classification 2A, the commonly used classification for general applications.

The standard metric screw thread series for general purpose equipment's threaded components and mechanical fasteners is a coarse thread series. The diameter-pitch combinations selected as preferred standard sizes in the coarse pitch series are listed in the following table. These combinations are in basic agreement with ISO standards.

The Metric M-Profile designation does not specify series of diameter/pitch combinations as does the Unified system, i.e., coarse, fine, etc. Although no indication of such grouping is given in the designation of a metric thread, series groupings are recommended. The coarse pitch series of diameter/pitch combinations shown in Table 11 is described as standard metric screw thread series for general purpose equipment's threaded components and mechanical fasteners.

A second series called Fine Pitch M-Profile metric screw threads, shown in Table 13, lists additional diameter/pitch combinations that are standard for general purpose equipment's threaded components. These combinations, in some instances, list more than one pitch for a nominal diameter. As with the coarse pitch series, they are in basic agreement with ISO screw standards.

Table 12. Standard Coarse Pitch M-Profile General Purpose and Mechanical Fastener Series

Nominal Size	Pitch	Nominal Size	Pitch
1.6	0.35	20	2.5
2	0.4	24	3.0
2.5	0.45	30	3.5
3	0.5	36	4.0
3.5	0.6	42	4.5
4	0.7	48	5.0
5	0.8	56	5.5
6	1.0	64	6.0
8	1.25	72	6.0
10	1.5	80	6.0
12	1.75	90	6.0
14	2.0	100	6.0
16	2.0		

Table 13. Standard Fine Pitch M-Profile Screw Threads

Nominal Size	Pitch		Nominal Size	Pitch	
8	1.0		40	1.5	
10	0.75	1.25	42	2.0	
12	1.0	1.25	45	1.5	
14	1.25	1.5	48	2.0	
15	1.0		50	1.5	
16	1.5		55	1.5	
17	1.0		56	2.0	
18	1.5		60	1.5	
20	1.0		64	2.0	
22	1.5		65	1.5	
24	2.0		70	1.5	
25	1.5		72	2.0	
27	2.0		75	1.5	
30	1.5	2.0	80	1.5	2.0
33	2.0		85	2.0	
35	1.5		90	2.0	
36	2.0		95	2.0	
39	2.0		100	2.0	

Table 14. M-Profile Metric Thread Tap Drill Sizes

Nominal Size	Pitch	Tap Drill Size	Inch Decimal
1.6	.35	1.25	.050
2.0	.4	1.6	.063
2.5	.45	2.05	.081
3.0	.5	2.5	.099
4.0	.7	3.3	.131
5.0	.8	4.2	.166
6.0	1.0	5.0	.198
8.0	1.0	7.0	.277
8.0	1.25	6.8	.267
10.0	.75	9.3	.365
10.0	1.25	8.8	.346
10.0	1.5	8.5	.336
12.0	1.0	11.0	.434
12.0	1.25	10.8	.425
12.0	1.75	10.3	.405
14.0	1.25	12.8	.503
14.0	1.5	12.5	.494
14.0	2.0	12.0	.474
16.0	1.5	14.5	.572
16.0	2.0	14.0	.553
18.0	1.5	16.5	.651
20.0	1.0	19.0	.749
20.0	1.5	18.5	.730
20.0	2.5	17.5	.692
22.0	1.5	20.5	.809
24.0	2.0	22.0	.868
24.0	3.0	21.0	.830
25.0	1.5	23.5	.927

Millimeter Drill Size for 75% Thread = Major Dia. – (.974 × Pitch)
Millimeters × .03937 = Inch Decimals

Thread Tapping

The cutting of internal screw threads with a hand tap is potentially a troublesome operation involving broken taps and time-consuming efforts to remove them. These troubles may be largely avoided by a better understanding of the tapping operating and the thread cutting tool. Fig. 87 illustrates a typical hand tap and gives the names of its various parts.

The threaded body of a tap is composed of lands, which are the cutters, and flutes or channels to let the chips out and permit cutting oil to reach the cutting edges. It is chamfered at the point to allow it to enter a hole and to spread the heavy cutting operation over several rings of lands or cutting edges.

The radial relief shown in Fig. 87 refers to material having been removed from behind the cutting edges to provide clearance and reduce friction. Another form of clearance provided on commercial taps is termed "back taper." This is done by a very slight reduction in thread diameter at the shank.

The three standard styles of hand taps, *taper*, *plug*, and *bottoming*, have varying amounts of point "chamfer." The taper taps have the longest (8 to 10 threads) and are usually used for starting a tapped hole. The plug tap has 3 to 4 threads chamfer and is used to provide full threads closer to the bottom of a hole than is possible with a taper tap. The bottoming tap has practically no chamfer, but when preceded by a plug tap and carefully used, can cut threads very close to the bottom of a hole.

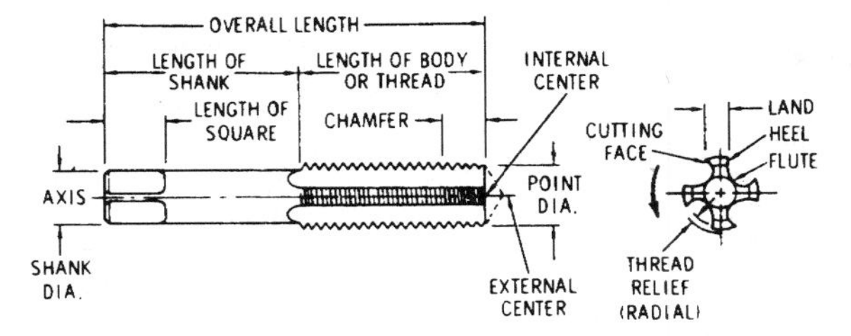

Fig. 87.

The hole for the tap must be made by a *tap drill* larger than the minor diameter of

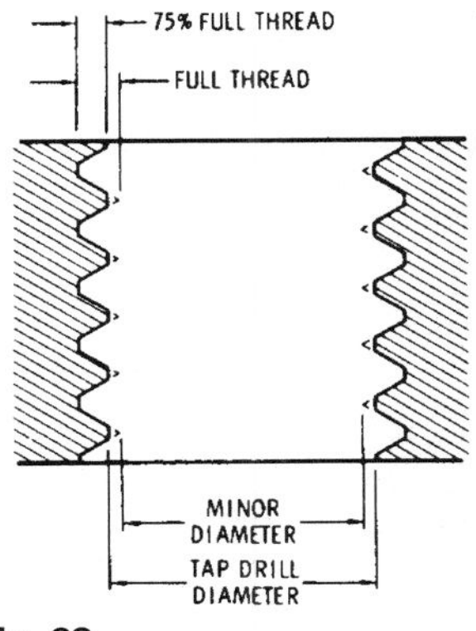

Fig. 88

the thread. The hole is made oversize to provide clearance between the wall of the hole and the roots of the tap threads. This gives chip space and allows free turning of the tap, reducing the tendency for the threads to tear.

The amount of hole oversize is such that the internal thread will be 75% of standard, as shown in Fig. 88. The 25% that is missing from the crest does not appreciably reduce its strength. Tap drill charts normally carry the notation: "Based on approximately 75% of full thread."

MECHANICAL FASTENERS

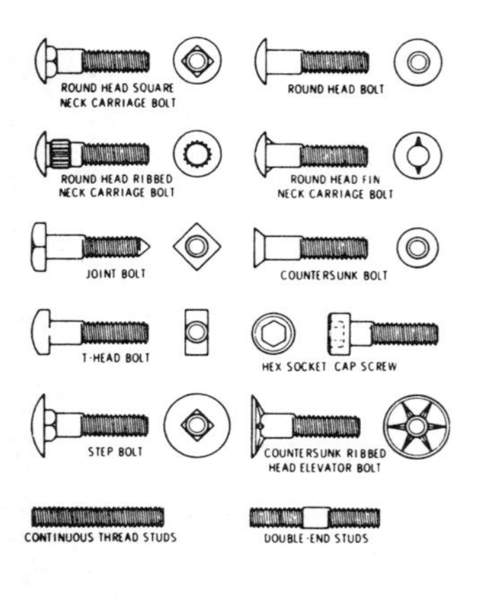

Fig. 89.

Machinery and equipment are assembled and held together by a wide variety of fastening devices. Threaded fasteners are by far the most widely used style, and the bolt, screw, and stud are the most common of the threaded fasteners.

The bolt is normally tightened and released by turning a mating nut. A screw differs from a bolt in that it is supposed to mate with an internal thread into which it is tightened or released by turning its head. Obviously these descriptions do not always apply, as bolts can be screwed into threaded holes and screws can be used with a nut. The stud is simply a cylindrical rod threaded on one or both ends or throughout its entire length. Some of the common bolts, screws, and stud types are shown in Fig. 89.

Threaded fasteners are furnished with either coarse threads conforming to *Unified National Coarse (UNC)* standards, or fine threads conforming to *Unified National Fine (UNF)* standards.

Coarse Threads

Fasteners with coarse threads are used for the majority of applications because they have the following advantages: they assemble easily and fast, providing a good start with little chance of cross threading; nicks and burrs from handling are not liable to affect assembly; the seizing probability in temperature applications and in joints where corrosion will form is low; they are not prone to strip when threaded into lower strength materials; and the coarse thread is more easily tapped in brittle materials and materials that crumble easily.

Fine Threads

The use of fine threads may provide superior fastening for applications where strength or specific qualities are required. Fine threads have the following advantages: they are 10% stronger than coarse threads because of greater cross-sectional area; fine threads tap more easily in very hard materials; they can be adjusted more precisely because of their smaller helix angle; they may be used with thinner wall thicknesses; and they are less liable to loosen from vibration.

Washers

Most threaded fasteners are installed where vibration occurs. This motion tends to overcome the frictional force between the threads, causing the fastener to back off and loosen. Washers are placed beneath the fastener head to help maintain frictional resistance to loosening.

Flat washers provide a bearing surface and spread the load over an increased holding area. Lockwashers tend to retard loosening of inadequately tightened fasten-

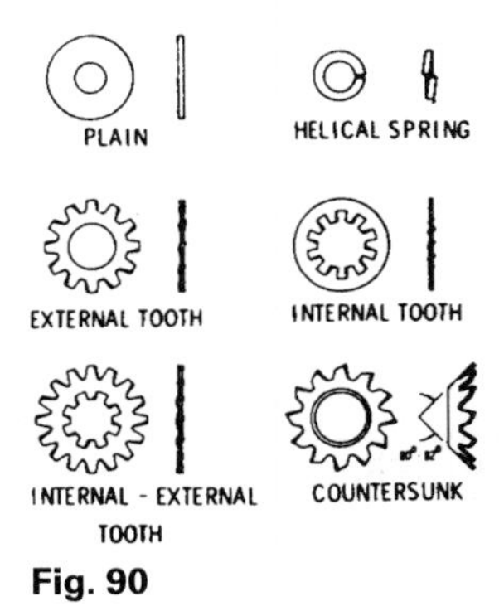

Fig. 90

ers. Theoretically, if fasteners are properly tightened, lockwashers would not be necessary. Multiple tooth locking washers provide a greater resistance to loosening because their teeth bite into the surface against which the head or nut bears. Their teeth are twisted to slide against the surface when tightened and hold when there is a tendency to loosen. They are most effective when the mating surface under the teeth is soft.

Nuts and Pins

The nuts is the mating unit used with bolt type fasteners to produce tension by rotating and advancing on the bolt threads. Nuts should be of equal grade of metal with the bolt to provide satisfactory service. Many types of nuts are available; those in common use are illustrated in Fig. 91.

Pins may be inserted through the nut and bolt after tightening to prevent the nut from turning. Some of the pins shown below are also used for other applications such as shear pins, locating and positioning parts, hinge applications, etc.

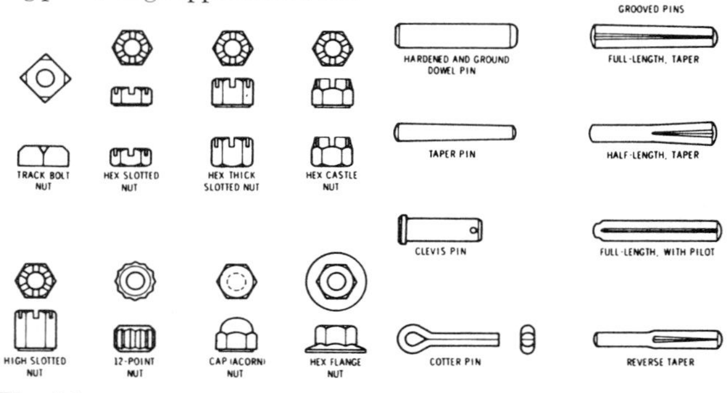

Fig. 91.

Characteristics

Mechanical fasteners are manufactured in a great variety of types and sizes to suit application requirements. However, their design

Chart 3. Characteristics of Standard Fasteners

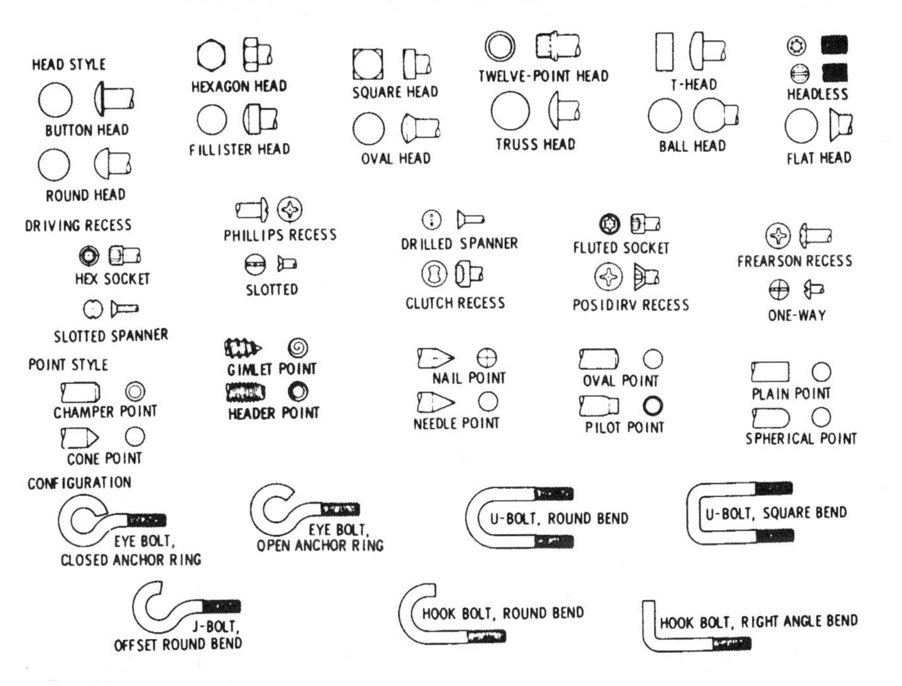

features, called *characteristics*, are standardized. Fasteners with almost any combination of these characteristics are commercially available. Chart 3 shows many of the characteristics that distinguish one fastener from another.

Measurements

Threaded fasteners are identified by their nominal diameter and one or more of the measurements illustrated in Fig. 92.

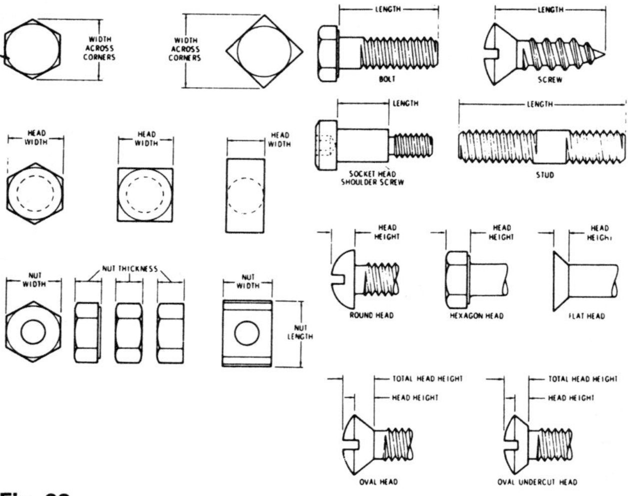

Fig. 92

Retaining Rings

Threaded fasteners are being replaced in an ever-increasing number of applications by metal "retaining rings." Because these rings are installed in grooves which often can be machined simultaneously with other production processes, they eliminate threading, tapping, drilling, and other machining operations. In addition to reducing manufacturing costs, the rings often provide a more compact and functional design, and in some cases make possible assemblies that would otherwise be impractical.

Retaining rings in a wide variety of styles serve as shoulders for accurately locating, retaining, or locking components on shafts and in boxes and housings. Assembly and disassembly is accomplished by expanding the external rings for assembly over a shaft, or by compressing the internal ring for insertion into a box or housing.

The names, size ranges, and series numbers for some of the commonly used *Waldes Truarc* retaining rings are listed in Chart 4.

Pliers are normally used for field installation and disassembly of retaining rings. They are designed to grasp the ring securely by the lugs and expand or compress it. Most standard pliers have straight tips; however, for applications where space is limited, bent tips are available.

PACKINGS AND SEALS

Stuffing Box Packings

The oldest and still one of the most widely used shaft seals in the mechanical arrangement called a *stuffing box*. It is used to control

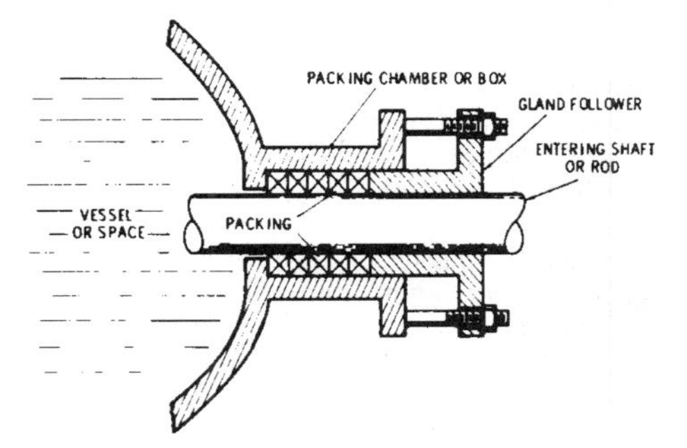

Fig. 93.

Chart 4. Retainer Ring Types

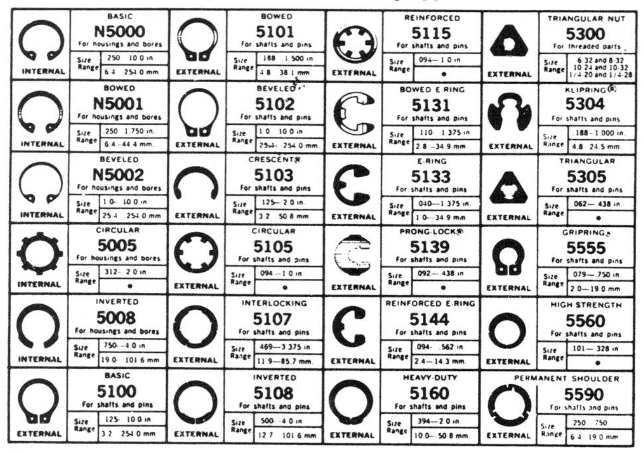

©1970 Waldes Kohinoor, Inc.
Reprinted with permission.

leakage along a shaft or rod. It is composed of three parts: the *packing chamber*, also called the *box;* the *packing rings*; and the *gland follower*, also called the *stuffing gland* (Fig. 93).

Sealing is accomplished by squeezing the packing between the throat or bottom of the box and the gland. The packing is subjected to compressive forces which cause it to flow outward to seal against the bore of the box and inward to seal against the surface of the shaft or rod.

Leakage along the shaft is controlled by the intimate contact of the packings on the surface of the shaft. Leakage through the packing is prevented by the lubricant contained in the packing. The packing material is called "soft" or "compression" packing. It is manufactured from various forms of fibers, and impregnated with binders and lubricant. The impregnated lubricant comprises about 30% of the total packing volume.

As the gland is tightened, the packing is compressed and wears; therefore it must have the ability to deform in order to seal. It must also have a certain ruggedness of construction so that it may be readily cut into rings and assembled into the stuffing box without serious breakage or deformation.

Packings require frequent adjustment to compensate for the wear and loss of volume that occurs continuously while they are subjected to operation conditions. A fundamental rule for satisfactory operation of a stuffing box is: *there must be controlled leakage*. This is necessary because in operation a stuffing box is a form of braking mechanism and generates heat. The frictional heat is held to a minimum by the use of smooth polished shaft surfaces and a continuous supply of packing lubricant to the shaft-packing interface. The purpose of leakage is to assist in lubrication and to carry off the generated heat. Maintaining packing pressures at the lowest possible level helps to keep heat generation to a minimum.

Stuffing Box Arrangement

The function of the multiple rings of packing in a stuffing box is to break down the pressure of the fluid being sealed so that it approaches zero atmospheric pressure at the follower end of the box. Theoretically a single ring of packing, properly installed, will seal. In practice the bottom ring in a properly installed set of packings does the major part of the sealing job. Because the bottom ring is the farthest from the follower, it has the least pressure exerted on it. Therefore, to perform its important function of a major pressure reduction, it is extremely important that it be properly installed.

Packing Installation

While the packing of a stuffing box appears to be a relatively simple operation, it is often done improperly. It is generally a hot, dirty, uncomfortable job that is completed in the shortest possible time with the least possible effort. Short packing life and damage to shaft surfaces can usually be traced to improper practices, rather than deficiencies in material and equipment.

For example, a common improper practice is to lay out packing material on a flat surface and cut it to measured lengths with square ends as shown in Fig. 94.

Fig. 94.

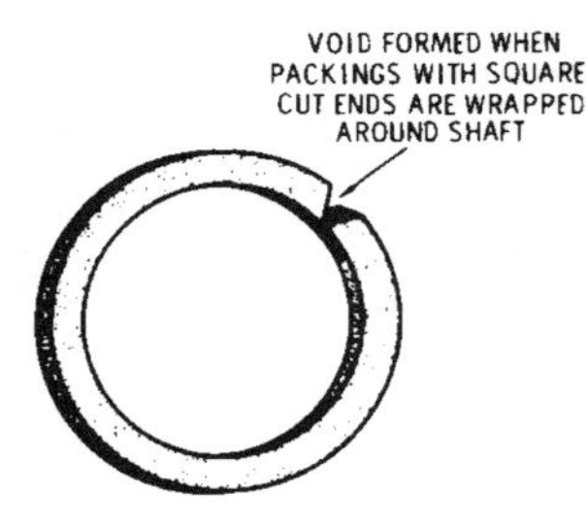

Fig. 95.

When lengths of packing with square ends are formed into rings they have a pie-shaped void at the ends, as shown in Fig. 95. The thicker the packing in relation to the shaft diameter the more pronounced the void will be, as the outside circumference of a ring is greater than the inside circumference. Such voids cause unequal compression and distortion of the packing. Overtightening is then required to accomplish sealing.

Correct practice is to cut the packing into rings while it is wrapped around a shaft or mandrel. A square-cut end to form a plain butt joint is as satisfactory as step, angle, or scive joints.

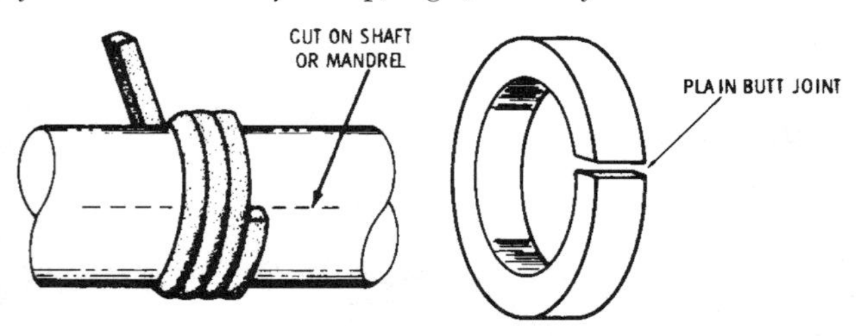

Fig. 96.

The manner of handling and installing compression packing has a greater influence on its service life than any other factor. Therefore, maximum packing life can only be realized when correct packing practices are followed. The basic steps in correct packing installation are as follows:

1. Remove all old packing and thoroughly clean the box.
2. Cut packings rings on shaft or mandrel as shown in Fig. 96. Keep the job simple and cut rings with plain butt joints.
3. Form the first packing gently around the shaft and enter the ends first into the box. The installation of this first packing ring is the most critical step in packing installation. The first ring should be gently pushed forward, keeping the packing square with the shaft as it is being seated. It must be seated firmly against the bottom of the box with the butt ends together before additional packing rings are installed. NEVER put a few rings into the box and try to seat them with the follower. The outside rings will be damaged and the bottom rings will not be properly seated.
4. Insert additional rings individually, tamping each one firmly into position against the preceding ring. Stagger the joints to provide proper sealing and support.
5. Install the gland follower.
6. Tighten follower snugly while rotating shaft by hand. When done in this manner it is immediately apparent if the follower becomes jammed due to cocking or if the packing is overtightened. Slack off and leave finger tight.
7. Open valves to allow fluid to enter equipment. Start equipment; fluid should leak from the stuffing box. If leakage is excessive, take up slightly on the gland follower. *Do not eliminate leakage entirely*; slight leakage is required for satisfactory service. During the first few hours of run-in operation the equipment should be checked periodically as additional adjustment may be required.

Stuffing Box Lantern Rings

Many stuffing box assemblies include a *lantern ring* or *seal cage*. Use of this device allows the introduction of additional lubricants or fluids directly to the interface of packing and shaft.

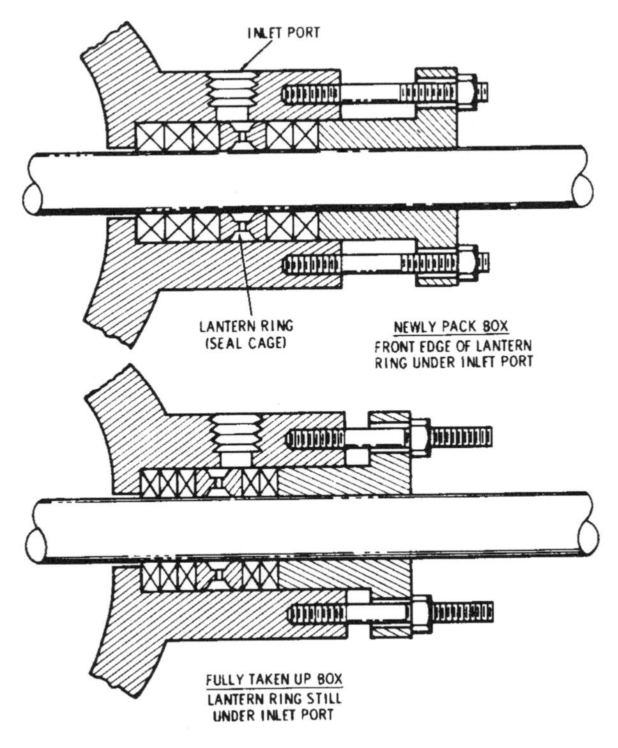

Fig. 97.

A common practice with pumps having a suction pressure below atmospheric pressure is to connect the pump discharge to the lantern ring. The fluid introduced through the lantern ring acts as both a seal to prevent air from being drawn into the pump, and as a lubricant for the packing and shaft.

Lantern rings are also commonly used in pumps handling slurries. In this case clear liquid from an external source, at a higher pressure than the slurry, is introduced into the stuffing box through the lantern ring.

Stuffing boxes incorporating lantern rings require special attention when packing. The lantern ring must be positioned between the packing rings so its front edge is in line with the inlet port at installation. As the packings wear and the follower is tightened, the lantern ring will move forward under the inlet port. When the packings have been fully compressed, the lantern ring should still be in a position open to flow from the inlet port. See Fig. 97.

While lantern rings may on occasion be troublesome to the mechanic, they should not be removed and discarded, as they are an important part of the stuffing box assembly.

Mechanical Seals

While the "stuffing box" seal is still widely used because of its simplicity and ability to operate under adverse conditions, it is used principally on applications where continuous slight leakage is not objectionable. The *mechanical seal* operates with practically no leakage and is replacing the stuffing box seal in an ever-increasing number of applications.

Principle of Operation — The mechanical seal is an *end-face* type designed to provide rotary seal faces that can operate with practically no leakage. This design uses two replaceable antifriction mating rings — one rotating, the other stationary — to provide sealing surfaces at the point of relative motion. These rings are statically sealed, one to the shaft, the other to the stationary housing. The mechanical seal therefore is made up of three (3) individual seals; two are static, having no relative movement, and the third is the rotary or dynamic seal at the end faces of the mating rings.

The mechanical seal also incorporates some form of self-contained force to hold the mating faces together. This force is usually provided by a spring-loaded apparatus such as a single coil spring, multiple springs, or wave springs, which are thin spring washers into which waves have been formed. The basic design and operating principle of the "inside" mechanical seal is diagrammatically illustrated in Fig. 98.

Mechanical-Seal Types

While there are many design variations and numerous adaptations, there are only three basic types of mechanical seals. These are the *inside, outside,* and *double* mechanical seals.

Inside Seal — The principle of the "inside" type is shown in Fig. 98. Its rotating unit is inside the chamber or box, thus its name. Because the fluid pressure inside the box acts on the parts and adds

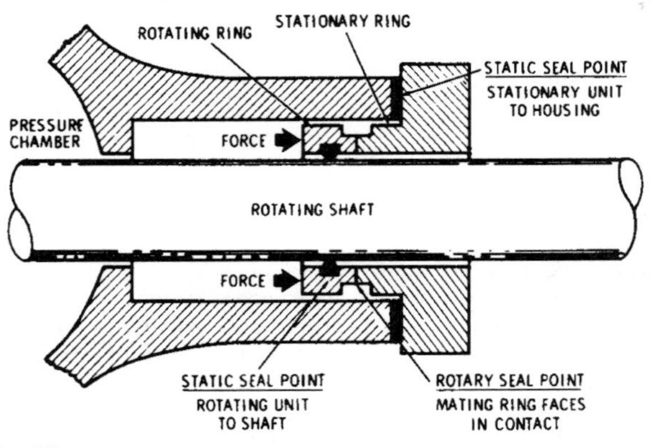

Fig. 98.

to the force holding the faces together, the total force on the faces of an inside seal will increase as the pressure of the fluid increases. If the force built into the seal plus the hydraulic force are high enough to squeeze out the lubricating film between the mating faces, the seal will fail.

While the mechanical seal is commonly considered to be a positive seal with no leakage, this is not true. Its successful operation requires that a lubricating film be present between the mating faces. For such a film to be present there must be a very slight leakage of the fluid across the faces. While this leakage may be so slight that it is hardly visible, if it does not occur the seal will fail. This is the reason a mechanical seal must never run dry. The stuffing box must be completely filled and the seal submerged in fluid before the equipment is started, and always while operating.

Outside Seal — This style of seal is also named for the location of the rotating unit, in this case outside of the stuffing box, as shown in Fig. 99. Because all the rotating parts are removed from the liquid being handled, it is superior for applications where corrosive or

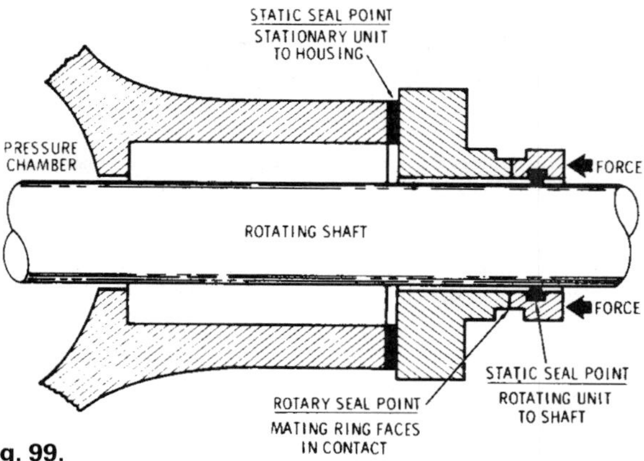

Fig. 99.

abrasive materials are present. Because the hydraulic pressure of the fluid is imposed on the sealing faces, tending to open them by overcoming the self-contained force, it is limited to moderate pressures.

Double Seal — A double seal is basically an arrangement of two single inside seals placed back to back inside a stuffing box. The double seal provides a high degree of safety when handling hazardous liquids. This is accomplished by circulating a non-hazardous liquid inside the box at higher pressure than the material being sealed. Any leakage therefore will be the non-hazardous lubricating fluid inward, rather than the hazardous material that is being sealed leaking outward. The basic principle of the "double mechanical seal" is illustrated in Fig. 100.

Lubrication of Seal Faces — The major advantage of the mechanical seal is it slow leakage rate. This is so low there is virtually no visible leakage. To operate satisfactorily in this manner requires

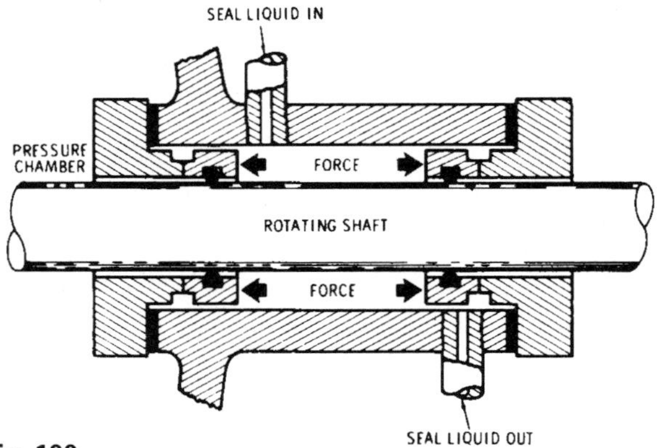

Fig. 100.

that the film of lubricant between the seal faces be extremely thin and uninterrupted. Maintenance of this extremely thin film is made possible by machining the seal faces to very high tolerances in respect to flatness and surface finish. To protect this high quality, precise surface finish requires that seal parts be carefully handled and protected. Mating faces should never be placed in contact without lubrication.

Operation of the mechanical seal depends on this thin film of lubricant furnished by leakage. If there is insufficient leakage to provide the lubricating film the faces will quickly overheat and fail. Liquid must always be present during operation, as running dry for a matter of seconds can destroy the seal faces.

Mechanical-Seal Installation

No one method or procedure for installation of mechanical seals can be outlined, because of the variety of styles and designs. In cases where the seal is automatically located in correct position by the shape and dimensions of the parts, installation is relatively simple and straightforward. In many cases, however, the location of the seal parts must be determined at installation. In such cases the location of these parts is critical, as their location determines the amount of force that will be applied to the seal faces. This force is a major factor in seal performance, as excessive face pressure results in early seal failure. Parts must be located to apply sufficient force to hold the mating rings together without exerting excessive face pressure.

The procedure or method of location and installation of outside seals is usually relatively obvious and easily accomplished. The inside style of mechanical seal is more difficult to install, as some parts must be located and attached while the equipment is disassembled. The location of these parts must be such that the proper force will be applied to the seal faces when the assembly is complete.

Seal designs and styles vary with manufacturers; however, the same basic principles apply to all when locating and installing inside mechanical seals. The following general procedure is applicable to most styles in common use.

Step #1 — Determine the compressed length of the seal component incorporating the force mechanism. This is its overall length when it is in operating position (springs properly compressed). Two widely used seal designs are shown in Fig. 101; one incorporates multiple springs and the other a single helical spring. In either case the spring or springs must be compressed the amount recommended by the seal manufacturer before the measurement to

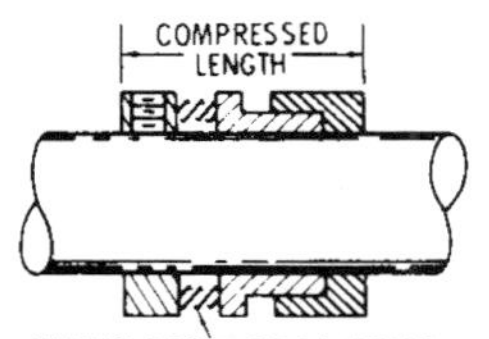

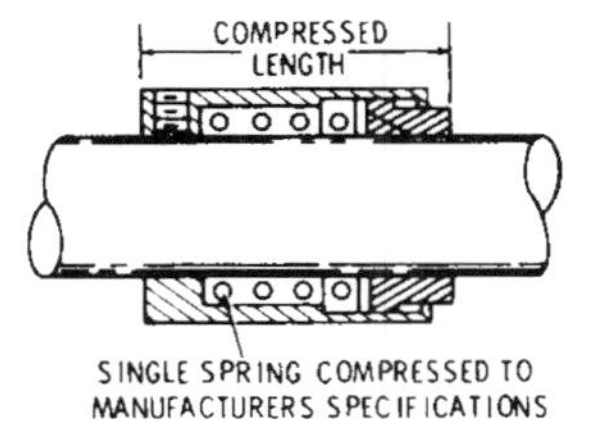

Fig. 101.

determine compressed length is made. This is vitally important, since the force exerted on the seal faces is controlled by the amount the springs are compressed.

Manufacturers' practices vary in the method of determining correct spring compression. In some cases the springs should be compressed to obtain a specific gap or space between sections of the seal assembly. In other cases it is recommended that spring or springs be compressed until alignment of lines or marks is accomplished. In any case, consult manufacturers' instructions and be sure the method used to determine compressed length is correct for the make and model of seal being installed.

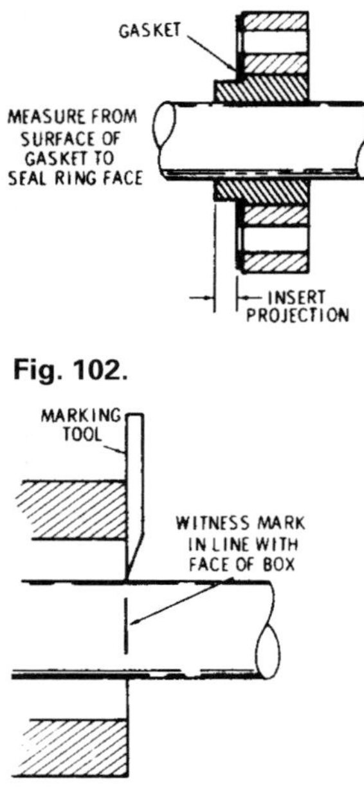

Fig. 102.

Fig. 103.

Step #2 — Determine the insert projection of the mating seal ring. This is the distance the seal face will project into the stuffing box when it is assembled into position. Care must be exercised to be sure the static seal gasket is in position when this measurement is made. Obviously the amount of projection can be varied by varying the thickness of the gasket. See Fig. 102.

Step #3 — Determine the "location dimension." This is done by simply adding the "compression-length" dimension found in *Step #1* to the "insert-projection" dimension found in *Step #2*.

Step #4 — "Witness-mark" the shaft in line with the face of the stuffing box (Fig. 103). A good practice is to blue the shaft surface in the area where the mark is to be made. A flat piece of hardened steel such as a tool bit, ground on one side only to a sharp edge, makes an excellent marking tool. The marker should be held flat against the face of the box and the shaft rotated in contact with it. This will provide a sharp clear witness-mark line that is exactly in line with the face of the box.

Step #5 — At this point the equipment must be disassembled in a manner to expose the area of the shaft where the rotary unit of the seal is to be installed. The amount and method of disassembly will vary with the design of the equipment. In some cases it may be necessary to completely remove the shaft from the equipment. In other cases, as with the back pull-out design pumps, it is only necessary to remove the back cover which contains the stuffing box chamber to expose the required area of the shaft.

Step #6 — With the shaft either removed or exposed, blue the area where the back face of the rotary unit will be located. From the witness mark, which was placed on the shaft in *Step #4*, measure the location-dimension distance and place a second mark on the shaft. This is called the "location mark," as it marks the point at which the back face of the rotary unit is to be located. The location dimension is the sum of the compressed length plus the insert projection (Fig. 104).

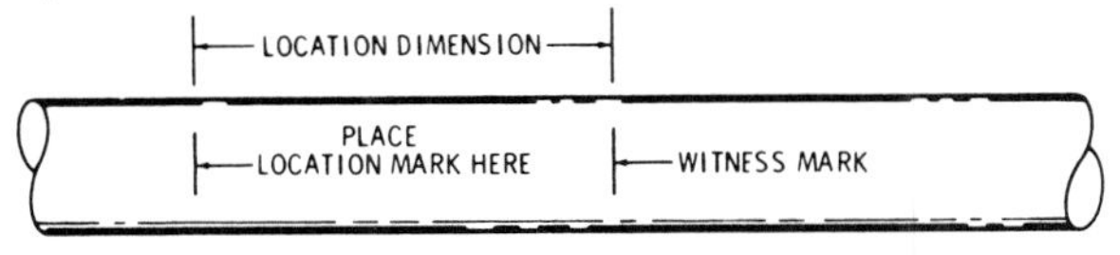

Fig. 104.

Step #7 — Assemble the rotary unit on the shaft with its back face on the location mark. Fasten the unit securely to the shaft at this location. Some seal designs allow separation of the rotary unit components. In such cases the back collar may be installed at this time and the other rotary unit parts later. Illustrated in Fig. 105 is a single-spring type rotary unit assembled on the shaft with its back face on the location mark. The spring is extended *and* will be compressed by tightening the stationary ring at assembly.

Step #8 — Reassemble the equipment with the rotary unit on the shaft inside the stuffing box chamber. Complete the seal assembly, BEING SURE THE SEAL FACES ARE LUBRICATED.

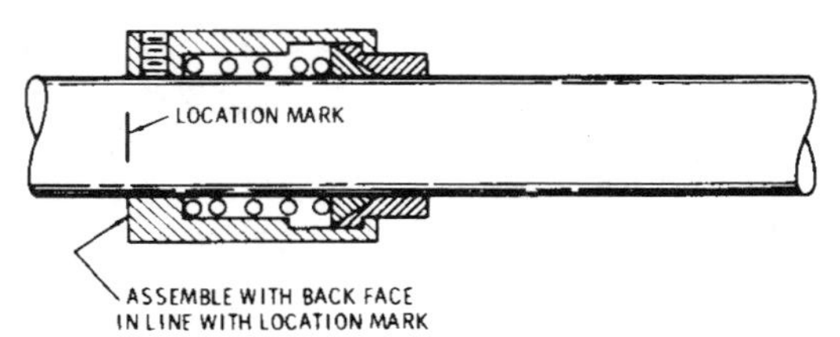

Fig. 105.

The sequence of parts assembled depends on the type and design of the seal and the equipment. Illustrated below is a completely installed inside seal of the multiple spring design.

The final assembly operation will be the tightening of the gland follower bolts or nuts. When this is done the lubricated seal faces should be brought into contact very carefully. When the faces initially contact there should be a space between the face of the box and the follower gland gasket (Fig. 106). This space should be the same amount as the springs were compressed in *Step #1*, when the compressed length of the rotary unit was determined. This should be very carefully observed, as it is a positive final check on correct location of the rotary unit of an inside mechanical seal.

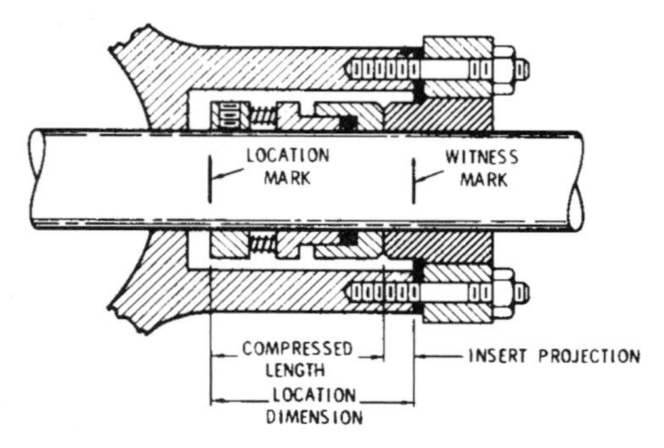

Fig. 106.

Installation Precautions

1. Check shaft with indicator for runout and end play. Maximum T.I.R. allowable .005 of an inch.
2. All parts must be clean and free of sharp edges and burrs.
3. All parts must fit properly without binding.
4. Inspect seal faces carefully. No nicks or scratches.
5. Never allow faces to make dry contact. Lubricate them with a good grade of oil or with the liquid to be sealed.
6. Protect all static seals such as O-rings, V-rings, V-cups, wedges, etc. from damage on sharp edges during assembly.
7. Before operating, *be sure proper valves are open and seal is submerged in liquid.* If necessary, vent box to expel air and allow liquid to surround seal rings.

O-Rings

The *O-ring* is a squeeze-type packing made from synthetic rubber compounds. It is manufactured in several shapes, the most common being the circular cross section from which it derives its name. The principle of operation of the O-ring can be described as controlled deformation. A slight deformation of the cross section, called a "mechanical squeeze," illustrated in Fig. 107A, deforms the ring and places the material in compression. The deformation squeeze flattens the ring into intimate contact with the confining surfaces, and the internal force squeezed into the material maintains this intimate contact.

Additional deformation results from the pressure the confined fluid exerts on the surface of the material. This in turn increases the contact area and the contact pressure as shown in Fig. 107B.

The initial mechanical squeeze of the O-ring at assembly should be equal to approximately 10% of its cross-sectional dimension. The general-purpose industrial O-ring is made to dimensions that, in effect, built the initial squeeze into the product. This is done by manufacturing the O-ring to a cross-sectional dimension 10% greater than its nominal size. The following are the nominal and actual cross-section diameter dimensions of the O-rings in general use:

Nominal —	1/32	3/64	1/16	3/32	1/8	3/16	1/4
Actual —	.040	.050	.070	.103	.139	.210	.275

Because the cross-section dimension of an O-ring is 10% oversize, the outside and inside dimensions of the ring must be proportionately larger and smaller. For example, a 2 × 2⅜ × 3/16 nominal size O-ring has actual dimensions of 1.975 × 2.395 × .210. Fig. 208 illustrates the relationship of such an O-ring to its groove.

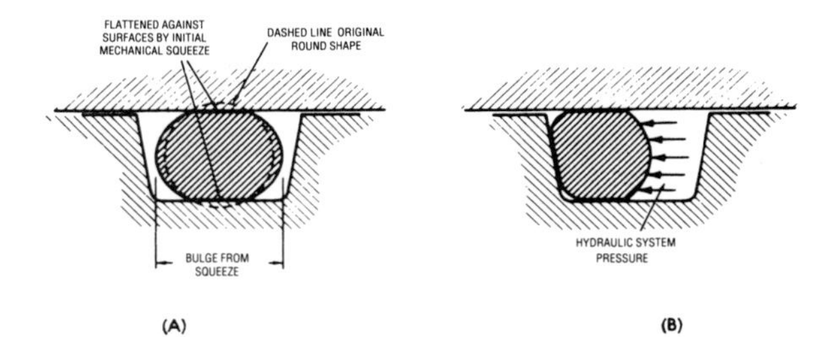

Fig. 107.

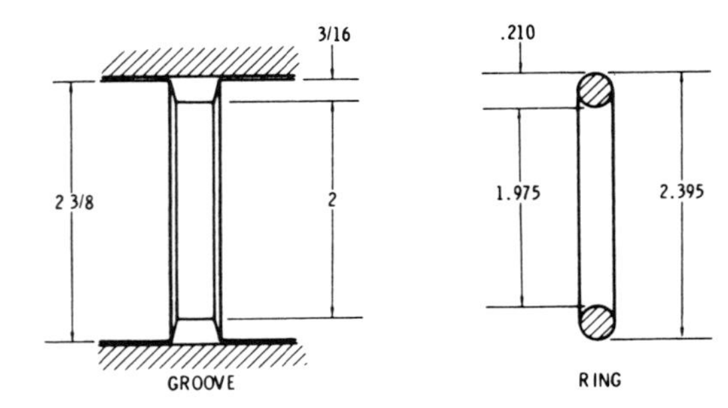

Fig. 108.

O-Ring Dash Number System

Many O-ring standards have been established by various agencies, societies, industrial groups, etc. The dimensions used in all of these are essentially the same; however, the numbering systems are not. In wide use by manufacturers of O-rings is what is called the *Uniform Dash Number* system, in which the numbers 001 to 475 are used to identify specific dimensions of O-rings.

Within the system of numbers the O-rings are grouped according to cross-section diameter as follows:

Dash Nos.—004 to—050 for 1/16" Diameter
Dash Nos.—110 to—170 for 3/32" Diameter
Dash Nos.—210 to—284 for 1/8" Diameter
Dash Nos.—325 to—395 for 3/16" Diameter
Dash Nos.—425 to—475 for 1/4" Diameter

Using a Uniform Dash Number table one can select an O-ring of correct size from the nominal dimensions. Simply select the dash number that corresponds to the nominal dimensions of the installation. Table 14 shows the dash numbers—270 to—329. Note that—281 is the largest O-ring having a 1/8" cross section and—325 is the smallest O-ring having a 1/16" cross section.

Formed and Molded Packings

The principle of operation of formed and molded packings is quite different from compression packings, in that no compression force is required to operate them. The pressure of the fluid being sealed provides the force which seats the packings against the mating surfaces. They are therefore often classed as automatic or hydraulic packings. Packings molded or formed in the shape of a "Cup," "Flange," "U-Shape," or "V-Shape" are classed as "lip"-type packings.

Lip-type packings are usually produced with lips slightly flared to provide automatic preload at installation. The fluid being sealed then acts against the lips, exerting the force that presses them against the mating surface. Lip-type packings are used almost exclusively for sealing during reciprocating motion. They must be installed in a manner that will allow the lips freedom to respond to the fluid forces.

Table 15.

	O-Ring Size							
	Actual					Nominal Size		
		Tolerance						
		Table I	Table II					
Dash No.	I.D.	±	±	I.D.	±	I.D.	O.D.	W.
—270	8.984	.030	.050	.139	.004	9	9 ¼	⅛
—271	9.234	.030	.055	.139	.004	9 ¼	9 ½	⅛
—272	9.484	.030	.055	.139	.004	9 ½	9 ¾	⅛
—273	9.734	.030	.055	.139	.004	9 ¾	10	⅛
—274	9.984	.030	.055	.139	.004	10	10 ¼	⅛
—275	10.484	.030	.055	.139	.004	10 ½	10 ¾	⅛
—276	10.984	.030	.065	.139	.004	11	11 ¼	⅛
—277	11.484	.030	.065	.139	.004	11 ½	11 ¾	⅛
—278	11.984	.030	.065	.139	.004	12	12 ¼	⅛
—279	12.984	.030	.065	.139	.004	13	13 ¼	⅛
—280	13.984	.030	.065	.139	.004	14	14 ¼	⅛
—281	14.984	.030	.065	.139	.004	15	15 ¼	⅛
——	15.955	.045	.075	.139	.004	16	16 ¼	⅛
——	16.955	.045	.080	.139	.004	17	17 ¼	⅛
——	17.955	.045	.085	.139	.004	18	18 ¼	⅛
—325	1.475	.010	.015	.210	.005	1 ½	1 ⅞	3⁄16
—326	1.600	.010	.015	.210	.005	1 ⅝	2	3⁄16
—327	1.725	.010	.015	.210	.005	1 ¾	2 ⅛	3⁄16
—328	1.850	.010	.015	.210	.005	1 ⅞	2 ¼	3⁄16
—329	1.975	.010	.018	.210	.005	2	2 ⅜	3⁄16

Packing Materials

Leather — One of the oldest packing materials and still the most satisfactory for rough and difficult applications. Has high tensile strength and resistance to extrusion. Absorbs fluids, therefore tends to be self-lubricating.

Fabricated — Made by molding woven duck, asbestos cloth, and synthetic rubber. Fabric reinforcement gives strength to withstand high pressures. Also resistant to acids, alkalis, and high temperature. Tends to wipe drier than leather packings, although the fabric will absorb some fluid and has slight lubricating action.

Homogenous — Compounded from a wide variety of synthetic rubbers. Low strength but high resistance to acids, alkalis, and high temperature. Requires a fine surface finish, close clearances, and clean operating conditions. Nonabsorbent, has no self-lubricating qualities, and usually wipes contact surfaces quite dry.

Plastic — Molded from various kinds of plastics for special applications. Inert to most chemicals and solvents but has little elasticity or flexibility. Some types have a slippery feel and resist adhesion to metal, but their friction under pressure is high.

Cup Packings — One of the most widely used styles of packing, simple to install and highly satisfactory for plunger end applications. Inside follower plate must not be overtightened, as the bottom of the packing will be crushed and cut through. The heel or shoulder is the point of greatest wear and usually the failure point. Clearances should be held to a minimum and the lips, protected from bumping (Fig. 109).

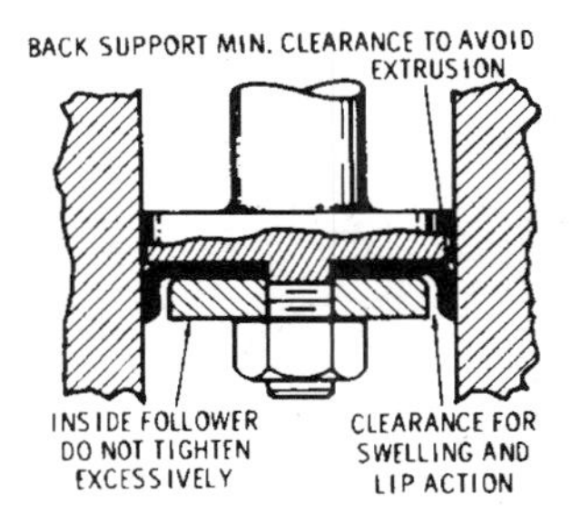

Fig. 109.

"U" Packings—A balanced packing, as sealing occurs on both the inside and outside diameter surfaces (Fig. 110). To support the lips, the recess of the U is filled with flax, hemp, rubber, fiber, etc. Metal rings called "pedestal" rings are also used for lip support. Fillers, if nonporous, must be provided with pressure equalization openings (holes) to allow equalization of pressure on all inside surfaces. Clearance between pedestal rings and inside packing wall is necessary to ac-

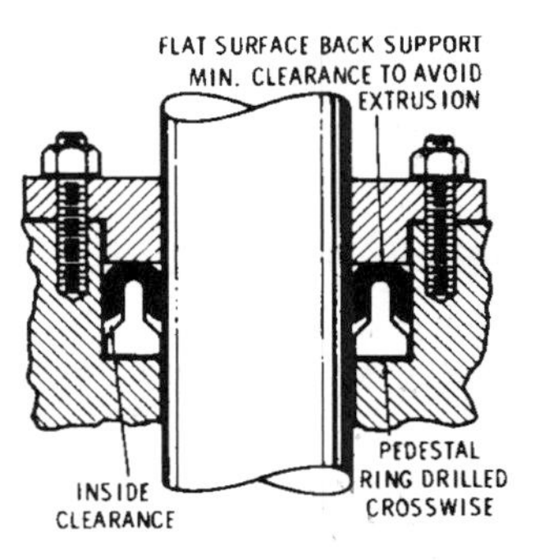

Fig. 110.

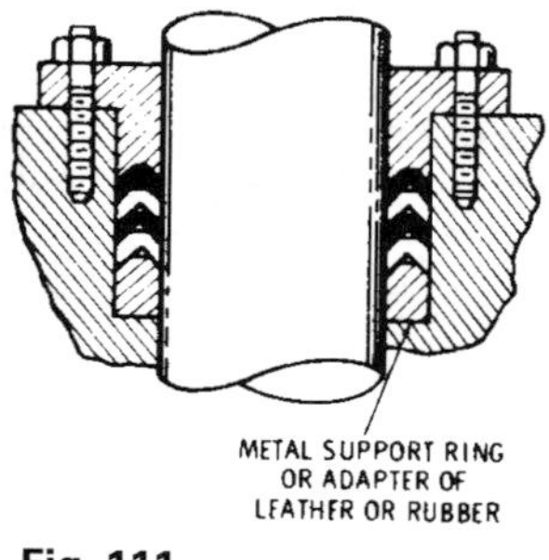

Fig. 111.

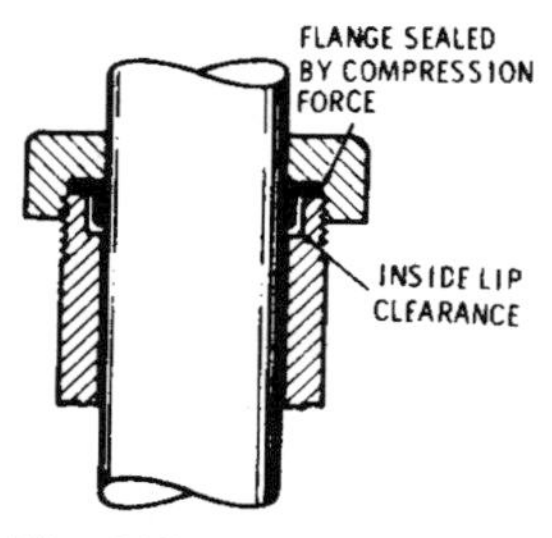

Fig. 112.

commodate swelling and allow the lips freedom to respond to the actuating fluid.

"V" Packings—Installed in sets, each set consisting of a number of V-rings and a male and female adapter. Have a small cross section and are suitable for both high and low pressures. Operate as automatic packings, but have the advantage of permitting taking up on the gland ring when excessive wear develops (Fig. 111).

Flange Packings — Seals on the inside diameter only; generally restricted to low pressures. Base sealed by an outside compressing force, usually a gland arrangement. Lip action same as cup packing and must be given same consideration for clearance, etc.

Packing Installation

The principle of operation of all lip-style packings is the same regardless of type or material. They must be installed in a manner that will allow them to expand and contract freely. They should not be placed under high mechanical pressure, as this transforms them to compression packings. Overtightening a lip-type packing improperly preloads it. While slight preload is needed for a tight fit and sealing at low pressure, it should occur automatically as a function of the size and shape of the packing.

Lip-type packings installed for one-directional scaling in glands or on pistons should be installed with the inside of the packing exposed to the actuating fluid. Proper installation of packings in this style of application is obvious if the principle of the lip-type automatic packing is understood. Double-acting applications require greater care and attention at installation to ensure that assembly conforms to operating principles. When mounted with the insides of the packing facing together, they are referred to as *face-to-face* mounted. When mounted with the bottoms toward one another and the insides facing away, they are referred to as *back-to-back* mounted.

The ideal arrangement for double-acting packing assembly is back-to-back with solid shoulders for back support (Fig. 113). Each packing is fully supported and no trapping of pressure between packings can occur. Because such an arrangement requires that packings be installed from both sides of the plunger head or end, it frequently is not practical. In such cases face-to-face mounting is used (Fig. 114). While this allows packing from one end, it also introduces an undesirable condition. The actuating fluid must pass the first packing and open the second packing, which is facing toward the fluid. When the second packing is expanded by the fluid, pressure tends to back up into the first packing, expanding it and locking pressure between the two packings.

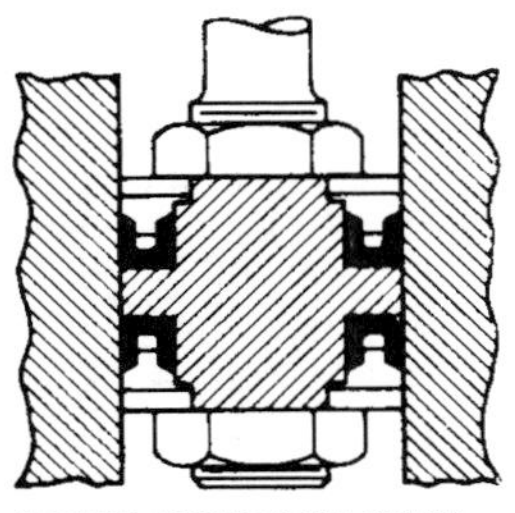

CORRECT: BACK TO BACK-PACKED FROM TWO SIDES

Fig. 113.

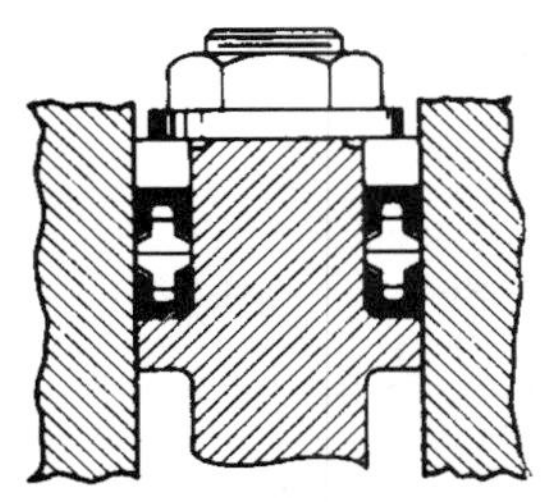

FACE TO FACE COMPROMISE TO ALLOW PACKING FROM ONE SIDE

Fig. 114.

Clearance Seals

Clearance seals limit leakage between a rotating shaft and housing by maintaining a small, closely-controlled annular clearance between shaft and housing. Because there is no rubbing contact, clearance seals offer many advantages: they do not wear, create heat due to friction, or consume power. The principal disadvantage of clearance seals is that they always have some leakage. The two general types are *bushing* and *labyrinth.* The bushing seal is simply a close-fitting stationary sleeve. Leakage from the high-pressure end to the low-pressure end is limited by the throttling action of the limited clearance. The labyrinth seal is probably the most common type of clearance seal. It is composed of a series of lands on one member which mate closely with a series of grooves on the other member. These lands and grooves retard leakage because of the tortuous escape path they provide. The labyrinth seal is not an off-the-shelf piece of equipment, but is usually custom designed. Slingers, which might be called clearance seals, are also used for shaft sealing. They are usually discs or rings on the revolving member which interrupt lubricant or foreign material from traveling laterally. Centrifugal force tends to throw the lubricant or foreign material from the slinger to the housing surface, from which it is collected in a collector or sump.

Oil Seals

Oil seals are more specifically called radial lip seals or radial contact seals. They are used primarily for sealing lubricants in and foreign material out. They are constructed so the lips must be expanded at assembly, which insures the positive rubbing contact necessary for their satisfactory operation. No external force is required as with compression packings, nor do they require internal fluid pressure to actuate them. They may, however, be made with a spring to improve contact of the sealing member with the shaft. The springs may be either coiled garter springs or bonded multiple leaf springs. The use of springs increases operating speed and allows some shaft run-out. This is possible because of the increased contact pressure of the lips with the shaft that is provided by the springs. Oil seals are primarily used on rotating shaft applications. The use of this type of seal in reciprocating assemblies is very infrequent. The oil seal is made in a wide variety of types, designs, and materials, most of which can be classified as *bonded*, *cased*, or *composition* types.

Bonded seals are constructed of synthetic rubber or other elastic materials, bonded to a metal case or washer. They are seldom used where pressures are encountered. The metal element to which the elastic material is bonded provides the support and stiffness that is necessary for mounting. The lip of the seal is made of elastic material, the tip of which provides the flexible sealing ring that contacts the shaft surface. This flexible ring edge interrupts the lubricant or foreign material, preventing it from traveling laterally on the surface of the shaft. Shown in Fig. 115 is a bonded-type seal with a coiled garter spring.

Garlock Inc. Mech. Packing Div.

Fig. 115. Bonded oil seal with garter spring.

Cased seals may use either elastic or leather sealing elements. The leather element is superior to elastic or synthetic rubber material where the seal is not constantly lubricated. As leather is porous and will absorb lubricant, it will operate for fairly long periods without additional lubrication. Leather does not require as smooth a shaft finish for satisfactory operation as synthetic rubber materials do. Leather, however, is limited to moderate temperatures and moderate surface speeds. Synthetic materials have greater resistance to acids, alkalis, etc., than leather and will withstand high temperatures. Another advantage of synthetic rubber is its ability to operate satisfactorily where run-out or eccentricity is present. Cased seals require more space than bonded seals, as they are larger in dimension for a given shaft size. Shown in Fig. 116 is a cased seal with multiple leaf or finger springs.

The composition seal is similar in basic design to both cased and bonded seals. The composition seal is constructed with a rigid asbestos heel to which the elastic material is molded. This gives a rigid, hard backing to support the seal and allows tight mounting in the housing. The flexible lips are made of the same variety of materials as are used in either the bonded or cases seals. The principal advantages of this type of construction are improved sealing in the bore and ability to conform to bores that may be somewhat out of round. Insofar as temperature and speed are concerned, the composition seal is satisfactory for the same conditions as bonded and cased seals of similar materials. Most split seals are of the composition type construction. Split seals can be installed by placing them around the shaft, instead of sliding the seal over the shaft as is necessary with solid seals of the conventional type.

Combination or dual element oil seals are designed for sealing where liquids are present on both sides of the seal. Other combinations provide a sealing action in one direction plus a wiping or an exclusion action in the reverse direction. The combination seal is similar in basic design to both cased and bonded seals except that it has double sealing lips, usually positioned in opposite directions. They are, however, made with both lips facing in the same direction for difficult sealing conditions or where leakage must be held to a minimum. The three functional installation modes of oil seal installation, shown in Fig. 117, are: *retention, exclusion*, and *combination*.

Garlock Inc. Mech. Packing Div.

Fig. 116. Cased oil seal with finger springs.

Installation

Sizes should be carefully checked before starting installation. The seal should be a light press fit into the housing, and the inside diameter of the seal such that its lips must be slightly expanded as they are assembled on the shaft. Surfaces should be checked carefully, as any sharp edges or roughness will soon damage the seal lips. The lips of the seal should be lubricated before assembly. This aids in installation and insures that the seal will not run dry when the machine is started. Leather seals, though frequently pre-lubricated,

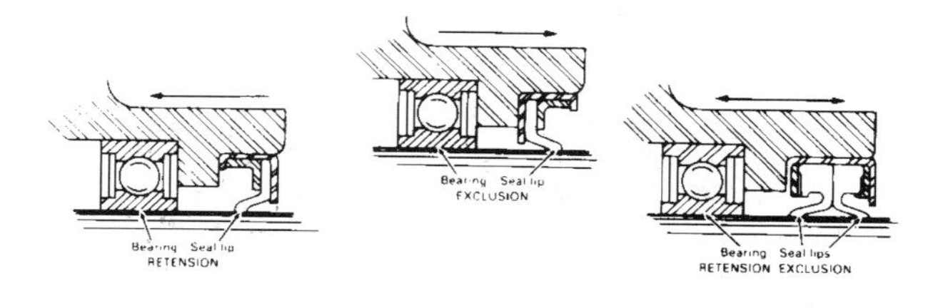

Garlock Inc. Mech. Packing Div.

Fig. 117. Oil seal installations.

should be oiled before installation. Synthetic seals can be dipped in oil before installing. This will also aid in pressing them into the housing. In some cases, particularly with cased seals, it may be advisable to coat the OD with some type of sealer. This will eliminate the possibility of leaks between the OD of the seal and the housing bore. Gasket cement, white lead, or some similar material may be used. In production operations special tools, such as thimbles to protect the seal lips and driving tools to properly apply the installation force, are usually provided. These are seldom available for field installation of oil seals. To protect the lips from sharp edges, such as keyways and splines, thin sheet material such as shimstock or plastic sheeting may be used to form a sleeve over which the seal can be assembled. In some cases it may be possible to cover these edges with one of the many adhesive tapes that are available, or even strong kraft paper.

The force to assemble the seal into the housing bore should be applied near the outer rim of the seal where it is stiff enough to withstand this treatment. If possible, a tubular tool or pipe that is slightly smaller in diameter than the OD of the seal should be used. Preferably this should be done in a press where the seal can be held square with the shaft.

In field installations where pressing is not possible, and driving tools are not available, the seal may be tapped into the housing bore with a wooden block and hammer. Care must be taken that the seal is at right angles to the shaft and not cocked and bound as it is being tapped in. The block should be moved from side to side and around the seal to accomplish this, using light blows and care that the block is squarely against the outer rim of the seal.

Wipers and Rod Scrapers

Two seals, very similar in appearance to the oil seal, but designed for reciprocating applications, are the wiper and the rod scraper. The wiper may be used either as a retention or exclusion seal, and its function is accomplished in the same manner as an oil seal, i.e., by pressure of the lip against the shaft surface. Wipers are easily confused with composition oil seals, as they are quite similar in appearance.

Rod scrapers are a special type of exclusion seal having a hard plastic or metallic lip or scraping element. They are used to scrape heavy or tenacious material from reciprocating shafts. Wipers are sometimes used behind them to catch any fine particles or fluid which pass the scraper. Shown in Fig. 118 are rod scrapers with polyurethane lip elements designed to exclude foreign materials such as dirt, mud, metal chips, etc.

BEARINGS

A bearing, in mechanical terms, is a support for a revolving shaft. In some cases this may also include an assembly of several components that hold or secure the member that supports the shaft.

Bearing Nomenclature

Journal — That part of a shaft, axle, spindle, etc., which is supported by and revolves in a bearing.

Axis — The straight line (imaginary) passing through a shaft on which the shaft revolves or may be supposed to revolve.

Garlock Inc. Mech. Packing Div.

Fig. 118. Rod scrapers with polyurethane lips.

Radial — Extending from a point or center in the manner of rays (as the spokes of a wheel are radial).

Thrust — Pressure of one part against another part, or force exerted end-wise or axially through a shaft.

Friction — Resistance to motion between two surfaces in contact.

Sliding Motion — Two parallel surfaces moving in relation to each other (plain bearings).

Rolling Motion — Round object rolling on mating surface with theoretically no sliding motion (antifriction bearing).

There are two broad classifications of bearings: *plain* and *antifriction*. The plain bearing operates on the principle of sliding motion, there being surface contact and relative movement between shaft and bearing surfaces. The antifriction bearing operates on the principle of rolling motion, there being a series of rollers or balls interposed between the shaft and the supporting member.

Plain Bearings

Plain bearings are simple in design and construction, operate efficiently, and are capable of supporting extremely heavy loads. During operation they develop an oil film between the journal and bearing surfaces that overcomes the friction of sliding motion. There are times, however, when this film is not present — during starting and stopping, during shock loading, or at misalignment, etc. For this reason the plain bearing is made of a material softer than the shaft material and having low frictional qualities. The most widely used bearing materials are bronze, babbit, cast iron, and plastics.

Another factor which has a major influence on plain bearing life is the surface finish of both the journal and the bearing. The rougher the surface, the thicker the film required to separate them. The high degree of surface smoothness which enables the plain bearing to operate with a very thin oil film is often achieved by a break-in or run-in period. During this period a wearing down or flattening of the peaks takes place which greatly reduces the maximum variations and smooths the surfaces.

The details of design and construction of plain bearings vary widely. They may be complete, self-contained units, or the bearing may be built into, and part of, a larger machine assembly. All, however, are variations of the following basic plain bearing types.

Fig. 119.

Solid Bearings — The sleeve bearing or bushing is the most common of all plain bearings. It provides the bearing surfaces for the shaft journal and is usually press-fitted into a supporting member (Fig. 119).

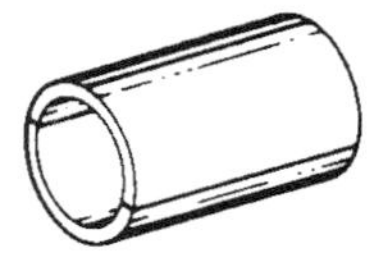

Fig. 120.

Split Bearings — A variation of the solid bearing, the split bearing is divided into two pieces to allow easy shaft assembly and removal (Fig. 120).

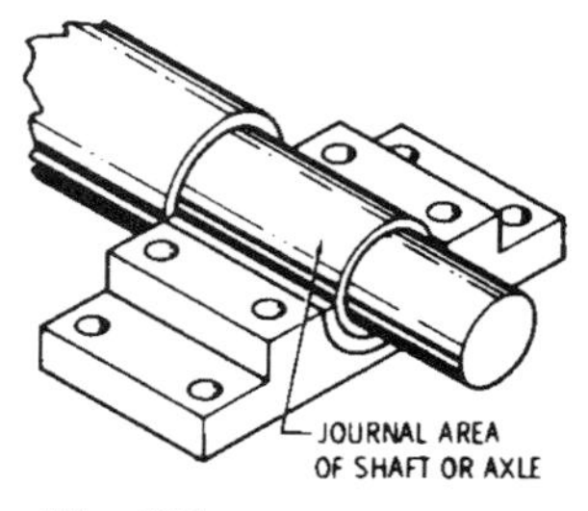

Fig. 121.

Journal Bearings — The plain bearing unit, used for support of radial loads, is commonly referred to as a "journal bearing." It takes its name from the portion of the shaft or axle that operates within the bearing (Fig. 121).

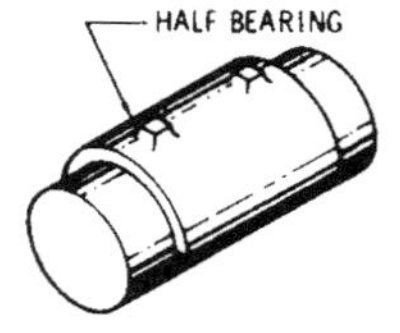

Fig. 122.

Part Bearings — The part or half bearing encircling only part of the journal is used when the principal load presses in the direction of the bearing. Its advantages are low material cost and ease of replacement (Fig. 122).

Thrust Bearings — The thrust bearing supports axial loads and/ or restrains endwise movement. One widely used style is the simple annular ring or washer. Two or more such rings of selected low-frictional materials, either hard or soft, are often combined. It is also common practice to support thrust loads on the end surface of journal bearings. If the area is too small for the applied load, a flange may be provided. The shaft surface which mates the thrust bearing surface and supports the axial load is usually provided by a shoulder (Fig. 123).

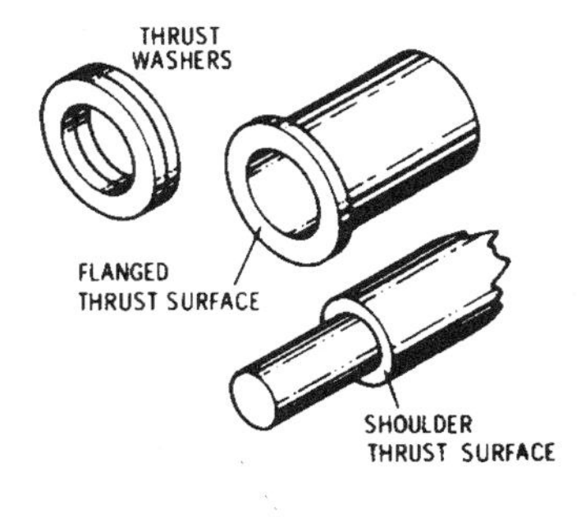

Fig. 123.

Lubrication

When bearing surfaces are separated with an oil film, the condition is described as *thick-film* lubrication. This film is a result of the hydraulic pressure that is generated by the rotation of the shaft. During rotation the oil is drawn into the clearance space between the journal and the shaft. The shaft is lifted on the film and separated from contact with the bearing. As the speed is increased, higher pressure develops and the shaft takes an eccentric position.

Clearance — To allow the formation of an oil film in a plain bearing, there must be clearance between the journal and the bearing. This clearance varies with the size of the shaft and the bearing material, the load carried, and the accuracy of shaft position desired. In industrial design of rotating machinery a diametral clearance of .001" per inch of shaft diameter is often used. This is a general base figure and requires adjustment for high speeds or heavy loading.

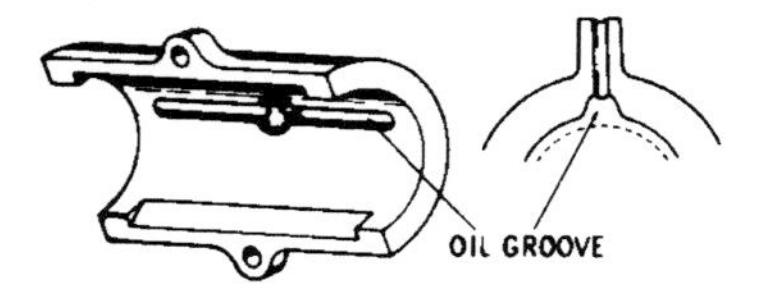

Fig. 124.

Lubrication Holes and Grooves — The simplest plain bearings have a hole in the top through which the lubricant travels to the journal. For longer bearings a groove or combination of grooves extending in either direction from the oil hole, as shown in Fig. 124, will distribute the oil.

The oil inlet hole is located in the center of the bearing, in the low-pressure region. It is important that the grooving be confined to the unloaded portion of the bearing surface. If grooving extends into the loaded or pressure region, the oil film will be disrupted, as the grooves will act as pressure relief passages.

Bearing Failures

Determination of the exact cause of a bearing failure is often a difficult matter. Usually the trouble lies in one or more of the following areas:

Unsuitable Materials — Under normal conditions babbit and bronze bearings may be used with soft steel journals. Harder bearing materials require a harder shaft surface.

Incorrect Grooving — If grooves are incorrectly located, interfering with oil-film formation, poor lubrication and bearing failure result.

Unsuitable Finish — The smoother the finish, the closer the surfaces may approach without metallic contact.

Insufficient Clearance — There must be sufficient clearance between journal and bearing to allow oil-film formation.

Operating Conditions — Speedup and overloading are the two most frequent causes of bearing failure.

Oil Contamination — Foreign material in the lubricant causes scoring and galling of the bearing surfaces.

Antifriction Bearing Types

Antifriction bearings are of two general types, *ball bearings* and *roller bearings.* They operate on the principle of rolling motion, using either balls or rollers between the rotating and stationary surfaces. Because of this, friction is reduced to a fraction of that in plain bearings, hence the name "antifriction."

Ball bearings may be divided into three main groups according to function: *radial, thrust,* and *angular contact.*

The radial bearing shown in Fig. 125A is designed primarily to carry a load in a direction perpendicular to the axis of rotation.

The thrust bearing shown in Fig. 125B can carry thrust loads only. That is a force parallel to the axis of rotation, tending to cause endwise movement of the shaft.

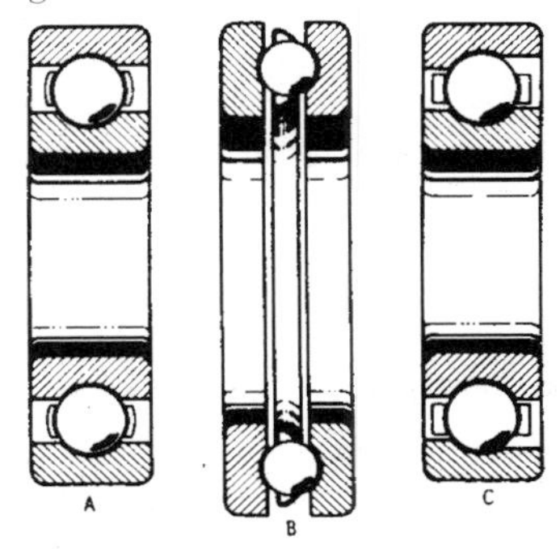

Fig. 125.

The angular contact bearing shown in Fig. 125C can support combined radial and thrust loads in one direction. Angular contact bearings must be installed in pairs to support thrust loads in both directions.

Roller bearings are also classified by their ability to support radial, thrust, and combination loads. In addition, they are further divided into styles according to the shape of their rollers. Widely used roller styles are: *cylindrical*, also known as *straight, taper, spherical,* and *needle*. Examples of bearings using these roller styles are shown in Fig. 126.

Combination load-supporting roller bearings are not called angular contact bearings, as they are quite different in design. The taper roller bearing, for example, is a combination load-carrying bearing by virtue of the shape of its rollers.

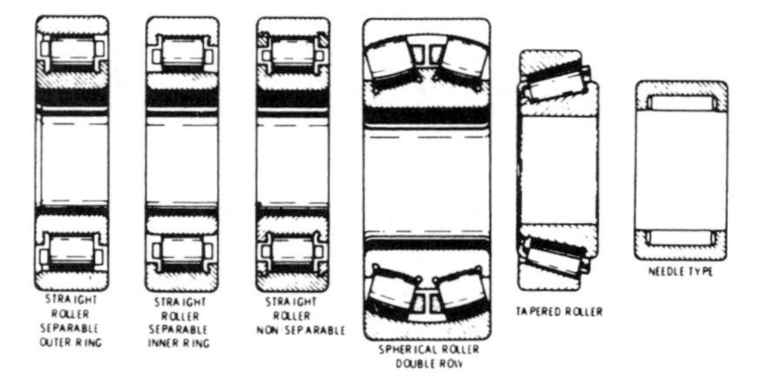

Fig. 126.

Self-Aligning Bearings

The *self-aligning* bearing is a specialized style of antifriction bearing which, as the name indicates, has the capability of angular self-alignment. This is usually accomplished by the use of a spherical raceway inside the outer ring of the bearing. The inner ring and the rolling elements rotate at right angles to the shaft center line in a fixed raceway. The position of the outer ring, because of its spherical raceway, may be misaligned within the limits of its width and still provide a true path for the rolling elements to follow. Bearings of this style are called *internal* self-aligning.

The *spherical double-row* roller bearing in Fig. 127A is a common self-aligning style of roller bearing. Another widely used style is the *self-aligning* ball bearing at (B). The angular movement this bearing allows is possible because the two rows of balls are rolling on the spherical inner surface of the outer ring. Another self-aligning ball bearing (single row) is shown at (C). It incorporates an additional outside ring with a spherical inner surface. The outside of the regular outer ring is made spherical to match the extra ring. This style of construction is used for single row self-aligning ball bearings. It is called an *external style* self-aligning ball bearing.

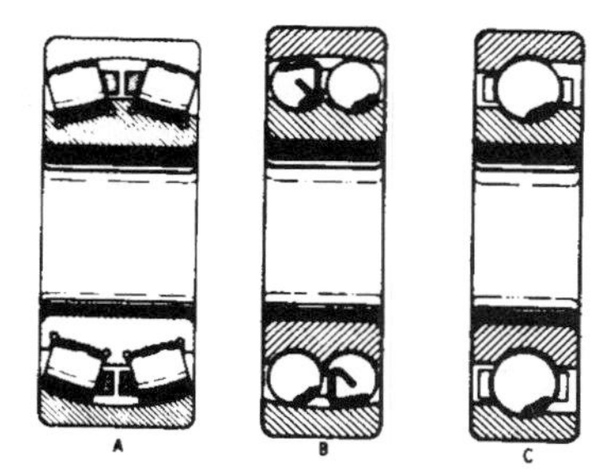

Fig. 127.

Another specialized style of bearing is the *wide inner ring* bearing. It is used primarily in pillow blocks, flange units, etc., termed transmission units. Wide inner ring bearings are commonly made in two types, rigid (type A), and self-aligning (type B). The rigid type has a straight cylindrical outside surface; the self-aligning type has a spherical outside surface. They are made to millimeter outside dimensions and inch bore dimensions.

They are an assembly of standard metric size ball bearing outer rings with special wide inner rings. The bore of the inner ring is made to inch dimensions to fit standard fractional-inch dimension shafting. As they are usually contained in an assembled unit, they are specified by their nominal fractional-inch bore size.

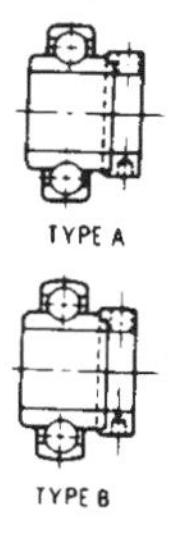

Fig. 128.

Antifriction Bearing Construction

All antifriction bearings consist of two hardened rings called the inner and outer rings, a separator, and hardened rolling elements which may be either balls or rollers. Bearing size is usually given in terms of what are called boundary dimensions. These are the *outside diameter*, the *bore*, and the *width*. The inner and outer rings provide continuous tracks or races for the balls or rollers to roll in. The separator or retainer properly spaces the rolling elements around the track and guides them through the load zone. Other words and terms used in describing antifriction bearings are the face, shoulders, corners, etc.

Terms used to describe roller bearings are a little different, in that what is normally the outer ring is called the *cup*, and the inner ring the *cone*. The word *cage* is standard for taper roller bearings, rather than separator or retainer.

The function of the separator, also called retainer or cage, is to properly space the rolling elements around the track and guide them through the load zone. As the bearing rotates and the rolling elements roll in the race, the separator rides with them. It is the weakest point in an antifriction bearing, as sliding friction is always present between the separator pockets and the rolling elements.

Ball Bearing Dimensions

Basic type bearings for general use in industry are manufactured to standardized dimensions of bore, outside diameter, and width (boundary dimensions). Tolerances for these critical dimensions and for the limiting dimensions for corner radii have also been standardized. Therefore, all types and sizes of ball bearings made to standardized specifications are satisfactorily interchangeable with other makes of like size and type.

Most basic ball bearings are available in for different "series" known as *extra light*, *light*, *medium*, and *heavy*. The names applied to each series are descriptive of the relative proportions and load-carrying capacities of the bearings. This means that there are as many as four bearings (one in each series) with the same bore size but with different widths, outside diameters, and load-carrying capacities.

It is also possible to select as many as four bearings with the same outside diameter (one in each series) with four different bore sizes, widths, and load-carrying capacities. Thus there is a choice of four different shaft sizes without changing the diameter of the housing.

Bearing manufacturers designate the various series by using numbers which they incorporate into their basic numbering systems. The extra light series is designated as the "100" series, the light as "200," the medium as "300," and the heavy as the "400" series.

Ball Bearing Numbering Systems

Antifriction ball bearings were first manufactured on a large scale in Europe, where the metric system of measurement is used. When the manufacture of ball bearings was started in America, the practice of using metric sizes was continued. Because of this, metric ball bearing sizes are interchangeable throughout the western world. However, the fact that they are made to even metric sizes results in dimensions that have no relation to inch fractions when converted to the inch system of measurement. For example, a ball bearing with a 60 millimeter bore measurement when converted to the inch system measures 2.3622 inches. Conversion of millimeter dimensions to inch dimensions is accomplished by multiplying the millimeter dimension by .03937, which is the equivalent of one millimeter in thousandths of an inch. For approximate conversion purposes, one millimeter is roughly 1/25 of an inch.

The basic ball bearing number is made up of three digits. The first digit indicates the bearing series, i.e., "100," "200," "300," or "400." The second and third digit, from 04 up, when multiplied by 5, indicates the bearing bore in millimeters. An example in each of the four duty series would be:

Basic Number	*Duty* Series	*Bore in* mm's	*Bore in* Inches
108	Extra Light	40	1.5748
205	Light	25	0.9843
316	Medium	80	3.1496
420	Heavy	100	3.9370

Ball bearings having a basic number under 04 have the following bore dimensions:

Basic Number	*Bore in* mm's	*Bore in* Inches
00	10	0.3937
01	12	0.4727
02	15	0.5906
03	17	0.6693

Cylindrical Roller Bearings (Numbers)

One of the most widely used types of roller bearings is one having rollers that are approximately equal in length and diameter. This specific type, which is called a "straight" roller bearing by some manufacturers, is made to the same dimensions as ball bearings. The same basic numbering system is used, although manufacture is limited to the light 200 series and the medium 300 series. This type of roller bearing is interchangeable with ball bearings of like size and series.

Standard Ball Bearing Sizes

The basic three digit ball bearing number indicates the bearing duty series and the bearing bore in millimeters. All standard ball bearings in any of the four duty series having the same last two digits in their number, have the same diameter bore. Listed in Table 16 are the standard bore sizes in nominal millimeters and equivalent decimal inches. Tables listing the boundary dimensions for standard size bearings in the 100, 200, 300, and 400, series are included in the appendix.

Spherical Roller Bearings

The double-row spherical roller bearing is a self-aligning bearing utilizing rolling elements shaped like barrels. The outer ring has a single spherical raceway. The double-shoulder inner ring has two spherical races separated by a center flange. The rollers are retained and separated by an accurately constructed cage.

This type of bearing is inherently self-aligning because the assembly of the inner unit (ring, cage, and rollers) is free to swivel

Table 16. Standard Ball Bearing Bore Sizes

Basic Bearing # Last 2 Digits	mm Bore Size	Inch Bore Size	Basic Bearing # Last 2 Digits	mm Bore Size	Inch Bore Size
00	10	0.3937	18	90	3.5433
01	12	0.4724	19	95	3.7402
02	15	0.5906	20	100	3.9370
03	17	0.6693	21	105	4.1339
04	20	0.7874	22	110	4.3307
05	25	0.9843	24	120	4.7244
06	30	1.1811	26	130	5.1181
07	35	1.3780	28	140	5.5118
08	40	1.5748	30	150	5.9055
09	45	1.7717	32	160	6.2992
10	50	1.9685	34	170	6.6929
11	55	2.1654	36	180	7.0866
12	60	2.3622	38	190	7.4803
13	65	2.5591	40	200	7.8740
14	70	2.7559	42	210	8.2677
15	75	2.9528	44	220	8.6614
16	80	3.1496	48	240	9.4488
17	85	3.3465			

within the outer ring. Thus there is automatic adjustment which allows successful operation under severe misalignment conditions. It will support a heavy radial load and heavy thrust loads in both directions.

Generally speaking, bearing applications have a rotating inner ring and a stationary outer ring. When correctly assembled, the inner ring is sufficiently tight on the shaft to ensure that both inner ring and shaft turn as a unit and "creeping" of the ring on the shaft does not occur. Should creeping occur, there will be overheating, excessive wear, and erosion between the shaft and the inner ring. For normal applications the inner ring is press-fitted to the shaft and/or clamped against a shoulder with a locknut. However, on applications subjected to severe shock or unbalanced loading, the usual press fit or locknut clamping does not grip tightly enough to prevent creeping. For such applications a design providing maximum grip of the inner ring on the shaft is required. In providing this tremendous grip two very important conditions must be controlled. First, the stress in the inner ring must remain below the elastic limit;

second, the internal bearing clearance must not be eliminated. Also, there must be a practical mounting method.

The taper bore self-aligning spherical roller bearing incorporates the features required. Size for size its capacity is greater than any other type of bearing. The taper bore provides a simple method of mounting which allows controlled stretching of the inner ring to obtain maximum gripping power. When mounting this style of bearing, the inner ring is forced on the taper by tightening a locknut. As the inner race stretches, and its grip increases, the internal bearing clearance is reduced. The bearing is manufactured with sufficient internal clearance to allow this stretching of the inner ring. The grip and the clearance are controlled by checking internal clearance before mounting, and by tightening the nut sufficiently to reduce the internal clearance by a specific amount.

Another feature that the taper bore bearing makes possible is the adapter-sleeve style of mounting. The use of a tapered adapter sleeve allows mounting the taper bore bearing on straight cylindrical surfaces. It also provides an easy means of locating the bearing. In many cases this feature is the reason for using the taper bore bearing for applications where loading is relatively light.

Installation of Taper-Bore Style — Before mounting spherical taper-bore roller bearings on taper-shaft fits or adapters, the internal clearance should be checked and recorded. The measurement is made on one side of the bearing only. Recommended practice is

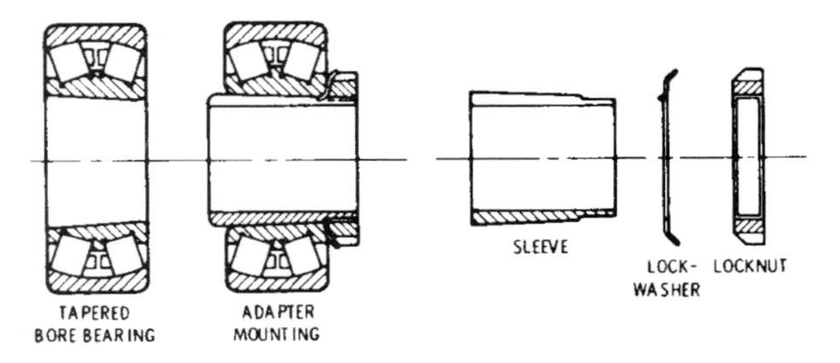

Fig. 129.

to rest the bearing upright on a table and insert the feeler gauge between the top roller and the inside of the outer ring.

Mounting Procedure

1. Check internal clearance between the rollers and the outer race on the open side. Feeler gauge must be inserted far enough to contact entire roller surface.
2. Lightly oil surface of bearing bore and install on taper shaft or taper adapter sleeve *without* lock washer.
3. Check internal clearance as nut is tightened. Nut must be tightened until internal clearance is reduced the amount shown in the table below.
4. Remove nut, install lock washer, and retighten nut. Secure lock washer.

Shaft Diameter	*Clearance Reduction*
1⅝ to 3½	.001 to .002
3½ to 6½	.002 to .0035
6½ to 10¼	.0035 to .0055

The above values are for most common applications. For special cases, such as high temperature equipment, bearings on hollow shafts with steam passing through, etc., consult manufacturers' specifications and instructions for special applications.

Taper Roller Bearings

Taper roller bearings are a separate group of roller bearings because of their design and construction. They consist fundamentally of tapered cone-shaped rollers operating between tapered raceways. They are so constructed, and the angle of all rolling elements so proportioned, that if straight lines were drawn from the tapered surfaces of each roller and raceway they would meet at a common point on the center line of the axis of the bearing, as illustrated in Fig. 130.

One major difference between taper roller bearings (separate types) and most antifriction bearings is that they are adjustable. In many cases this is a distinct advantage, as it permits accurate control

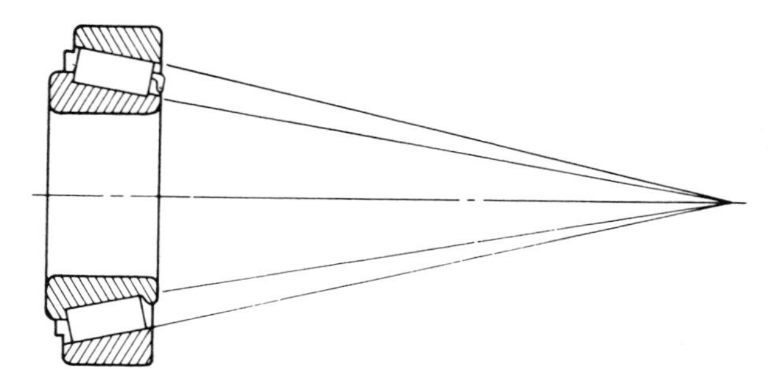

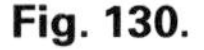

Fig. 130.

of bearing running clearance. The proper amount of running clearance may be assembled into the bearing at installation to suit the specific application. This feature also permits preloading the bearing for applications where extreme rigidity is required. This adjustable feature also requires that proper procedure be followed at assembly to ensure correct setting. Generally the best setting is one of minimum clearance, allowing free running with no appreciable end play.

The basic parts of a taper roller bearing are the *cone* or inner race, the *taper rollers*, the *cage* (which is called the retainer or separator), and the *cup* or outer race.

The most widely used taper roller bearing is a single-row type with cone, rollers, and cage factory-assembled into one unit, with the cup independent and separable (Fig. 131). Many additional styles of single and multiple row taper roller bearings are made, including unit assemblies factory preadjusted.

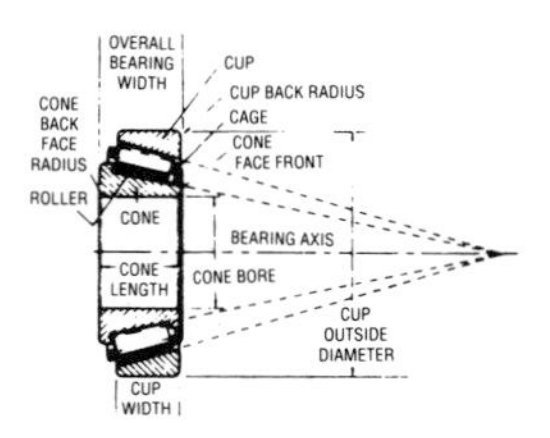

Fig. 131.

Taper Roller Bearing Adjustments—While each single row taper bearing is an individual unit, their construction is such that they must be mounted in pairs so that thrust may be carried in either direction. Two systems of mounting are employed, *direct* and *indirect*, as shown in Fig. 132.

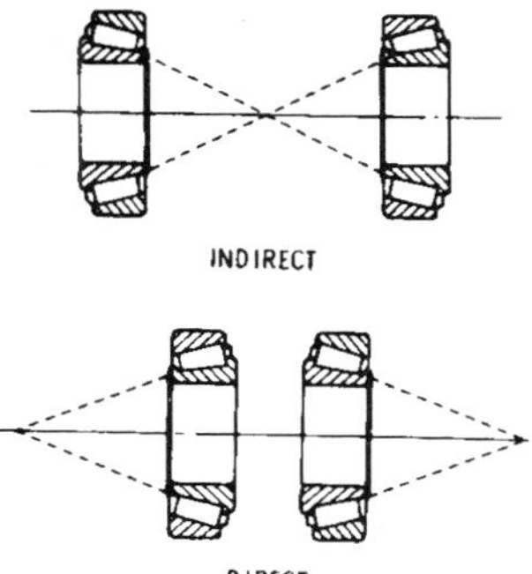

Fig. 132.

Indirect mounting is used for applications where maximum stability must be provided in a minimum width.

Direct mounting is used where this assembly system offers mounting advantages and maximum stability is not required or desired.

The terms *face-to-face* and *back-to-back* may also be used to describe these systems of mounting. If these terms are used they must be qualified by stating whether the terms are used in regard to the cup or the cone. Indirect and direct mounting may more appropriately be referred to as *cone-clamped* and *cup-clamped*. Cone-clamped describes indirect mounting which is normally secured by clamping against the cone. Cup-clamped describes direct mounting which is normally secured by clamping against the cup.

Many devices are used for adjusting and clamping taper roller

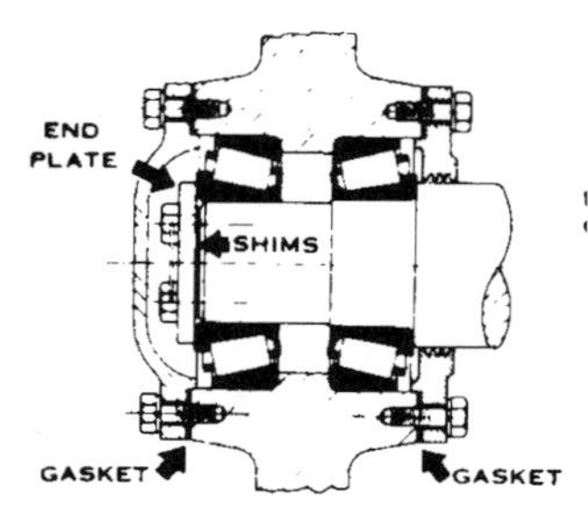

Fig. 133.

1. The slotted hex nut and the cotter pin are used to adjust bearings in implement wheels and automotive front wheels. Simply tighten the nut while rotating the wheel until a slight bind is obtained in the bearing. That insures proper seating of all parts. Then back off the nut one slot, lock with a cotter pin. Result is free-running clearance in the bearing. Bearing is rotated in order to seat rollers against the cone rib — should be done when making any bearing adjustment.

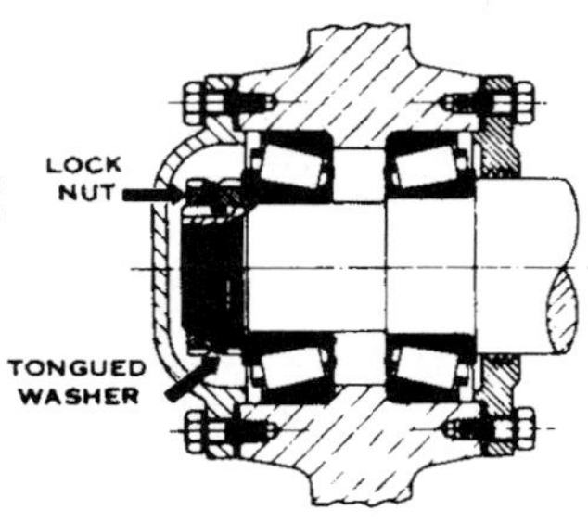

Fig. 134.

2. Here two standard lock nuts and a tongued washer are used to adjust the bearing. They provide a much finer adjustment than a slotted hex nut. Pull up the inner nut until there's a slight bind on the bearings; then back off just enough to allow running clearance after the outer or "jamb" nut is tightened. This type of adjustment device is used in full floating rear axles and industrial applications where slotted hex nuts are not practicable or desirable.

bearing assemblies. Figs. 133 to 138 are the basic ones; others are, in general, variations of these.

Fixed and Floating Bearings

Temperature variations will expand and contract the components of any machine. Because of this, it is essential that such parts be permitted to expand and contract without restriction. For that reason, only one bearing on any one shaft should be fixed axially in the housing (called a *fixed* or *held* bearing) to prevent axial or

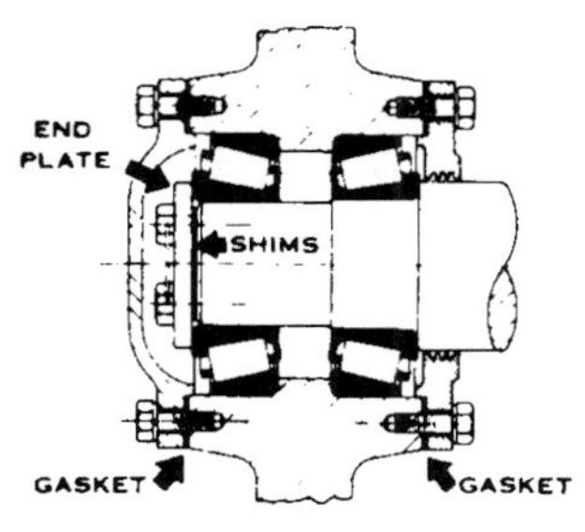

Fig. 135.

3. In this case, shims are used between the end of the shaft and the end plate. The shim pack is selected to give the proper setting of the bearing. This will vary with the unit in which the bearings are used. The end plate is held in place by cap screws. The cap screws are wired together for locking. A slot may be provided in the end plate to measure the shim gap.

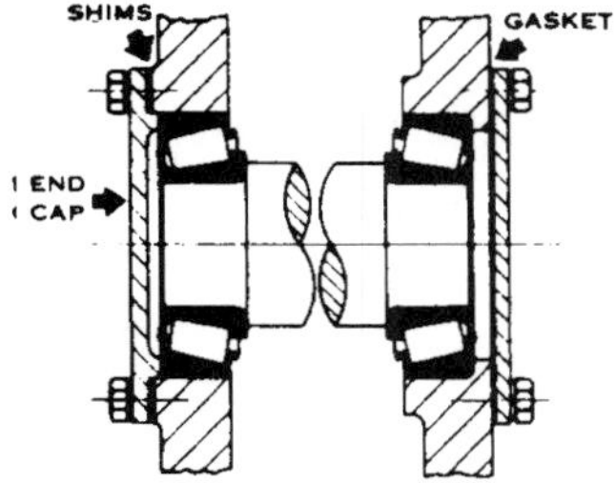

Fig. 136.

4. Here is another adjustment using shims. They are located between the end cap flange and the housing. Select the shim pack which will give the proper bearing running clearance recommended for the particular application. End cap is held in place by cap screws, which can be locked with lock washer (as shown) or wired. This easy method of adjustment is used with loose-fitted cups.

endwise motion. All other bearings on that same shaft should have adequate axial clearance in the housing (referred to as *floating* or *free* bearings).

The illustration in Fig. 139 shows a typical fixed and floating bearing assembly. The fixed bearing is clamped securely in its housing. The floating bearing has clearance on both sides within the housing. Thus movement is allowed as changes in temperature cause the shaft to increase or decrease in length.

Generally it is preferred to hold the bearing at the drive end. However, consideration is sometimes given to bearing load, and the

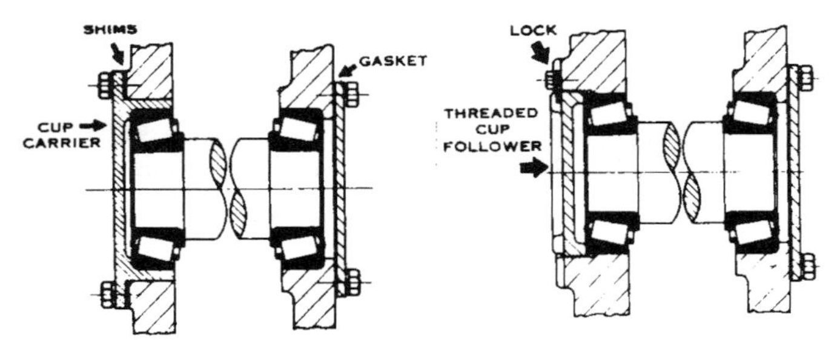

Fig. 137.

5. Shims are also used in this adjustment, as shown in No. 4. The difference here is that one cup is mounted in a carrier. This adjusting device is commonly used in gear boxes and drives. It is also used for industrial applications for ease of assembly or disassembly.

Fig. 138.

6. The threaded cup follower shown above is another common adjusting device. It is used in gear boxes, drives, or automotive differentials. The follower is locked by means of a plate and cap screws. The plate fits between the lugs of the follower.

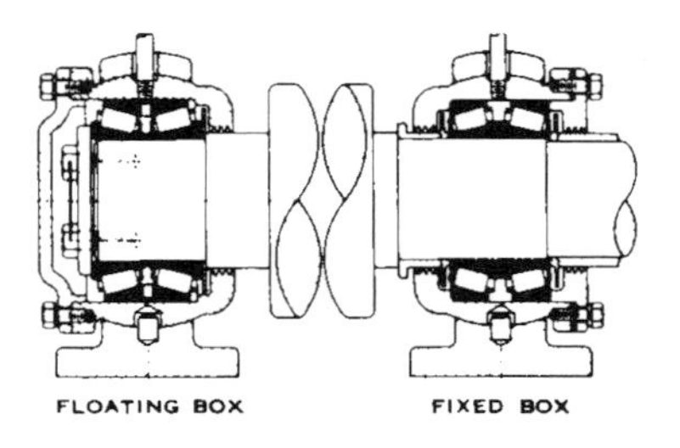

Fig. 139.

bearing carrying the smaller radial load is fixed. This is usually in cases where the fixed bearing is subjected to a thrust load of some magnitude and distributing the load in this manner tends to equalize it.

Another very important consideration in the determination of which bearing should be fixed is operating clearances. When the clearances of components rotating with the shaft must be held to close tolerances, the closest bearing to the close clearance point should be fixed.

In gear reducers where herringbone gears are used, only one bearing on one of the shafts should be fixed, usually the output shaft. In this style of assembly the "V" shape of the gear teeth will locate the mating gear and shaft axially.

When a flexible coupling is used to connect two shafts, a fixed bearing is required on each end of the shafts, as a flexible coupling permits endwise motion of both shafts.

Antifriction Bearing Care and Handling — The principal reason for antifriction bearing failure is the entrance of dirt or grit. Second is mechanical damage from improper handling. The two most important rules, therefore, in handling antifriction bearings are:

1. Keep bearings and parts clean.
2. Apply force to the tight fitting ring only.

PUMPS

Pumps are broadly classified with respect to their construction, or the service for which they are designed. The three groups into which most pumps in common use fall are: *centrifugal*, *reciprocating*, and *rotary*.

Centrifugal Pumps

Centrifugal force, from which this pump takes its name, acts upon a body moving in a circular path, tending to force it farther from the center of the circle (Fig. 140).

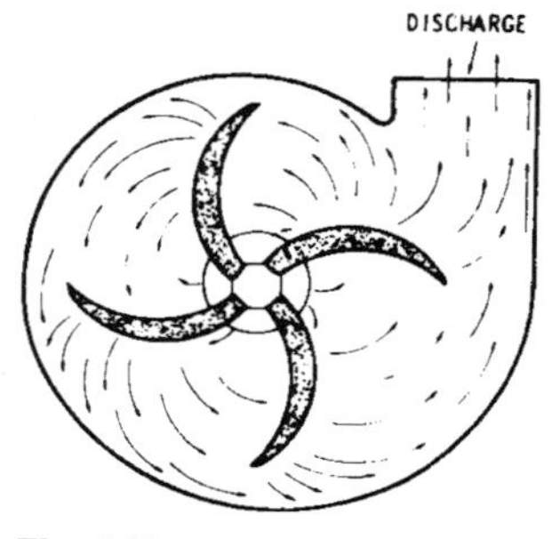

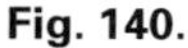
Fig. 140.

Inside the body of a centrifugal pump the impeller forces the liquid to revolve and generate centrifugal force. The impeller blades, as shown at right, are usually curved backward with reference to the direction of rotation. The liquid is drawn in through the center or "eye" of the impeller, whirled around by the blades and thrown outward by the centrifugal force, and passes through the discharge outlet.

There are two principal classes of centrifugal pumps, single-stage and multistage. The single-stage pump has a single impeller. By arranging a number of centrifugal pumps in series so the discharge of one is led to the suction of the succeeding pump, the head or pressure may be multiplied as required. Multistage pumps are made with a common housing, and internal passages so arranged that liquid flows from the discharge of one stage to the inlet of the next.

Reciprocating Pumps

The reciprocating pump has a back-and-forth motion as the pumping element alternately moves forward and backward. It moves liquid by displacing the liquid with a solid, usually a piston or plunger. The principle of operation is called *positive displacement.*

The *piston* pumping element is a relatively short cylindrical part that is moved back and forth in the pump chamber, or cylinder. The distance that the piston travels back and forth, called the *stroke*, is generally greater than the length of the piston. Leakage past the piston is usually controlled by packings or piston rings. The piston in normal operation moves back and forth within the cylinder.

A *plunger* pumping element is generally longer than the stroke of the pump. In operation the plunger moves into and withdraws from the cylinder. To prevent leakage past the plunger, packings are contained in the end of the cylinder through which the plunger moves.

As the pumping element in a reciprocating pump travels to and fro, liquid is alternately moved into the pump chamber and moved out. The period during which the element is withdrawing from the chamber and liquid is entering is called the *suction*, or *intake* stroke. Travel in the opposite direction during which the element displaces the liquid is called the *discharge* stroke. Check valves are placed in the suction and discharge passages to prevent backflow of the liquid. The valve in the suction passage is opened and the discharge passage valve is closed during the suction stroke. Reversal of liquid flow on the discharge stroke causes the suction valve to close and the discharge valve to open.

Fig. 141 illustrates the position of the valves during travel in each direction. At one end of each cylinder the suction valves are open to admit liquid and the discharge valves closed to prevent backflow from the discharge passage. On reversal of direction, the suction valves are closed to prevent backflow into the suction passage, and liquid moves out through the open discharge valves.

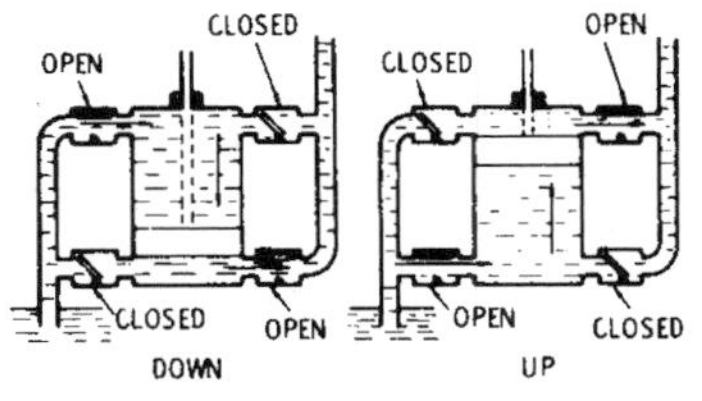

Fig. 141.

Rotary Pumps

Rotary pumps are also positive-displacement type pumps in operation. As their flow is continuous in one direction, no check valves are required. Different designs make use of such elements as vanes, gears, lobes, cams, etc., to move the material. The principle of operation is

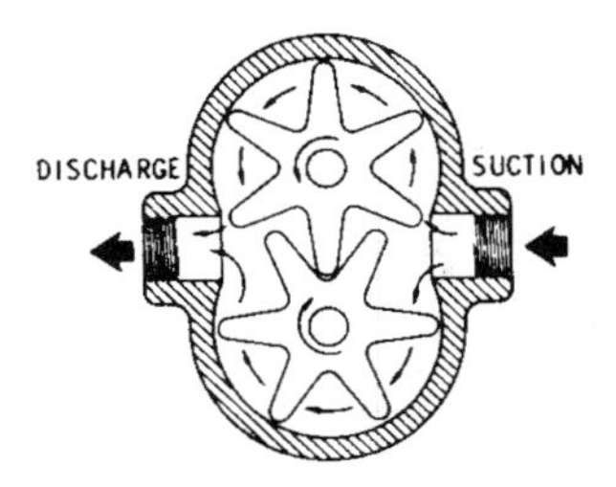

Fig. 142.

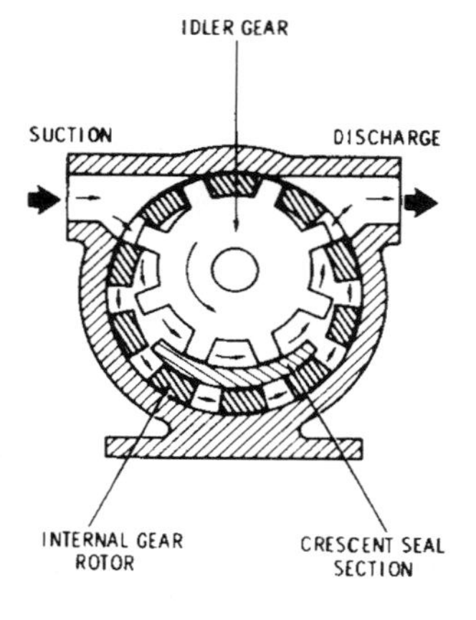

Fig. 143.

similar with all of these elements, in that the element rotates within a close-fitting casing which contains the suction and discharge connections (Fig. 142).

At the pump suction port the liquid enters chambers formed by spaces in the elements, or between the surface of the elements and the internal chamber surface. The liquid is carried with the elements as they rotate, and it is literally squeezed out the discharge as the elements mesh or the volume of the chambers is reduced to practically zero (Fig. 143).

Rotary pumps have close running clearances and generally are self-priming. In operation they produce a very even continuous flow with almost no pulsation. The delivery capacity is constant regardless of pressure, within the limits of operating clearances and power.

Table 17. Troubleshooting

Trouble Symptom	Key to Causes
Pump fails to discharge	1, 2, 3, 4, 6, 7, 12
Pump discharges then stops	4, 8, 9, 10, 11
Pump not up to capacity	1, 2, 4, 7, 8, 12, 13, 15
Pump noisy or vibrates	7, 11, 12, 14, 16, 17, 18, 19, 23
Pump takes too much power	5, 14, 17, 20, 21, 22, 24

Trouble Causes

1. Suction or discharge valves closed
2. Direction of rotation wrong
3. Lift too high
4. Supply level low

5. Supply level high
6. Pump not primed
7. Blocked lines
8. Leaks in stuffing box
9. Leaks in suction line
10. Vents blocked
11. Air or vapor in liquid
12. Impeller or rotor damaged
13. Speed too low
14. Misalignment
15. Pump worn or excessive clearances
16. Shaft bent
17. Binding or rotating elements
18. Cavitation
19. Worn or defective pump or motor bearings
20. Motor undersize
21. Specific gravity
22. Viscosity of material changed
23. Relief valve chattering
24. Stuffing box tight

STRUCTURAL STEEL

American Standard Angles

The symbol used to indicate an angle shape is (∠). The usual method of billing is to state the symbol, then the long leg, the short leg, the thickness, and finally the length. For example, ∠ 6 × 4 × 3/8 × 12'4". When the legs are equal, both lengths are stated (Fig. 144).

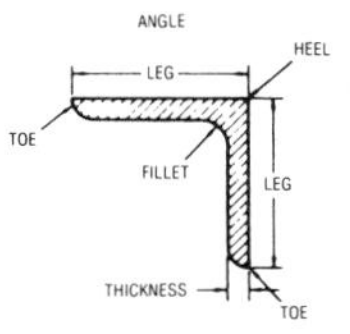

Fig. 144.

American Standard Beams

Called "I" beams because of their resemblance to the capital letter "I." The symbol used to indicate the beam shape is the letter (I). The usual method of billing is to state the depth, the symbol, the weight per foot, and finally the length. For example, 15 I 42.9 X 18'4½" (Fig. 145).

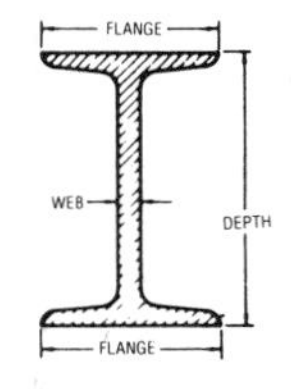

Fig. 145.

American Standard Channels

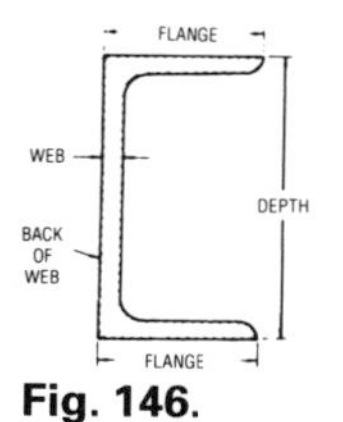

Fig. 146.

May be compared to an I beam that has been trimmed on one side to give a flat back web. The symbol used to indicate the standard channel is ([). The usual method of billing is to state the depth, the symbol, the weight per foot, and finally the length. For example, 10 [15.3 - 16'16" (Fig. 146).

American Standard Wide Flange Beams or Columns

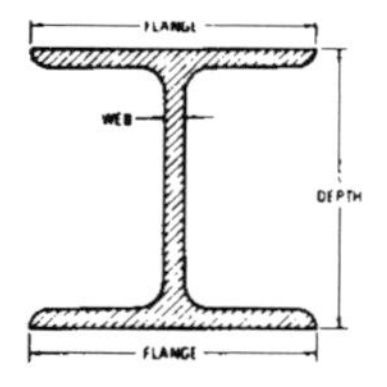

Fig. 147.

Also referred to as "B," "CB," or "H" shapes. May also be compared to an I beam with extra wide flanges. The symbol used to indicate the standard wide flange shape is (WF). The usual method of billing is to state the depth, the symbol, the weight per foot, and finally the length. For example, 12 (WF) 45 — 24'8" (Fig. 147).

Structural steel is produced at rolling mills in a wide variety of standard shapes and sizes. In this form it is referred to as "plain material." In addition to the plates and bar stock, the four shapes illustrated above are the most widely used. Other standard shapes produced are *Tee's, Zee's, Rails,* and various special shapes. All standard structural shapes are made to a standardized series of nominal sizes. Within each size group there is a wide range of weights and dimensions.

Simple Square-Framed Beams

Square-framed beams are, as the name implies, beams that intersect or connect at right angles. This is the most common type of steel construction. Two types of connections may be used in framed construction — *framed* and *seated*. In the framed type, shown in Fig. 148A, the beam is connected by means of fittings

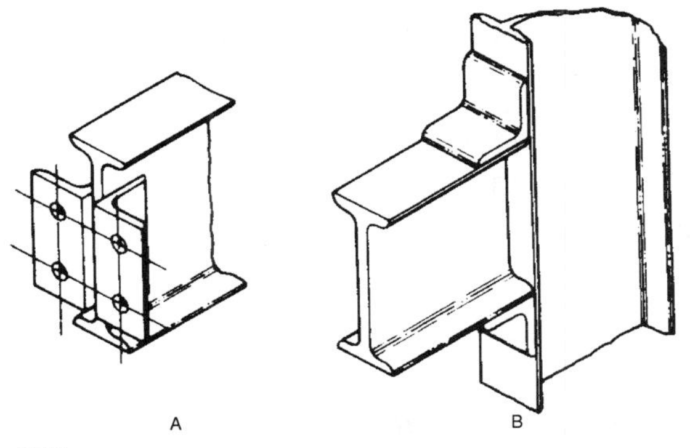

Fig. 148.

(generally a pair of angle irons) attached to its web. With the seated connection (Fig. 148B) the end of the beam rests on a ledge or seat.

Clearance Cuts

When connecting one member to another, it is often necessary to notch or cut away both flanges of the entering member to avoid flange interference. Such a notch is called a *cope*, a *block*, or a *cut*. The term "cope" is usually used if the cut is to follow closely the shape of the member into which it will fit. When the cut is rectangular in shape with generous clearance, it is usually called a "block-out." Unless there is some reason for a close-matching fit, the block-out is recommended, as it is the easiest and most economical notch to make (Fig. 149).

When making block-out cuts, the dimensions of the rectangular notch may be obtained from tables of structural steel dimensions. The tables list the "K" and "a" dimensions for various size and weight members. The "K" and "a" values determine the dimensions of the cut, as they indicate the maximum points of interference. While the notch is made in the entering member, the values from the table for the supporting member determine the notch dimensions. The steel must be cut to length before the block-out cut is made.

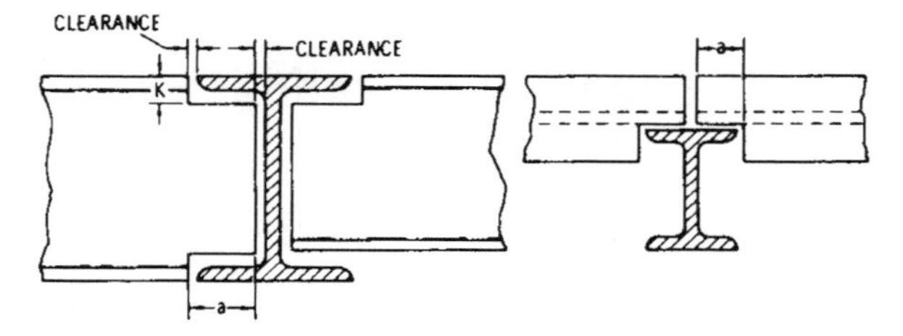

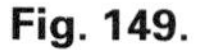
Fig. 149.

Square-Framed Connections (Two-Angle Type)

"Standard" connections should be used when fabricating structural steel members to ensure proper assembly with supporting members at installation. Some of the terms, dimensions, etc., of standardized connections are the following:

Spread — The distance between hole centers in the web of the supporting member, and the holes in the *attached* connection angles, is called the *spread* of the holes. The spread dimension is standardized at 5½ inches, as shown in Fig. 150.

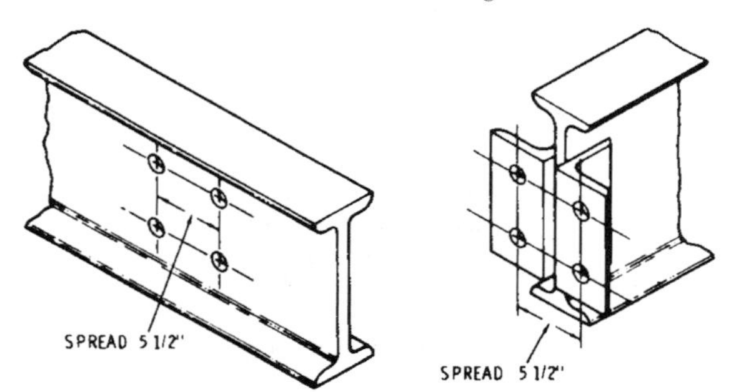

Fig. 150.

Angle Connection Legs

The legs of the angles used as connections are specified according to the surface to which they are connected, as shown in Fig. 151. The legs which attach to the entering steel to make the connection are termed *web* legs. The legs of the angles that attach to the supporting member are termed *outstanding* legs. The lines on which the holes are placed are called *gauge* lines. The distances between gauge lines, or from a gauge line to a known edge, are called *gauges*.

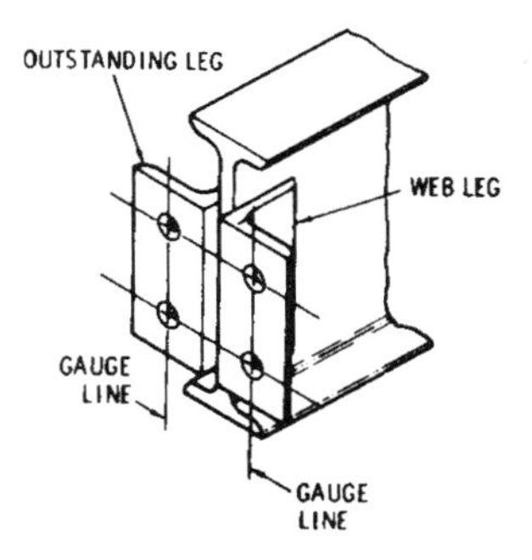

Fig. 151.

Fabrication Terms

Commonly used structural steel fabrication terms are illustrated in Fig. 152. Use of these terms, because they are descriptive, aids in understanding of steel fabrication and reduces the probability of errors.

Steel — Various structural steel shapes and forms.

Member — An assembly of a length of steel and its connection fittings.

Center-to-Center — The distance from the center line of one member to the center line of another member.

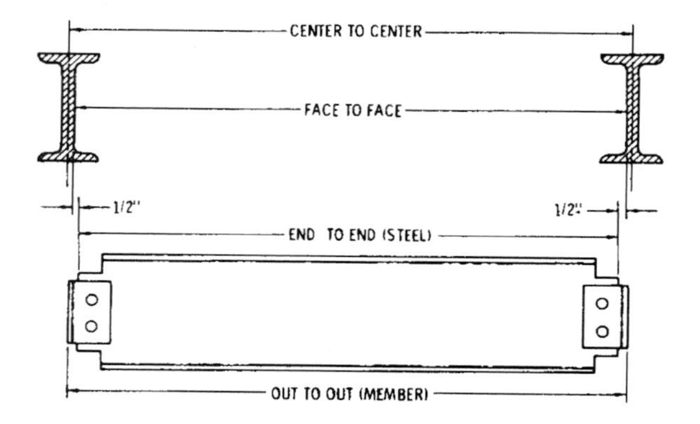

Fig. 152.

Face-to-Face—The distance between the facing web surfaces of two members. It is the center-to-center distance minus the ½-web thickness of each member.

End-to-End—The overall length of the steel. It should be 1 inch shorter than the opening into which it will be placed. The 1-inch clearance is provided for assembly and as an allowance for inaccuracies.

Out-to-Out—The overall length of the assembled member. To provide assembly clearance, the connection angles are positioned on the member so the out-to-out distance before assembly is slightly less than the face-to-face distance.

Connection Hole Locations

The various terms and the constant dimension for standard square-framed two-angle type connections are illustrated in Fig. 153.

Web-Leg Gauge—The distance from the heel of the angle to the first gauge line on the web leg is called "web-leg gauge." This dimension is constant, as it is standardized at 2¼ inches.

Outstanding-Leg Gauge — The distance from the heel of the angle to the first gauge line on the outstanding leg is called the "outstanding-leg gauge." This dimension varies, as the thickness of the web of the member varies in order to maintain a constant 5½-inch spread dimension. The outstanding leg-gauge dimension is determined by subtracting the web thickness from 5½ inches and dividing by two. A simpler way to determine the outstanding-leg dimension is to subtract the ½-inch web thickness from 2¾ inches, which is one-half the spread.

Gauges — The distance between gauge lines, or from a gauge line to a known edge, are called "gauges." When more than one row of holes is used, the gauge is 2½ inches. This dimension is constant.

Pitch — The distance between holes on any gauge line is called "pitch." This dimension is standardized at 3 inches.

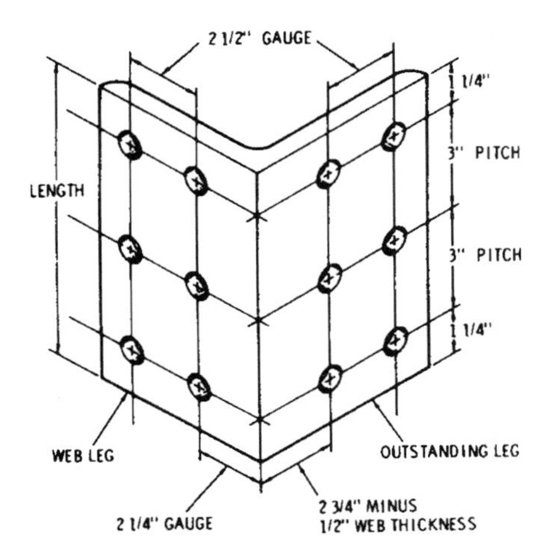

Fig. 153.

End Distance — The end distance is equal to one-half of the remainder left after subtracting the sum of all pitches from the length of the angle. By common practice the angle length is selected to give a 1¼-inch end distance.

Dimensions of Two-Angle Connections

Detail drawings for fabrication of structural steel members specify dimensions, hole locations, etc. A typical detail drawing for a two-angle connection is illustrated in Fig. 154. Note that the angles are slightly above center, the uppermost hole being 3 inches below the top of the beam. Whenever practicable, the uppermost holes are set at 3 inches below the top of the beam, as it makes for standardization and tends to reduce errors in matching connections.

The selection of steel sizes is usually done by persons competent to make strength calculations based on anticipated loadings. Connection details, however, may not be provided, and the selection of connections may be left to the fabricator. To aid in the selection of connections, Table 18 gives two-angle connections and two-angle connection dimensions.

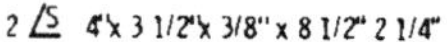

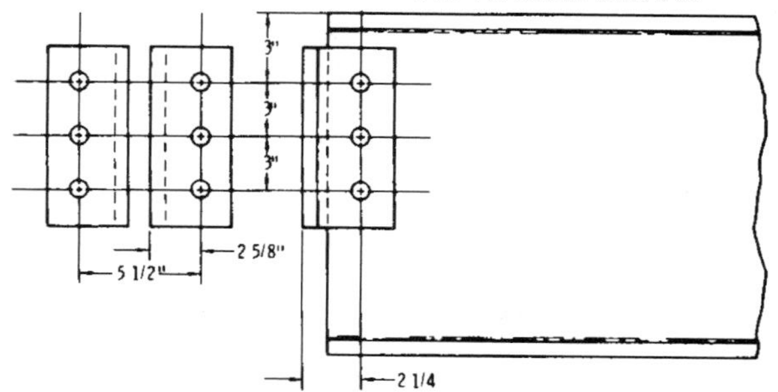

Fig. 154.

For purposes of simplification, the connections have been given numbers 1 through 6. While these are not standard connection numbers, dimensions conform to connection standards. Note also that only connections using ¾-inch diameter bolts are shown. This is also for the purpose of simplification, as ¾-inch bolts are adequate for use with the steel sizes for which these connections are used.

To select angle connections, find the size in the first column of the appropriate table, and the weight per foot in the second column. Follow along the weight line to the column that corresponds to the span length. The number in this column indicates the connection angle to use. The dimensions to which the selected angles should be fabricated are given in Fig. 155.

Table 18. Two-Angle Connections for Uniformly Loaded Beams and Channels

Size	Weight	Span in Feet												
Inches	Lbs/Ft	4	6	8	10	12	14	16	18	20	22	24	26	28
24	120									6	6	6	6	6
	100							6	6	6	6	6	6	6
	79.9							6	6	6	6	6	6	6
20	95							5	5	5	5	5	5	5
	65.4					5	5	5	5	5	5	5	5	5
18	70					4	4	4	4	4	4	4	4	4
	54.7					4	4	4	4	4	4	4	4	4
15	50			4	4	4	4	4	4	4	4	4	4	4
	42.9			4	4	4	4	4	4	4	4	4	4	4
12	50			3	3	3	3	3	3	3	3	3	3	3
	31.8			3	3	3	3	3	3	3	3	3	3	3
10	35			2	2	2	2	2	2	2	2	2	2	2
	25.4		2	2	2	2	2	2	2	2	2	2	2	2
8	23	2	2	2	2	2	2	2	2	2	2	2	2	2
	18.4	2	2	2	2	2	2	2	2	2	2	2	2	2
7	20	1	1	1	1	1	1	1	1	1	1	1	1	1
	15.3	1	1	1	1	1	1	1	1	1	1	1	1	1
6	17.2	1	1	1	1	1	1	1	1	1	1	1	1	1
	12.5	1	1	1	1	1	1	1	1	1	1	1	1	1
5	14.7	1	1	1	1	1	1	1	1	1	1	1	1	1
	10	1	1	1	1	1	1	1	1	1	1	1	1	1

Table 19. Two-Angle Connections for Uniformly Loaded Wide Flange Beams

Size	Weight	Span in Feet												
Inches	Lbs/Ft	4	6	8	10	12	14	16	18	20	22	24	26	28
24	120												6	6
	76							6	6	6	6	6	6	6
21	96							6	6	5	5	5	5	5
	62						5	5	5	5	5	5	5	5
18	70						5	5	5	5	5	4	4	4
	50						4	4	4	4	4	4	4	4
16	96									4	4	4	4	4
	64						4	4	4	4	4	4	4	4
	36				4	4	4	4	4	4	4	4	4	4
14	38						3	3	3	3	3	3	3	3
	34					3	3	3	3	3	3	3	3	3
	30					3	3	3	3	3	3	3	3	3
12	36				3	3	3	3	3	3	3	3	3	3
	31					3	3	3	3	3	3	3	3	3
	27				3	3	3	3	3	3	3	3	3	3
10	29			2	2	2	2	2	2	2	2	2	2	2
	25			2	2	2	2	2	2	2	2	2	2	2
	21		2	2	2	2	2	2	2	2	2	2	2	2
8	20	2	2	2	2	2	2	2	2	2	2	2	2	2
	17	2	2	2	2	2	2	2	2	2	2	2	2	2

Note: Connections must not be used for shorter spans than indicated as their capacity does not equal that of the steel in shorter spans.

TWO-ANGLE CONNECTION DIMENSIONS

13⁄16" Holes For 3⁄4" Bolts

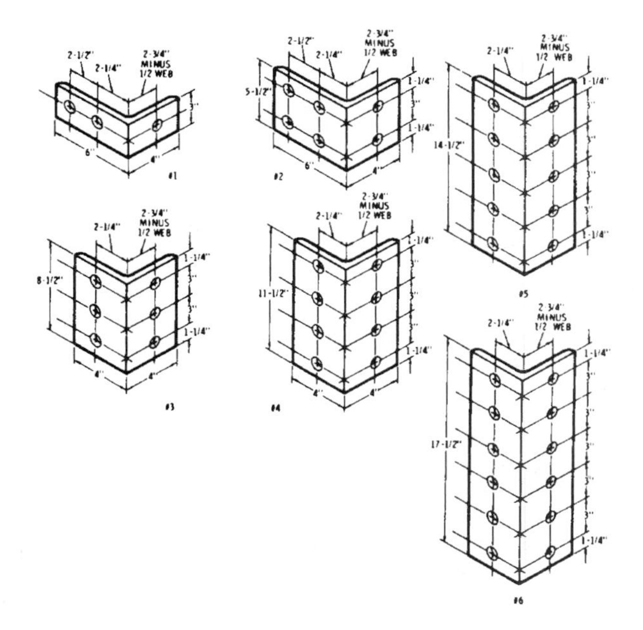

Fig. 155.

Steel Elevation

Unless otherwise stated on the drawing, all square framed beam members are presumed to be parallel or at right angles to one another. It is also presumed that their webs are in a vertical plane and that they are in a level position end to end. Elevation information is usually given by note, stating the vertical distance above some established horizontal plane.

Table 20. Standard Channels

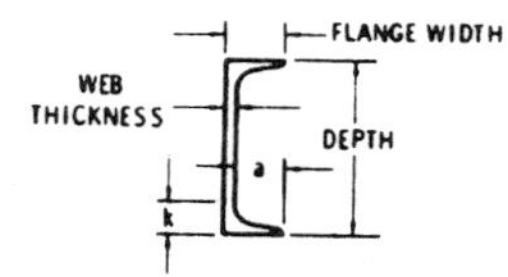

Depth	Weight Per Ft.	Flange Width	Web Thick.	a	k
3"	4.1 5.0 6.0	1⅜ 1½ 1⅝	3/16 ¼ ⅜	1¼ 1¼ 1¼	⅝ ⅝ ⅝
4"	5.4 7.25	1⅝ 1¾	3/16 5/16	1⅜ 1⅜	⅝ ⅝
5"	6.7 9.0	1¾ 1⅞	3/16 5/16	1½ 1½	11/16 11/16
6"	8.2 10.5 13.0	1⅞ 2 2⅛	3/16 5/16 7/16	1¾ 1¾ 1¾	¾ ¾ ¾
7"	9.8 12.25 14.75	2⅛ 2¼ 2¼	¼ 5/16 7/16	1⅞ 1⅞ 1⅞	13/16 13/16 13/16
8"	11.5 13.75 18.75	2¼ 2⅜ 2½	¼ 5/16 ½	2 2 2	13/16 13/16 13/16
9"	13.4 15.0 20.0	2⅜ 2½ 2⅝	¼ 5/16 7/16	2¼ 2¼ 2¼	⅞ ⅞ ⅞
10"	15.3 20.0 25.0 30.0	2⅝ 2¾ 2⅞ 3	¼ ⅜ 9/16 11/16	2⅜ 2⅜ 2⅜ 2⅜	15/16 15/16 15/16 15/16
12"	20.7 25.0 30.0	3 3 3⅛	5/16 ⅜ ½	2⅝ 2⅝ 2⅝	1 1/16 1 1/16 1 1/16
15"	33.9 40.0 50.0	3⅜ 3½ 3¾	7/16 9/16 ¾	3 3 3	1 5/16 1 5/16 1 5/16
18"	42.7 45.8 51.9 58.0	4 4 4⅛ 4¼	7/16 ½ ⅝ 11/16	3½ 3½ 3½ 3½	1 5/16 1 5/16 1 5/16 1 5/16

Table 21. Standard Beams

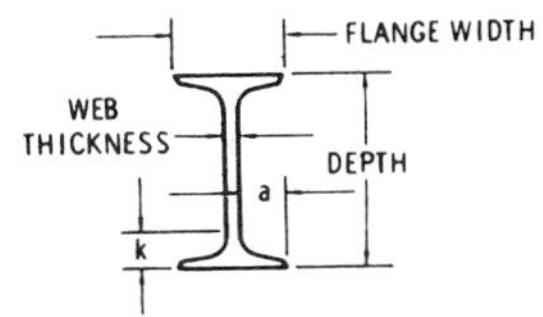

Depth	Weight Per Ft.	Flange Width	Web Thick.	a	k
3"	5.7 7.5	2⅜ 2½	3⁄16 ⅜	1⅛ 1⅛	9⁄16 9⁄16
4"	7.7 9.5	2⅝ 2¾	3⁄16 5⁄16	1¼ 1¼	⅝ ⅝
5"	10.0 14.75	3 3¼	¼ ½	1⅜ 1⅜	11⁄16 11⁄16
6"	12.5 17.75	3⅜ 3⅝	¼ ½	1½ 1½	¾ ¾
7"	15.3 20.0	3⅝ 3⅞	¼ 7⁄16	1¾ 1¾	13⁄16 13⁄16
8"	18.4 23.0	4 4⅛	5⁄16 7⁄16	1⅞ 1⅞	⅞ ⅞
10"	25.4 35.0	4⅝ 5	5⁄16 ⅝	2⅛ 2⅛	1 1
12"	31.8 35.0	5 5⅛	⅜ 7⁄16	2⅜ 2⅜	1⅛ 1⅛
12"	40.8 50.0	5¼ 5½	½ 11⁄16	2⅜ 2⅜	1 5⁄16 1 5⁄16
15"	42.9 50.0	5½ 5⅝	7⁄16 9⁄16	2½ 2½	1¼ 1¼
18"	54.7 70.0	6 6¼	½ ½	2¾ 2⅞	1⅜ 1 9⁄16
20"	65.4 75.0	6¼ 6⅜	⅝ ¾	2⅞ 2¾	1 9⁄16 1⅜
20"	85.0 95.0	7 7¼	11⁄16 13⁄16	3¼ 3¼	1¾ 1¾

Table 22. Wide Flange — CB Sections

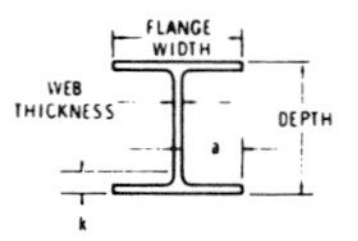

Nom. Depth	Weight Per Ft.	Flange Width	Web Thick.	a	k
4"	10	4	¼	1 ⅞	7/16
5"	16	5	¼	2 ⅜	⅝
6"	12	4	¼	1 ⅞	9/16
	15.5	6	¼	2 ⅜	9/16
8"	13	4	¼	1 ⅞	9/16
	17	5 ¼	¼	2 ½	⅝
	20	5 ¼	¼	2 ½	11/16
	24	6 ½	¼	3 ⅛	13/16
	28	6 ½	5/16	3 ⅛	13/16
	31	8	5/16	3 ⅞	13/16
	35	8	5/16	3 ⅞	⅞
	40	8 ⅛	⅜	3 ⅞	15/16
	48	8 ⅛	7/16	3 ⅞	1 1/16
	58	8 ¼	½	3 ⅞	1 3/16
	67	8 ¼	9/16	3 ⅞	1 5/16
10"	15	4	¼	1 ⅞	9/16
	21	5 ¾	¼	2 ¾	11/16
	25	5 ¾	¼	2 ¾	13/16
	29	5 ¾	5/16	2 ¾	⅞
	33	8	5/16	3 ⅞	15/16
	39	8	5/16	3 ⅞	1 1/16
	45	8	⅜	3 ⅞	1 ⅛
	49	10	⅜	4 ⅞	1 1/16
	60	10 ⅛	7/16	4 ⅞	1 3/16
	72	10 ⅛	½	4 ⅞	1 5/16
	100	10 ⅜	11/16	4 ⅞	1 ⅝
12"	27	6 ½	¼	3 ⅛	13/16
	36	6 ⅝	5/16	3 ⅛	15/16
	40	8	5/16	3 ⅞	1 ⅛
	50	8 ⅛	⅜	3 ⅞	1 ¼
	58	10	⅜	4 ⅞	1 ¼
	65	12	⅜	5 ¾	1 3/16
	106	12 ¼	⅝	5 ¾	1 9/16
14"	30	6 ¾	5/16	3 ¼	⅞
	48	8	⅜	3 ⅞	1 3/16
	68	10	7/16	4 ¾	1 5/16
	84	12	7/16	5 ¾	1 ⅜

TWIST DRILLS

Drill Terms

For general-purpose drilling of steel, a twist drill point angle of 118 degrees is generally recommended. For hard and tough materials, and for field work using a drill motor, a point angle of 135 degrees is recommended.

The heel or clearance angle for the 118-degree point should be about 8 to 12 degrees. The 135-degree point should have a clearance angle of about 6 to 9 degrees.

A twist drill cuts by wedging under the material and raising a chip. The steeper the point and the greater the clearance angle, the easier it is for the drill to penetrate. The blunter the point and the smaller the lip clearance, the greater is the support for the cutting edges. Thus, the greater point angle and lesser lip clearance is suitable for hard and tough materials, and the decreased point angle and increased lip clearance for softer materials.

When drilling some of the nonferrous metals, there is a tendency for the drill point to bite in, or penetrate too rapidly. To overcome this, the cutting edge of the drill is slightly flattened in front for brass and copper.

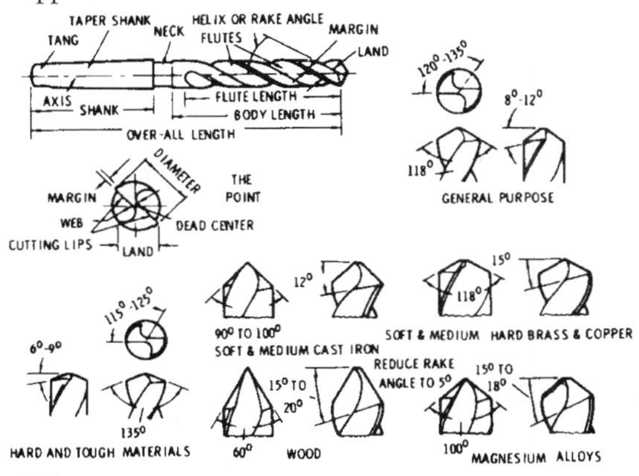

Fig. 156.

Drill Sharpening

Fig. 157

A new twist drill starts to wear as soon as it is placed in operation. Wear starts as a dulling along the cutting edges or lips and also a slight rounding of the corners, as shown at left. These dulled edges result in heat generation and a faster rate of wear which tends to extend back along the margins. When regrinding a twist drill all of this worn section must be removed. Sharpening the edges or lips only, without removing the worn margins, will not properly recondition a twist drill.

Most drills are made with webs which increase in thickness toward the shank. Therefore, after several sharpenings and shortenings of a twist drill the web thickness at the point increases, resulting in a longer chisel edge as shown at left. When this occurs it is necessary to reduce the web so that the chisel edge is restored to its normal length. This operation is called *web-thinning.*

Several different types of web-thinning are in common use. The method shown in Fig. 157 is perhaps the most common. The length A is usually made about ½ to ¾ the length of the cutting lip.

After the worn portion of the drill has been removed and the web thinned if necessary, the surfaces of the point must be reground. These two conical surfaces intersect with the faces of the flutes to form the cutting lips, and with each other to form the chisel edge. As in the case of any other cutting tool, the surface back of these cutting lips must not rub on the work, but must be relieved in order to permit the cutting edge to penetrate. Without such relief the drill could not penetrate the metal, but would only rub around and around.

In addition to grinding the conical surfaces to give the correct point angle and cutting clearance, both surfaces must be ground alike. Regardless of the point angle, the angles of the two cutting lips (A1 and A2) must be equal. Similarly, the lengths of the two lips (L1 and L2) must be equal. Drill points of unequal angles or lips of unequal lengths will result in one cutting edge doing most of the cutting. This type of point will cause oversized holes, excessive wear, and short drill life.

To maintain the necessary accuracy of drill point angles, lip-lengths, lip-clearance angle, and chisel edge angle, the use of machine point-grinding is recommended. However, the lack of a drill-point grinding machine is not sufficient reason to excuse poor drill points. Drills may be pointed accurately by hand if proper procedure is followed and care exercised.

Freehand Drill Point Grinding

1. Adjust grinder tool rest to a convenient height for resting back of forehand on it while grinding drill point.
2. Hold drill between thumb and index finger of left hand. Grasp body of drill near shank with right hand.
3. Place forehand on tool rest with centerline of drill making desired angle with cutting face of grinding wheel (Fig. 158A) and slightly lower end of drill (Fig. 158B).
4. Place heel of drill lightly against grinding wheel. Gradually raise shank end of drill while twisting drill in fingers in a counterclockwise rotation and grinding conical surfaces in the

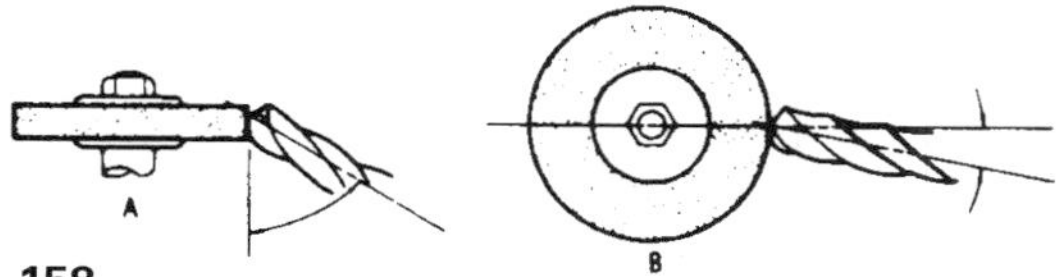

Fig. 158.

direction of the cutting edges. Exert only enough pressure to grind the drill point without overheating. Frequently cool drill in water while grinding.
5. Check results of grinding with a gauge to determine if cutting edges are the same lengths and at desired angle, and that and that adequate lip clearance has been provided.

PORTABLE POWER TOOLS

The field of portable power tools covers everything from small electric hand tools to heavy duty drilling, grinding, and driving tools. Most mechanics have experience with the more common types and are knowledgeable in their correct usage and safe operation. There are, however, some more specialized types which mechanics may use only occasionally. Because these specialized tools, as well as all other power tools, are relatively high speed, using sharp-edged cutters, safe and efficient operation requires knowledge and understanding of both the power unit and the auxiliary parts, tools, etc.

Magnetic Drilling

The electromagnetic drill press (Fig. 159) is the basic equipment used for magnetic drilling. It can be described as a portable drilling machine incorporating an electromagnet, with a capability of fastening the machine to metal work surfaces. The magnetic drill allows bringing the drilling equipment to the work, rather than bringing the work to the drilling machine. A major advantage over the common portable drill motor is that it is secured in positive position electromagnetically, rather than depending on the strength and steadiness of an operator. This makes possible drilling holes with a greater degree of precision in respect to size, location, and direction, with little operator fatigue.

Magnetic drilling is limited to flat metal surfaces large enough to accommodate the magnetic base in the area where the hole or holes are to be drilled. In operation the work area is cleaned of chips, dirt, etc., to ensure good mating of the magnetic base to the work surface. The unit is placed in appropriate position and the drill point aligned with the center point location. When proper alignment has been

Milwaukee Electric Tool Corp.

Fig. 159. Electromagnetic drill press.

established, the magnet is energized to secure the unit. A pilot hole is recommended for drilling holes larger than ½" diameter. Operation proceeds in a manner similar to a conventional drill press. Enough force should be applied to produce a curled chip. Too little force will result in broken chips and increased drilling time; too great a force will cause overheating and shorten drill life.

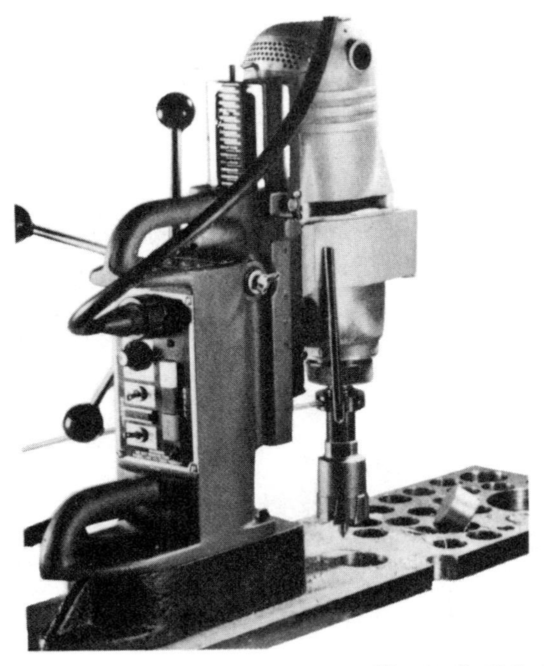

Milwaukee Electric Tool Corp.

Fig. 160. Hole cutting with carbide tipped hole cutter.

The capability of magnetic drilling equipment can be expanded through the use of carbide-tipped hole cutters. These are tubular shaped devices with carbide-tipped multiple cutting edges which are highly efficient. The alternating inside and outside cutting edges are ground to cut holes rapidly with great precision. Having hard carbide cutting tips, they will outlast regular high-speed steel twist drills. They are superior tools for cutting large diameter holes, because their minimal cutting action is fast and the power required is less than when removing all the hole material. Approximately a ⅛"-wide kerf of material is removed and a cylindrical plug of material is ejected upon completion of the cut. When used in conjunction with the magnetic drill, they enable the mechanic to cut large-diameter holes with little effort and great accuracy. Fig. 160 shows the hole cutter's ability to notch or cut material away from the edge of a workpiece or from the edge of an existing hole.

The hole cutter makes possible operations in the field that could otherwise only be performed on larger fixed machinery in the shop. The arbor center pin which is visible in Fig. 160 allows accurate alignment for pre-marked holes. As with most machining operations, a cooling and lubricating fluid should be used when cutting holes with this type of cutter. Ideally it would be applied with some pressure to furnish a flushing action as well as cooling and lubricating. This may be accomplished by applying the fluid directly to the cutter and groove, or by use of an arbor lubricating mechanism. With the special arbor shown in Fig. 160, it is possible to introduce fluid under pressure through the stationary ring which is part of the arbor assembly. The fluid is forced-fed by a hand pump which is part of hand-held fluid container. By introducing the fluid on the inside surfaces of the cutter a flushing action across the cutting edges occurs, tending to carry the chips away from the cutting area and up the outside surface of the cutter.

Hammer Drilling

The *hammer drill* combines conventional or rotary drilling with hammer or percussion drilling. Its primary use is the drilling of holes in concrete or masonry. It incorporates two-way action in that it may be set for hammer drilling with rotation or for rotation only. When drilling in concrete or masonry, special carbide-tipped bits are required. These are made with alloy steel shanks for durability, to which carbide tips are brazed to provide the hard cutting edges necessary to resist dulling. The concrete or masonry is reduced to granules and dust by this combined hammering and rotating action, and shallow spiral flutes remove the material from the hole.

Fig. 161 shows a hammer drill having a two-speed gear shift drive. Other models may incorporate trigger control variable speed drives.

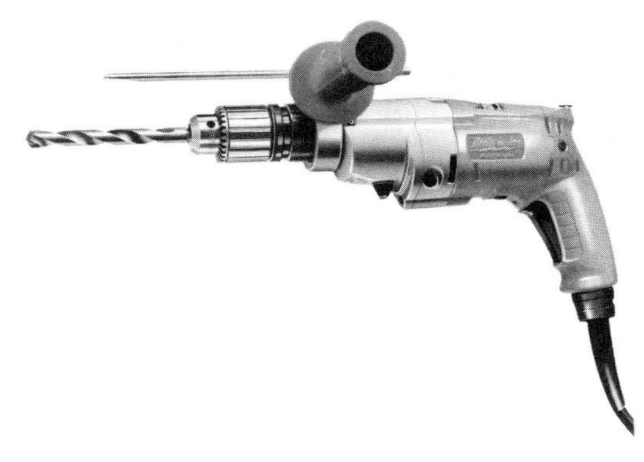

Milwaukee Electric Tool Corp.

Fig. 161. Hammer Drill.

Diamond Concrete Core Drilling

Holes in concrete structures, prior to the introduction of diamond concrete core drilling, required careful planning and form preparation, or breaking away sections of hard concrete and considerable patch-up. The tool that has changed these practices is the *diamond core bit.* It is basically a metal tube, on one end of which is a matrix crown imbedded with industrial-type diamonds. One type has the diamonds distributed throughout the crown in a carefully controlled manner. During use, the matrix is worn away, assuring that new sharp diamonds are exposed for constant cutting away of the concrete. A second type of diamond bit has all the diamonds set on the surface of the bit crown and arranged in a predetermined pattern for maximum cutability and exposure. Bits are made in two styles of construction, the *closed back* and the *open back,* shown in Fig. 162.

There is a cost advantage in the use of the open back bit, in that the adapters are reusable, offering savings in cost on each bit after the first. Another sometimes appreciable advantage is that should a

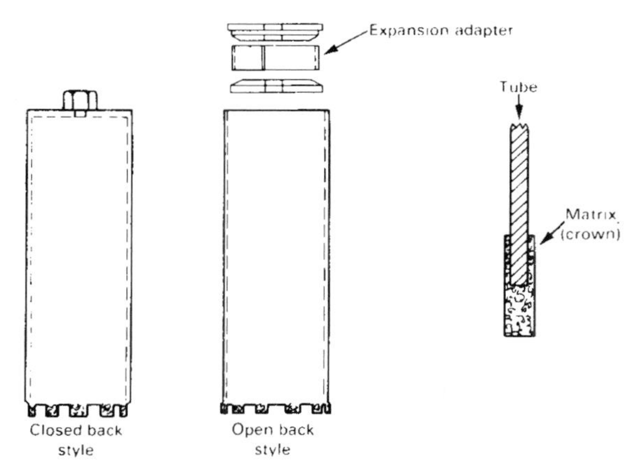

Fig. 162. Diamond Core Bits.

core become lodged in the bit, removing the adapter makes core removal easier. The closed back bit offers the advantage of simplicity. Installation requires only turning the bit onto the arbor thread, with no problem of positioning or alignment. As the bit is a single, complete unit, there is no problem with mislaid, lost, or damaged parts.

Successful diamond core drilling requires that several very important conditions be maintained. These are: rigidity of the drilling unit, adequate water flow, and uniform steady pressure. Understanding the action which takes place when a diamond core bit is in operation will result in better appreciation of the importance of these conditions being maintained. Fig. 163 illustrates the action at the crown end of a diamond core bit during drilling. Arrows indicate the water flow inside the bit, down into the kerf slot, and around the crown as particles are cut free and flushed up the outside surface of the bit.

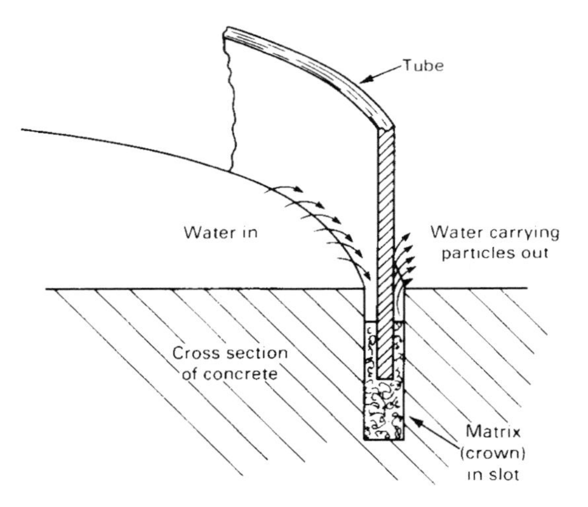

Fig. 163. Diamond bit cutting action.

The diamond concrete core drilling machine is in effect a special drill press. The power unit is mounted in a cradle which is moved up and down the column by moving the operating handles. The handles rotate a pinion gear which meshes with a rack attached to the column. To secure the rig to the work surface, the top of the column is provided with a jack screw to lock the top of the column against the overhead with the aid of an extension. Fig. 164 shows a concrete core drilling rig equipped with a vacuum system which makes possible attachment of the rig directly to the work surface.

The power unit is a heavy-duty electric motor with reduction gears to provide steady rotation at the desired speed. The motor spindle incorporates a water swivel which allows introduction of water through a hole in the bit adapter to the inside of the core bit.

Milwaukee Electric Tool Corp.

Fig. 164. Diamond concrete core drilling rig.

The power unit shown in Fig. 165 attaches to the cradle of the rig shown in Fig. 164. The diamond core bits (both open and closed back syles) fit the threaded end of the motor spindle.

The rigidity of the drilling rig plays a very important part in successful diamond core drilling. The rig must be securely fastened to the work surface to avoid possible problems. Slight movement may cause chatter of the drill bit against the work surface, fracturing the diamonds. Greater movement will allow the bit to drift from location, resulting in crowding of the bit, binding in the hole, and possible seizure and damage to the bit.

An easy way to anchor the unit is with the jack screw provided at the top of the column. A telescoping extension may be used, or a pipe, 2×4, or other material, cut to appropriate length. Through the use of such an extension the rig may be braced against an overhead object, or if used on a wall it may be braced against the opposite wall.

The versatility of a diamond core drilling rig may be greatly increased with a vacuum system. With this device it is possible to anchor the unit directly to the work surface, eliminating the need for extensions, braces, or other securing provisions. The rig in Fig. 164 is equipped with such a system. It consist of a vacuum pad to which the base may be attached, a vacuum pump unit, and auxiliary hose, gauge, fittings, etc.

To use the vacuum system of attachment, the area where the work is to be performed should first be cleaned to remove any loose dirt or material that might cause leakage at the seal of the pad to the work surface. The rig, with the diamond core bit on the spindle, should be placed in the desired location. The pad nuts should be loosened to allow the pad to contact the work surface without restraint. The vacuum pump should be started to evacuate the air from inside the pad. This produces what is commonly called the "suction," which holds the pad in place. A vacuum gauge indicates the magnitude of the vacuum produced. The graduated gauge face is marked to show the minimum value required for satisfactory operation. A clean, relatively smooth surface should make possible building the vacuum value to the maximum. Should the gauge register a value below the minimum required, drilling should not be attempted. A check should be made for dirt, porous material, cracks in the surface, or any other condition which might allow air to leak

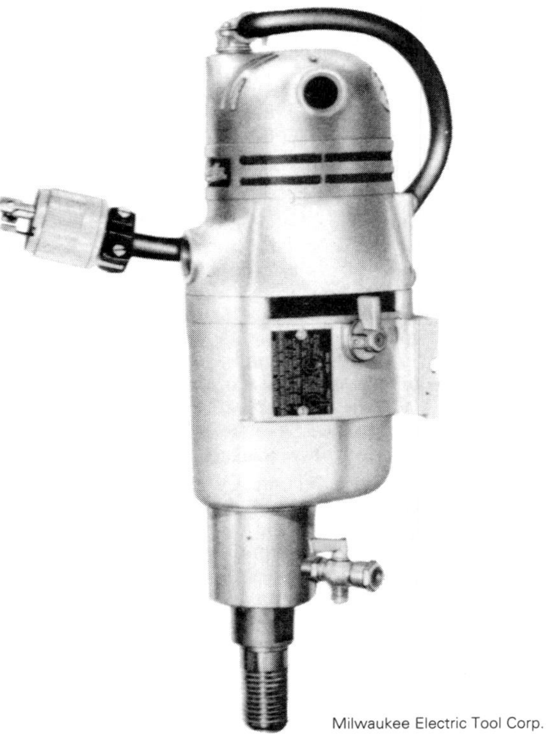

Milwaukee Electric Tool Corp.

Fig. 165. Diamond drill power unit.

past the pad seal. When the gauge reading indicates the pad is secure, the pad nuts should be tightened to fasten the rig base firmly to the vacuum pad. Standing on the base of the rig is not a substitute for good pad fastening. While additional weight will add a little to the downward force, it will not prevent the rig from floating or shifting out of position.

Water, which is vital to the success of diamond core drilling, is introduced through the water swivel, which is a component part of the lower motor housing. The swivel has internal seals which prevent leakage as the water is directed into the spindle and into the inside of the bit. The preferred water source is a standard water hose (garden hose), which provides dependable flow and pressure. When not possible or practical, a portable pressure tank, such as used for garden sprayers, or a gravity feed tank or similar arrangement could be used. Whatever the arrangement, precautions should be taken to ensure adequate flow and pressure to handle the job. Depending on the operating conditions, provisions may be necessary to dispose of the used water. On open, new construction, it may be permissible to let the water flow freely with little or no concern for runoff. In other situations it will be necessary to contain runoff, and find a way to dispose of used water. A water collector ring and pump are made for this purpose, or the common wet-dry vacuum cleaner may be employed by building a dam with rags or other material.

When the rig is secured, and water supply and removal provisions are made, drilling may commence. Starting the hole may present a problem because the bit crown has a tendency to wander, particularly when starting in hard materials and on irregular and inclined surfaces. At the start the crown may contact only one spot, and thrust tends to cause the bit to walk. Often it is sufficient to guide the bit lightly with a board notched at the end. To start, a light pressure on the bit is recommended. After the crown has penetrated the material, pressure may be increased, and when the crown is well entered into the material, pressure may be fully applied. The feed force must be uniformly applied in a steady manner, not jerky or intermittent. Too little pressure will cause the diamonds to polish; too much can cause undue wear. To aid the operator in maintaining a steady pressure of the proper magnitude, most diamond drilling rigs incorporate an ammeter to indicate motor loading. In addition to the regular calibrations, the meter dial also has a green area, which is the working range, and a red area to indicate when too much pressure is being applied. Enough force should be exerted on the operating handles to keep the ammeter needle in the green area, indicating the proper bit pressure and drilling speed is being maintained, preventing overload and providing optimum bit life.

The importance of flowing water during drilling cannot be overstressed. A constant, adequate water flow must be maintained at all times. The water pressure and flow must be sufficient to wash cuttings from under the bit crown and up the outside of the bit, as illustrated in Fig. 163. Another important function of the water is to act as a coolant, carrying away the heat that might otherwise cause the diamonds to polish, or in an extreme case cause the bit to burn (to turn blue). All water connections must be tight, and water flow steady (1 to 2 gallons per minute). Water flow and bit rotation must *not* stop while the bit is in the hole. The bit should be raised out of the hole while turning and then the water and power shut off. Stopping and starting the bit while in the hole could cause binding and damage to the bit. When the bit is cutting freely the operator can feel the movement into the concrete. The outflowing water will have a slightly sludgy appearance as it carries away the concrete particles.

Diamond core bits are capable of drilling all masonry materials: concrete, stone, brick, tile, etc., as well as steel imbedded in concrete, such as reinforcing rod and structural steel. They are not, however, capable of continuous steel cutting and so experience considerable wear and/or deterioration of the diamonds if required to do extensive steel cutting. When the bit encounters imbedded steel, the operator will notice an increased resistance to bit feed. The feed rate should be decreased to accommodate the slower cutting action which occurs when cutting steel. When cutting steel at reduced feed rate, the water becomes nearly clear and small gray metal cuttings will be visible in the off flow water. Diamond core bits will also cut through electrical conduit buried in concrete. This can be a major consideration when cutting holes for changes or revisions in operating areas. Not only is there a possibility of shutting down operations if power lines are cut, but there is a possible electrical shock hazard for the operator. Common sense dictates that the possibility of buried conduit be carefully considered when doing revision work. If the location of buried lines cannot be determined, drilling should not be attempted unless all power to lines that might be severed is disconnected.

When the concrete being drilled is not too thick, the hole may be completed without withdrawal of the bit. With greater thicknesses, drilling should proceed to a depth equal to at least two times the diameter of the hole. The bit should then be withdrawn and the core broken out. This may be done with a large screwdriver or pry bar inserted into the edge of the hole and rapped firmly with a hammer. The first section of the core can usually be removed with two screwdrivers, one on each side of the hole, by prying up and lifting. If imbedded steel is encountered, after it is cut through, the bit should be withdrawn and the core and any loose pieces of steel removed. When deeper cores must be removed, a little more thought and effort is required. Larger holes may permit reaching into them, and if necessary drilling a hole to insert an anchor to aid in removal. If the core cannot be snapped off and removed, it may be necessary to break it out with a demolition hammer. A situation which must be given consideration when drilling through a concrete floor is that of escaping water and the possibility of the concrete core falling from the hole as the bit emerges from the underside of the floor. Provisions must be made to contain the water and catch the falling core to prevent damage and avoid injury.

Portable Band Saw

The portable band saw, while not in common use by mechanics, has the capability to perform many cutting operations efficiently and with little effort. Many field operations that require cutting in place consume excessive time and effort if done by hand. If acetylene torch cutting were used, the resulting rough surfaces might not be acceptable and/or the flame and flying molten metal might not be permissible. There are many times when the object being worked on cannot be taken to another location to perform a sawing operation. In such instances a portable band saw can do the work in the field. The portable band saw is in effect a lightweight, self-contained small version of the standard shop band saw. It uses two rotating rubber-tired wheels to drive a continuous saw blade. Power is supplied by a heavy-duty electric motor which transmits its power to the wheels through a gear train and worm wheel reduction. In this manner speed reduction is accomplished and powerful torque is developed to drive the saw band at proper speed. Blade guides, similar in design to the standard shop machine, are built into the unit, as well as a bearing behind the blade to handle thrust loads.

Milwaukee Electric Tool Corp.

Fig. 166. Portable Band Saw

Blades for portable band saws are available in a tooth pitch range from 6 to 24 teeth per inch. The rule of thumb for blade selection is to always have three (3) teeth in the material at all times. Using too coarse a blade will cause thin metals to hang up in the gullet between two teeth, and to tear out a section of teeth. Too fine a blade will prolong the cutting job, as only small amounts of metal will be removed by each tooth. *Do not use cutting oil.* Oil and chips transfer and stick to rubber tires, causing the blade to slip under load. Chips build up on tires, causing misalignment of the blade. Fig. 166 shows a portable band saw attached to a table to increase its versatility.

Portable Fluid Power Tools

The driving power for a great many portable power tools is what is known as *fluid power*. While many materials capable of flowing, either liquid or gas, may be classified as fluid, the mediums most widely used to transmit power in the portable power field are: air (pneumatic) or oil (hydraulic).

All manner of pneumatic tools are in wide use, both in manufacturing and in service and maintenance. Probably the most common of the pneumatic power tools are those that fall into the impact category. These range from the heaviest of demolition hammers to lightweight impact tools. All deliver a high-speed series of impacts or blows to accomplish their purpose. Another large group of pneumatic tools are those that deliver rotary motion, most of which are powered by high-speed air turbine. The common nut runner and pneumatic drills are examples of this type of tool.

The second medium, hydraulic power, is the activating force in a wide variety of portable tools that have become a recognized necessity in mechanical operations. The two basic components of most of these tools and accessories are the hydraulic pump, both hand and power operated, and the hydraulic cylinder. The development of synthetic packing materials, improved metal alloys, and the advanced techniques for fine metal finishing has made possible the production of portable power tools that are highly dependable and capable of withstanding severe service conditions. The principle that is utilized in tools involving the hydraulic pump and cylinder is the multiplication of force by hydraulic leverage. In operation a force is exerted on a small surface to generate hydraulic pressure, which in turn is transmitted to a greater surface to produce a force magnified as many times as the ratio of the power cylinder surface area to the output cylinder surface area. The principle is illustrated in Fig. 167.

This multiplication of force is offset by a corresponding decrease in travel; that is, the small plunger in the diagram must travel 40 inches to displace enough fluid to move the large plunger 1 inch.

The primary component in a hydraulic power system is the hydraulic pump. The widely used hand-operated, single-speed hydraulic pump is a single piston type designed to develop up to

10,000 pounds per square inch (psi) pressure. It is commonly constructed with a reservoir body, at one end of which is the power head. The head contains the power cylinder and piston, usually about ⅜ to ½ inch in diameter. Force is developed by use of a lever handle, which enables the effort applied to the handle end to be multiplied many times. Thus a force of 80 lbs., as shown in Fig. 168, applied to the handle end, would result in a force of 1200 lbs. acting on the hydraulic piston when under load. Under this loaded condition there would develop in the hydraulic system a pressure of 8000 psi.

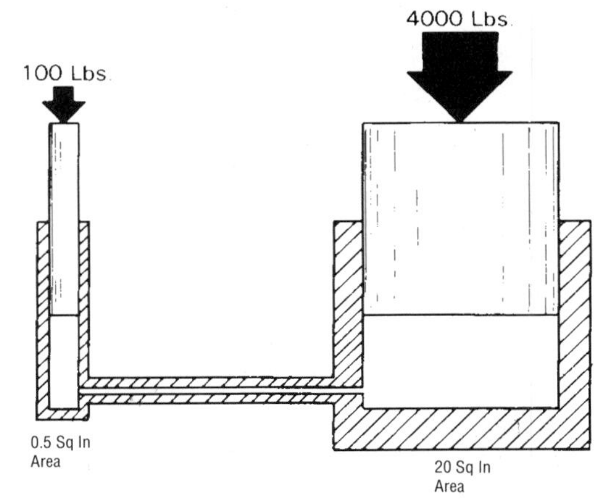

Fig. 167. Multiplication of force.

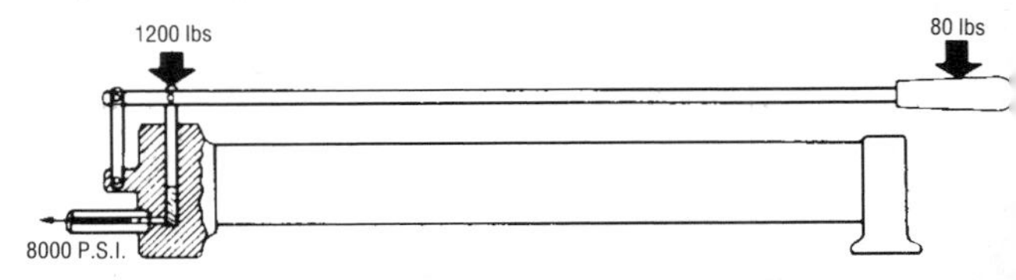

Fig. 168. Single speed hand operated hydraulic pump.

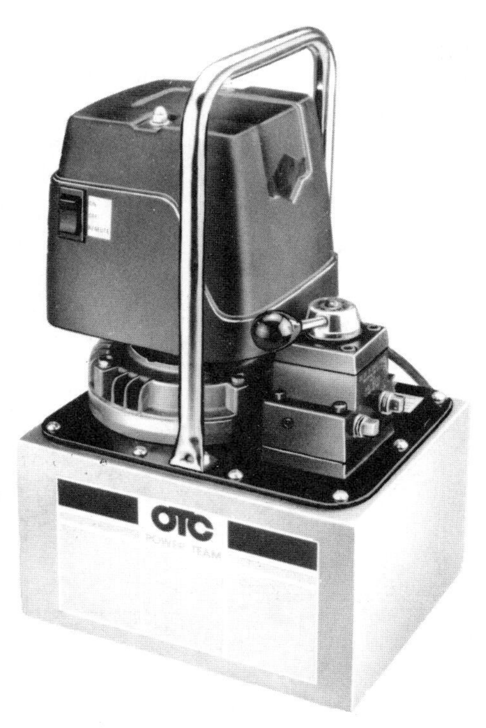

OTO Power Team

Fig. 169. Electric Driven Hydraulic Pump.

Also widely used, where time or speed is a factor, is the two-speed hand pump. The principle of operation is the same as the single speed, except that it has the feature of a second piston of larger diameter. The two-speed operation provides high oil volume at low pressure for rapid ram approach, and when switched to the small piston provides a high pressure, low volume stage for high force operation. For applications of greater frequency, or where a large volume of hydraulic fluid may be required, such as operation of several cylinders simultaneously, a power-driven pump unit may be more suitable.

Power-driven pump units—electric motor, gasoline engine, or air—are usually of the two-stage type. Many manufacturers use a design which combines a roller-vane or a gear-type pump, with an axial-piston or radial-piston type. The roller-vane or gear pump provides the low-pressure, high-volume stage, assuring fast ram approach and return. When pressure builds to approximately 200 psi, the high-pressure, low-volume stage provided by the axial-piston or radial-piston takes over to handle high-pressure requirements. Power pumps come equipped with various size reservoirs to provide hydraulic fluid for particular applications.

Shown in Fig. 169 is an electric driven, two-stage hydraulic pump for use with single- or double-acting rams. It is equipped with a 3-position 4-way manual valve for driving double-acting rams or multiple single-acting rams. Compressed air driven power units as well as gasoline engine driven units in a great variety of designs are also in wide use.

Shop presses, both hand pump and power operated, are not normally considered as portable tools; however, they are included in this grouping because they utilize portable hydraulic power pumps. While they are often used in production operations, they are almost essential for mechanical work involving controlled application of force to assemble or disassemble machinery. Shown in Fig. 170 is a press with an electric motor driven power unit. While presses with additional features are available, most shop work can be handled adequately with the type shown. It provides a large work area under the ram, a winch and cable mechanism to quickly raise or lower the press bed, and rapid ram advance and return.

An accessory vital to the operation of portable hydraulic tools is the high-pressure hose. Most hose manufactured for this purpose is designed for an operating pressure of 10,000 psi, with a bursting strength of 20,000 to 30,000 psi. The lightweight hoses are constructed with a nylon core tube and polyester fiber reinforcement, while those designed for heavy-duty, more severe service have one

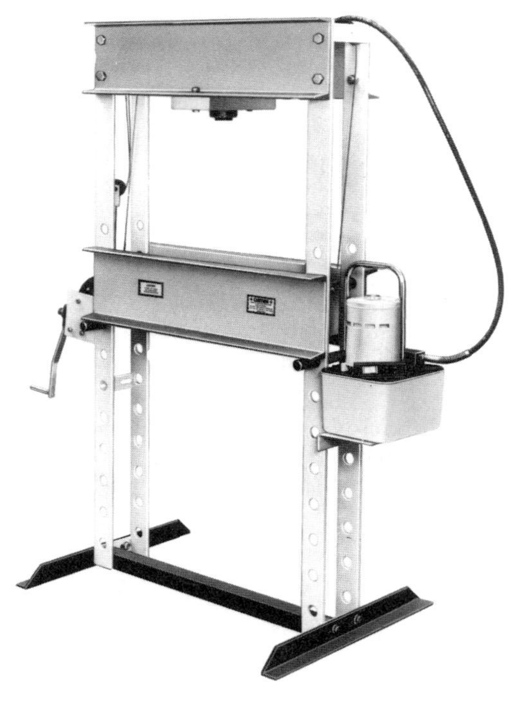

OTO Power Team

Fig. 170. Power Driven Shop Press.

or more layers of braided steel webbing to help withstand the internal hydraulic pressure, as well as the external abuse which may be encountered in some applications.

Another accessory, while not a necessity, is a valuable component to include in the hydraulic power system — the pressure gauge. Its principal advantage is that it provides a visual indication of the pressure generated by the pump, and thus can assist in preventing overloading of the hydraulic system as well as the equipment upon which work is being done. Gauges, in addition to being graduated in psi, may also include a scale calibrated in tonnage. This style of gauge must be matched to the cylinder diameter to obtain correct tonnage values. Correct practice with gauge manufacturers is to show a danger zone on the gauge face. This is done with a red background coloring in the area over 10,000 psi.

The second major basic component in a hydraulic power system is the hydraulic cylinder. Hydraulic cylinders operate on either a single- or double-acting principle, which determines the type of "return" or piston retraction. Single-acting cylinders have one port, and in simplest form retract due to weight or force of the load (load return). They are also made with an inner spring assembly which enables positive retraction regardless of the load (spring return). Double-acting cylinders have two ports, and the fluid flow is shifted from one to the other to achieve both hydraulic cylinder lifting and retraction (hydraulic return). Both single-acting and double-acting cylinders are manufactured with solid pistons or with center-hole pistons. Center-hole pistons allow insertion of pull rods for pulling applications. The single-acting, push-type, load-return style hydraulic cylinder is shown in Fig. 171.

The single-acting, either load or spring return, is the most commonly used cylinder for hydraulic power tool operation performed by industrial mechanics. There are some operations, however, where power is required in both the lift and retract directions. The double-acting cylinder provides hydraulic function in both the lifting and lowering modes. Fig. 172 shows the principle of operation of a double-acting cylinder.

Double-acting cylinders should be equipped with a four-way value to prevent trapping of hydraulic fluid in the retract system. Most manufacturers build a safety value into the retract system to prevent damaging the ram if the top hose is inadvertently left unconnected and the ram is actuated.

All manner of pumps, cylinders, accessories, and special attachments are available to adapt hydraulic power to a multitude of uses. Perhaps the most frequent application the mechanic is concerned with is the disassembly of mechanical components. This usually

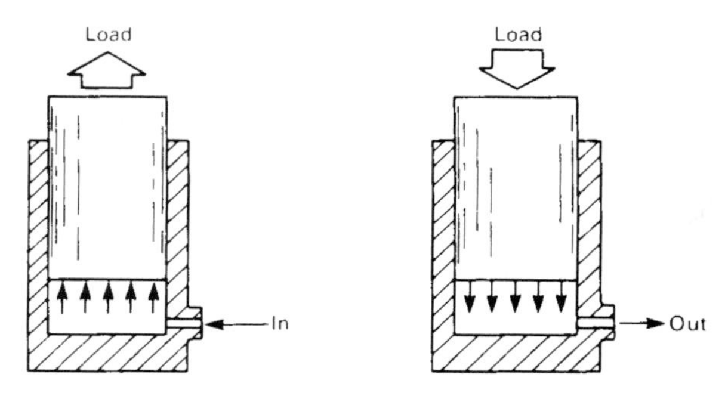

Fig. 171. Single-acting, load-return cylinder.

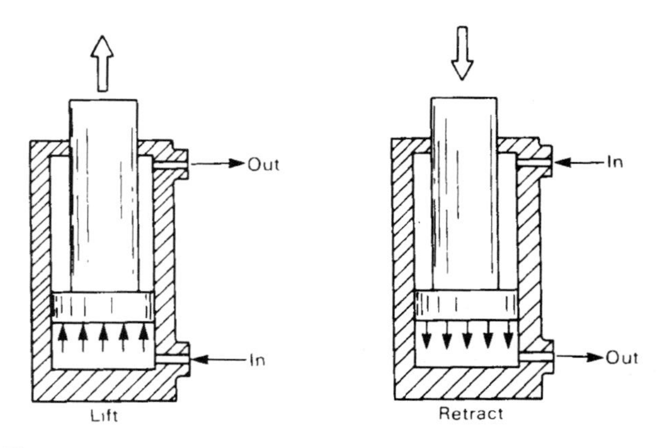

Fig. 172. Double-acting cylinder in lift and retract modes.

requires the use of considerable force to remove one closely fitted part from another. One of the great advantages of hydraulic power over striking or driving to accomplish the operation is the manner in which force is applied. Instead of heavy blows which might cause

Fig. 173. A jaw puller, hydraulically actuated, is used to push a shaft from a sheave.

distortion or damage to components, the force is applied in a steady controlled manner. Fig. 173 shows a hydraulic power unit used with a jaw-type mechanical puller. Note this has the advantage of mechanical positioning and adjustment plus controlled hydraulic power.

A great variety of hydraulic power accessories and puller sets, including all types of jaw pullers, bar pullers, puller legs, adapters, etc., are available for all manner of pulling and pushing operations. Fig. 174 shows the use of a hydraulic unit in combination with a bar-type puller and extension legs.

Hydraulic power equipment is widely used in manufacturing operations of all kinds, from simple holding and clamping operations to bending, punching, assembling, etc. While the mechanic may not be concerned with the operation of the manufacturing machines, he may be called upon to install, adjust, and maintain them. Although in some cases they may differ in appearance from those in the illustrations, the principle of operation is the same.

Powder Actuated Tools

The principle of the powder actuated tool is to fire a fastener into material and anchor or make it secure to another material. Some applications are: wood to concrete, steel to concrete, wood to steel, steel to steel, and numerous applications of fastening fixtures and special articles to concrete or steel. Because the tools vary in design details and safe handling techniques, general information and descriptions of the basic tools and accessories are given, rather than specific operation instructions. Because the principle of operation is similar to that of a firearm, safe handling and use must be given the highest priority. A powder actuated tool in simple terms is a "pistol." A pistol fires a round composed of two elements:

1. Cartridge with firing cap and powder
2. Bullet

When you pull the trigger on a pistol, the firing pin detonates the firing cap and powder and sends a bullet in free flight to its destination. The direct-acting type powder-actuated tool can be described in the same manner, with one small change. The bullet is loaded as two separate parts. Fig. 175 shows the essential parts of a powder-actuated tool system—the tool and the two-part bullet (the combination of the fastener and the power load).

In operation the fastener and power load are inserted in the tool, the tool is pressed against the work surface, the trigger is pulled, and the fastener travels in free flight to its destination.

While different manufacturers' tools may vary in appearance, they are all similar in principle and basic design. They all have, within a suitable housing, a chamber to hold the power load, a firing

Owatonna Tool Co.

Fig. 174. Hydraulically-powered bar-type puller removes generator rotor assembly from shaft.

mechanism, a barrel to confine and direct the fastener, and a shield to confine flying particles. There are two types of tools:

Direct acting, shown in Fig. 176, in which the expanding gas acts directly on the fastener to be driven into the work.

Indirect acting, shown in Fig. 177, in which the expanding gas of a powder charge acts on a captive piston, which in turn drives the fastener into the work.

Fasteners

Fasteners used in powder-actuated tools are manufactured from special steel and heat treated to produce a very hard yet ductile fastener. These properties are necessary to permit the fastener to penetrate concrete or steel without breaking. The fastener is equipped with some type of tip, washer, eyelet, or other guide member. This guide aligns the fastener in the tool, guiding it as it is being driven. Some of the alignment tips in use are shown in Fig. 178.

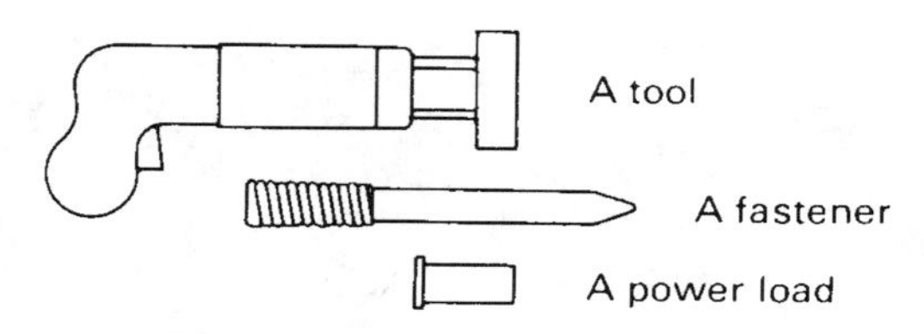

Fig. 175. Basic parts of powder-actuated tool.

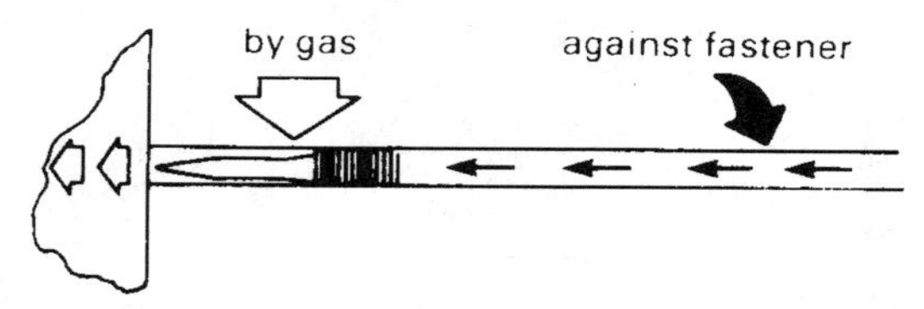

Fig. 176. Direct-acting operating principle.

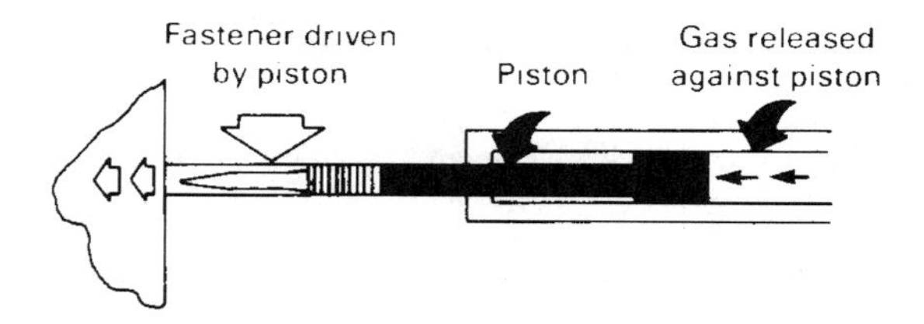

Fig. 177. Indirect-acting operating principle.

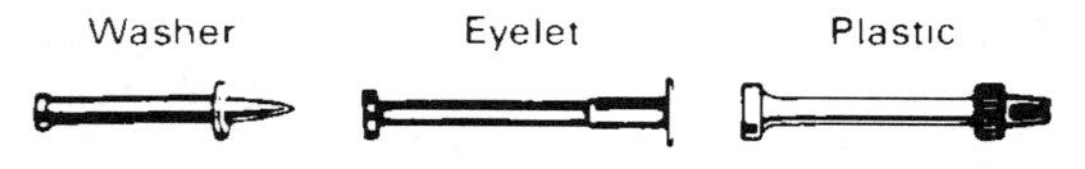

Fig. 178. Typical alignment tips.

Two basic types of fasteners are in common use: the *drive pin* and the *threaded stud.* The drive pin is a special nail-like fastener designed to attach one material to another, such as wood to concrete or steel. Head diameters are generally ¼", 5⁄16", or ⅜" in diameter. For additional head bearing in conjunction with soft materials, washers of various diameters are either fastened through or made a part of the drive pin assembly. Typical examples of drive pins are shown in Fig. 179.

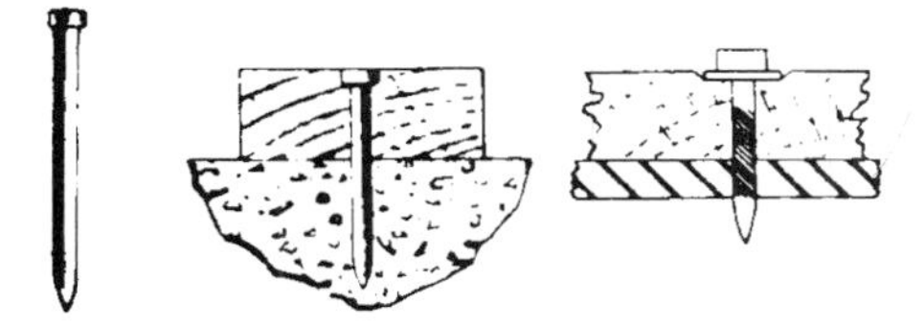

Fig. 179. Typical drive pins.

The *threaded stud* is a fastener composed of a shank portion which is driven into the base material and a threaded portion to which an object can be attached with a nut. Usual thread sizes are 8-32, 10-24, ¼-20, 5⁄16-18, ⅜-16. Typical examples of threaded studs are shown in Fig. 180.

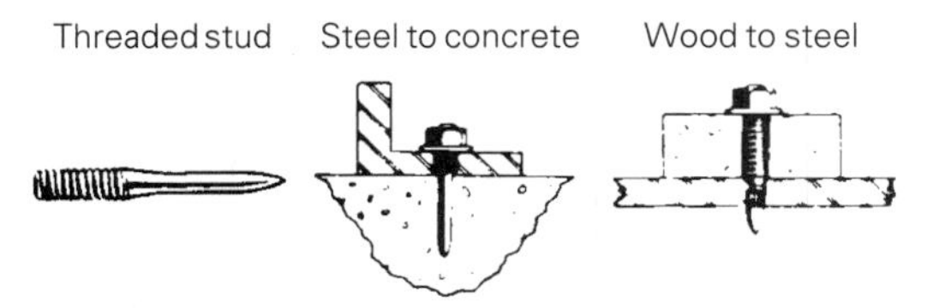

Fig. 180. Typical threaded studs.

A special type of drive pin with a hole through which wires, chains, etc., can be passed for hanging objects from a ceiling is shown in Fig. 181.

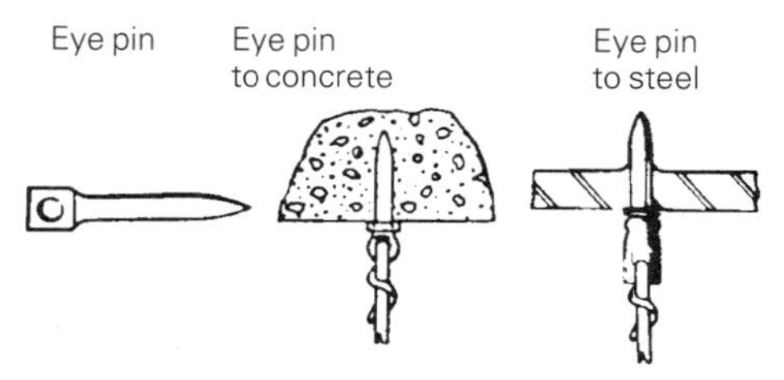

Fig. 181. Special type "eye" pins.

Another special type of fastener, in this case a variation of the threaded stud, is the *utility stud.* This is a threaded stud with a threaded collar which can be tightened or removed after the fastener has been driven into the work surface. Fig. 182 shows the stud and its use to fasten wood to concrete.

Power Loads

The power load is a unique, portable, self-contained energy source used in powder-actuated tools. These power loads are supplied in two common forms: cased or caseless. As the name indicates, the propellant in a cased power load is contained in a metallic case. Some of these power loads and their construction are illustrated in Fig. 183. The caseless power load does not have a case and the propellant is in a solid form.

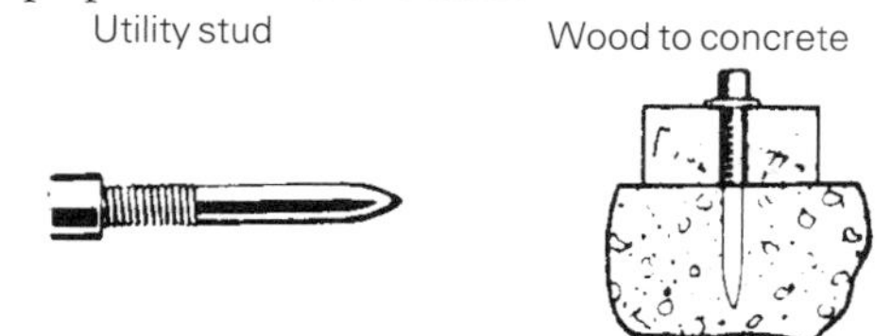

Fig. 182. Special utility studs.

Whatever the type, caliber, size, or shape, there is a standard number and color code used to identify the power level or strength of all power loads. The power loads are numbered 1 through 12, with

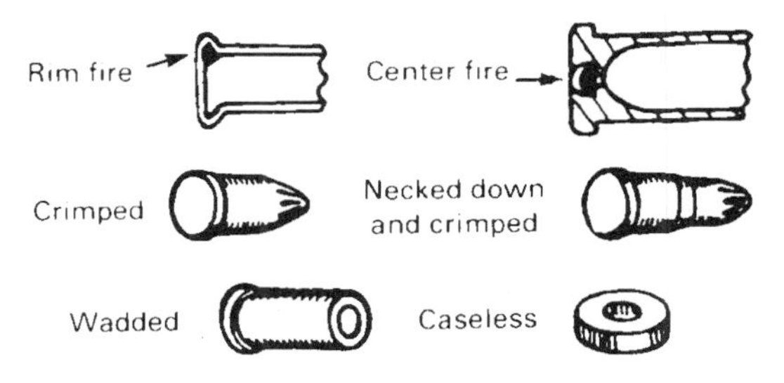

Fig. 183. Power load construction.

the lightest being No. 1 and the heaviest being No. 12. In addition, because there are not twelve readily distinguishable permanent colors, power loads No. 1 through 6 are in brass-colored cases and No. 7 through 12 are in nickel-colored cases. It is a combination of the case color and load color that defines the load level or strength. The number and color identification code is shown in the following listing:

Power Level	Case Color	Load Color
1	Brass	Gray
2	Brass	Brown
3	Brass	Green
4	Brass	Yellow
5	Brass	Red
6	Brass	Purple
7	Nickel	Gray
8	Nickel	Brown
9	Nickel	Green
10	Nickel	Yellow
11	Nickel	Red
12	Nickel	Purple

In selecting the proper power load to use for an application, it is important to start with the lightest power level recommended for the tool being used. If the first test fastener does not penetrate to the desired depth, the next higher load should be tried, until the proper penetration is obtained.

Shields and special fixtures are important parts of the powder-actuated fastening system and are used for safety and tool adaption to the job. The shield should be used whenever fastening directly into a base material, such as driving threaded studs or drive pins into steel or concrete. In addition to confining flying particles, the shield also helps hold the tool perpendicular to the work. Medium and high velocity class tools are designed so that the tool cannot fire unless a shield or fixture is attached. One of the most important acts that a conscientious user of a powder-actuated fastener system can perform is to see that the tool being used is equipped with the proper safety shield to assure both safety and good workmanship.

The material into which the fastener shank is driven is known as the base material. In general, base materials are metal and masonry of various types and hardness. Suitable base materials, when pierced by the fastener, will expand and/or compress and have sufficient hardness and thickness to produce holding power and not allow the fastener to pass completely through. Unsuitable materials may be put into three categories: too hard, too brittle, too soft. If the base material is too hard, the fastener will not be able to penetrate and could possibly deflect or break. Such things as hardened steel, welds, spring steel, marble, natural rock, etc., fall in this category. If the base material is too brittle it will crack or shatter, and the fastener could deflect or pass completely through. Such things as glass, glazed tile, brick, slate, etc., fall in this category. If the base material is too soft it will not have the characteristics to produce holding power, and the fastener will pass completely through. Such things as wood, plaster, drywall, composition board, plywood, etc., fall in this category.

Because masonry is one of the principal base materials suitable for powder-actuated tool fastening, it is important to understand what happens when a fastener is driven into it. The holding power of the fastener results primarily from a compression bond of the masonry to the fastener shank. The fastener, on penetration, displaces the masonry, which tries to return to its original form and exerts a squeezing effect. Compression of the masonry around the fastener shank takes place, with the amount of compression increasing in relation to the depth of penetration and the compressive strength of the masonry. Fig. 184 illustrates the bond of a fastener driven into masonry material.

When an excessive direct pullout load is applied to a fastener driven into masonry material, failure will occur in either of two ways: the fastener will put out of the masonry, as shown in Fig. 185A, or failure of the masonry will occur, as shown in Fig. 185B.

This illustrates an important relationship between the depth of penetration and the strength of bond of the fastener shank, and the strength of the masonry itself. When the depth of penetration produces a bond on the fastener shank equal to the strength of the masonry, the maximum holding power results.

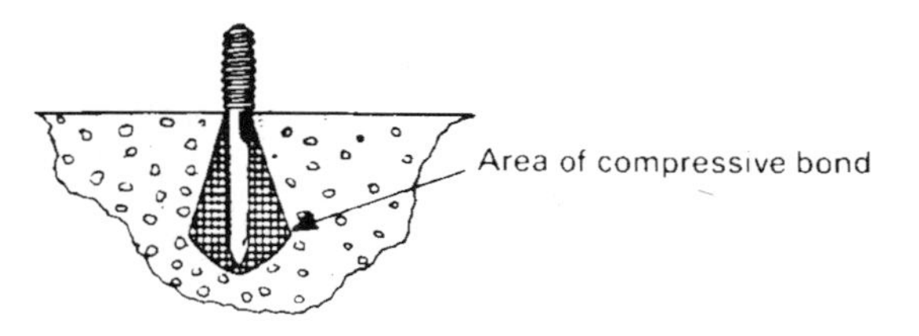

Fig. 184. Fastener bond in masonry.

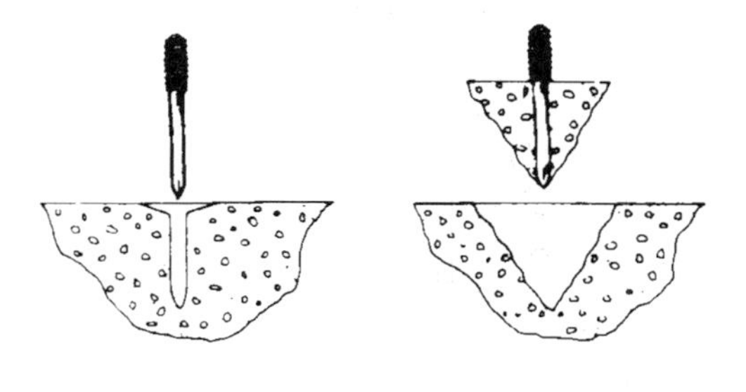

Fig. 185. Pullout from masonry.

Because the tensile strength of masonry is relatively low, care should be taken not to place it under high tension when driving fasteners. This may occur if fasteners are driven too close to the edge, as is illustrated in Fig. 186.

Do not fasten closer than 3" from the edge of the masonry. If the masonry cracks, the fastener will not hold and there is a chance a chunk of masonry or the fastener could escape in an unsafe way. Setting fasteners too close together can also cause the masonry to crack. Spacing should be at least 3" for small diameter fasteners, 4" for medium size, and 6" for the larger diameters.

The potential hazards associated with the use of powder-actuated tools exceed those commonly encountered with other portable tools. The care and attention devoted to safe operation must therefore be proportionally greater. Manufacturers stress safe operation in their instructional material, and it is very important that their instructions be carefully studied before using the tool. In some areas there are regulations that require instruction by a qualified instructor, with examination and issuance of a permit before the tool may be used without supervision. Perhaps the greatest hazard involved is that of release of the fastener. Careful thought and effort must be exerted to ensure that the fastener is at all times under control and prevented from escaping into free flight.

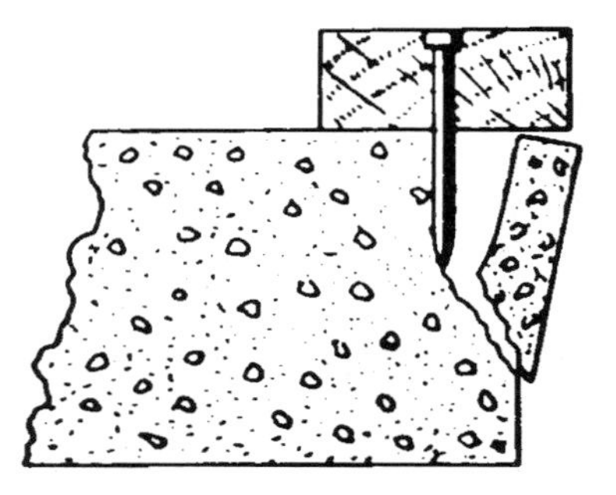

Fig. 186. Edge failure.

WELDING

Because of the development of numerous welding processes, and also because of the development of new steels and other metals that can be welded, welding has become the most important metal joining process. The following is an explanation, in concise form, of the most widely used of these processes, giving basic information that should be of value in understanding the welding processes in general use. This includes information regarding welding equipment, procedures, filler metals, etc., that are involved in the operation called welding.

The commonly used welding processes may be grouped into four (4) general categories: *gas oxyacetylene welding, shielded metal-arc welding, gas metal-arc welding,* and *gas tungsten-arc welding.* The word "shielding" is used to described the creation of an environment of controlled gas or gases around the weld zone to protect the molten weld metal from contamination by the oxygen and nitrogen in the atmosphere.

Oxyacetylene Welding

Oxyacetylene gas welding is perhaps the oldest of the gas welding processes. It came into use about one hundred years ago and is still being used in much the same manner. The process is extremely flexible and one of the most inexpensive as far as equipment is concerned. Today, its most popular application is in maintenance welding, small-pipe welding, auto body repairs, welding of thin materials, and sculpture work. The high temperature generated by the equipment is used for soldering, hard soldering, or brazing. Also, using a special torch, a variation of the process allows for flame cutting. This is accomplished by bringing the metal to a high temperature and then introducing a jet of oxygen which burns the metal apart. It is a primary cutting tool for steel.

In the oxyacetylene gas welding process, coalescence is produced by heating with a gas flame obtained from the combustion of acetylene with oxygen, with or without the use of filler metal. An oxyacetylene flame is one of the hottest of flames—6300°F. This hot flame melts the two edges of the pieces to be welded and the filler

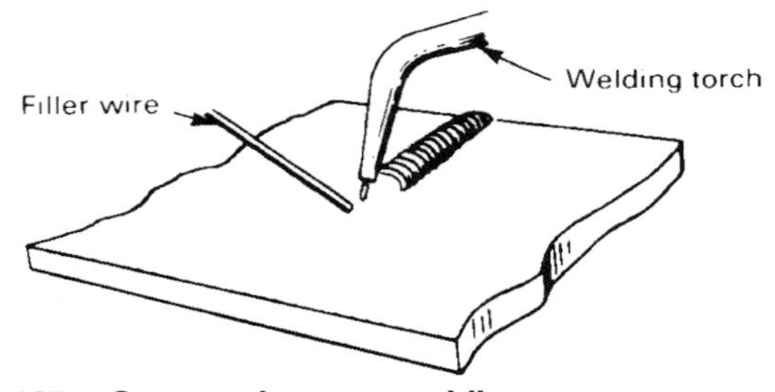

Fig. 187. Oxyacetylene gas welding process.

metal (added to fill the gaps or grooves) so the molten metal can mix rapidly and smoothly. The acetylene and the oxygen gases flow from separate cylinders to the welding torch, where they are mixed and burned at the torch tip. Fig. 187 shows the oxyacetylene gas welding process.

The proportions of oxygen and acetylene determine the type of flame. The three basic types are *neutral, carburizing,* and *oxidizing.* The neutral flame is generally preferred for welding. It has a clear, well-defined white cone indicating the best mixture of gases and no gas wasted. The carburizing flame has an excess of acetylene, a white cone with a feathery edge, and adds carbon to the weld. The oxidizing flame, with an excess of oxygen, has a shorter envelope and a small pointed white cone. This flame oxidizes the weld metal and is used only for specific metals. Flame-cutting is accomplished by addition of an extra oxygen jet to burn the metal being cut. The equipment required for oxyacetylene welding is shown in Fig. 188.

The standard torch can be a combination type used for welding, cutting, and brazing. The gases are mixed within the torch. A thumb screw needle valve controls the quantity of gas flowing into a mixing chamber. A lever type valve controls the oxygen flow for cutting with a cutting torch or attachment. Various types and sizes of tips are used with the torch for specific applications of welding, cutting, brazing, or soldering. The usual welding outfit has three or more tips. Too small a tip will take too long or will be unable to melt the base metal. Too large a tip may result in burning the base metal.

The gas hoses may be separate or molded together. The green (or blue) hose is for oxygen, and the red (or orange) for acetylene. The hose fitting are different to prevent hooking them up incorrectly. Oxygen hose has fittings with right-hand threads and acetylene hose has fittings with left-hand threads.

Gas regulators keep the gas pressure constant, insuring steady volume and even flame quality. Most regulators are dual stage and have two gauges; one tells the pressure in the cylinder and the other shows the pressure entering the hose. Gases for the process are oxygen and, primarily, acetylene. Other gases, including hydrogen, city gas, natural gas, propane, and mapp gas, are used for specific applications. However, with its higher burning temperature, acetylene is the preferred gas in most instances.

Gas cylinders for acetylene contain porous material saturated with acetone. Since acetylene cannot, with safety, be compressed over 15 psi, it is dissolved in the acetone which keeps it stable and allows pressure of 250 psi. Because of the acetone in the acetylene cylinders, they should always stand upright. The oxygen cylinder capacities vary from 60 to 300 cu. ft. with pressures up to 2400 psi. The maximum charging pressure is always stamped on the cylinder.

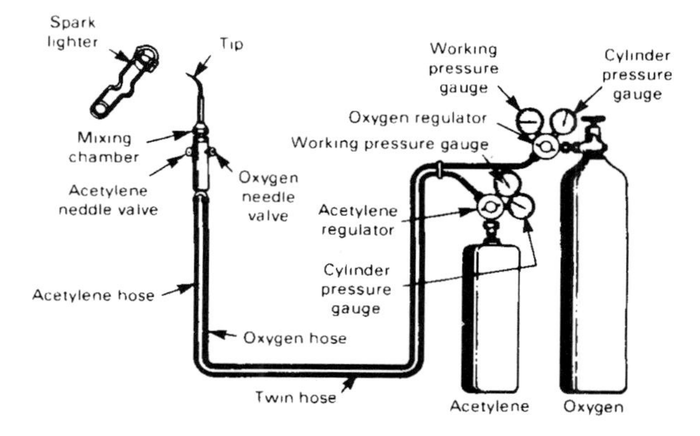

Fig. 188. Oxyacetylene welding equipment.

Shielded Metal-Arc Welding

This is perhaps the most popular welding process in use today. The high quality of the metal produced by the shielded arc process, plus the high rate of production, has made it a replacement for other fastening methods. The process can be used in all positions and will weld a wide variety of metals. The most popular use, however, is the welding of mild carbon steels and the low-alloy steels.

Shielded metal-arc welding is an arc welding process wherein coalescence is produced by heating with an arc between a covered metal electrode and the work. Shielding is obtained from decomposition of the electrode covering. The electrode also supplies the filler metal. Fig. 189 shows the covered electrode, the core wire, the arc area, the shielding atmosphere, the weld, and solidified slag.

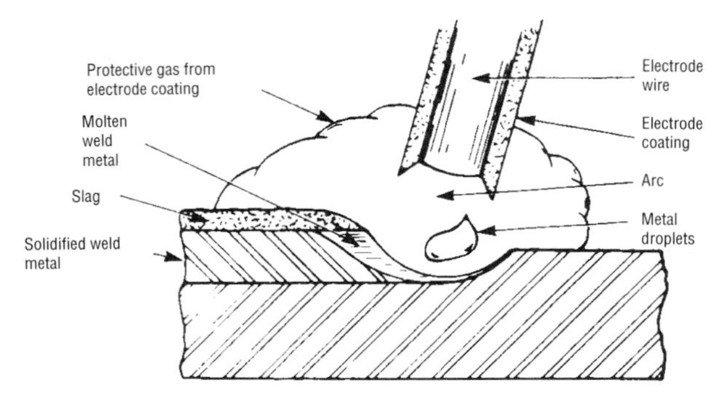

Fig. 189. Shielded metal-arc welding.

This manually controlled process welds all nonferrous metals ranging in thickness from 18 gauge to the maximum encountered. For material thicknesses of ¼", a beveled edge preparation is used and the multipass welding technique employed. The process allows for all position welding. The arc is under the control of, and is visible to, the welder. Slag removal is required. The major components required for shielded metal-arc welding are shown in Fig. 190.

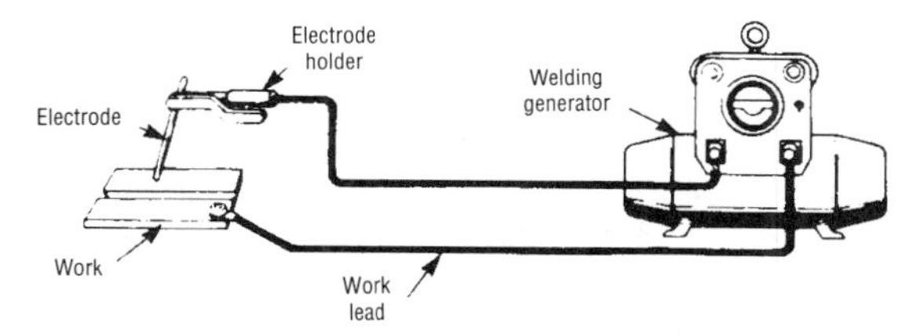

Fig. 190. Shielded metal-arc welding equipment.

The welding machine (power source) is a major item of required equipment. Its primary purpose is to provide electric power of the proper current and voltage sufficient to maintain a welding arc. Shielded metal-arc welding can be accomplished by either alternating current (AC) or direct current (DC). Direct current can be employed straight (electrode negative) or reverse (electrode positive). A variety of welding machines are used, each having its specific advantages or special features. The AC transformer is simple, inexpensive, and quiet. The transformer-rectifier type machine converts AC power to DC power, and provides direct current at the arc. There is also the AC-DC transformer-rectifier type machine which combines the features of both the transformer and the rectifier. Probably the most versatile welding power source is the direct current generator. The conventional dual-control single-operator generator allows the adjustment of the open-circuit voltage and the welding current. When electric power is available, the generator is driven by an electric motor. Away from power lines, the generator can be driven by a gasoline internal combustion engine or a diesel engine.

The electrode holder is held by the operator and firmly grips the electrode, carrying the welding current to it. The insulated pincer-type holders are the most popular. Electrode holders come in various sizes and are designated by their current-carrying capacity.

The welding circuit consists of the welding cables and connectors used to provide the electrical circuit for conducting the welding current from the machine to the arc. The electrode cable forms one side of the circuit and runs from the electrode holder to the electrode terminal of the welding machine. Welding cable size is selected based on the maximum welding current used. Sizes range from AWG No. 6 to AWG No. 4/0 with amperage ratings from 75 amperes upward. The work lead is the other side of the circuit and runs from the work clamp to the work terminal of the welding machine.

Covered electrodes, which become the deposited weld metal, are available in sizes from 1⁄16" to 5⁄16" diameter and from 9" to 18" in length, with the 14" length the most popular. The covering on the electrode dictates the usability of the electrode and provides the following:

1. Gas shielding
2. Deoxidizers for purifying the deposited weld metal
3. Slag formers to protect weld metal from oxidation
4. Ionizing elements for smooth operation
5. Alloying elements to strengthen deposited weld metal
6. Iron powder to improve the productivity of the electrode

The usability characteristics of different types of electrodes are standardized and defined by the American Welding Society. The AWS identification system indicates the characteristics and usability by classification numbers printed on the electrodes. Color code markings used in the past for this purpose are no longer employed.

Gas Metal-Arc Welding

Gas metal-arc (MIG) welding is an arc welding process wherein coalescence is produced by heating with an arc between a continuous filler-metal (consumable) electrode and the work. The electrode is in the form of a wire which is continuously and automatically fed into the arc to maintain a steady arc. This electrode wire, melted into the heat of the arc, is transferred across the arc and becomes the deposited weld metal. Shielding is obtained entirely from an externally supplied gas mixture. Fig. 191 shows the electrode wire, the gas shielding envelope, the arc, and the deposition of the weld metal.

Some of the outstanding features of gas metal-arc welding are: the top-quality welds in almost all metals and alloys, little after-weld cleaning, relatively high speed, no slag production. Some of the variations of the process involve Microwire for thin gauge materials, CO_2 for low-cost high-speed welding, and argon/oxygen for stainless steels. The major components required for gas metal-arc welding are shown in Fig. 192.

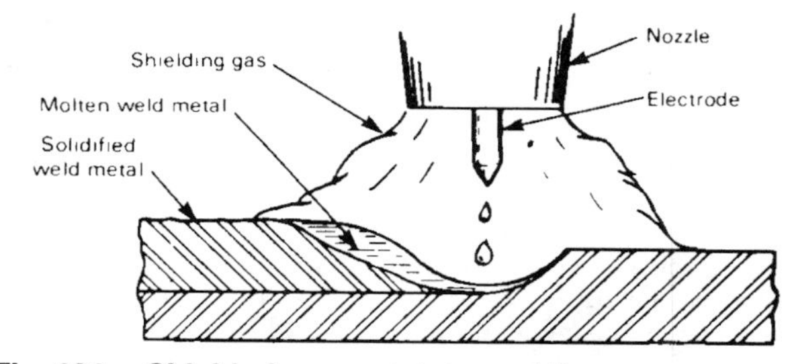

Fig. 191. Shielded gas metal-arc welding.

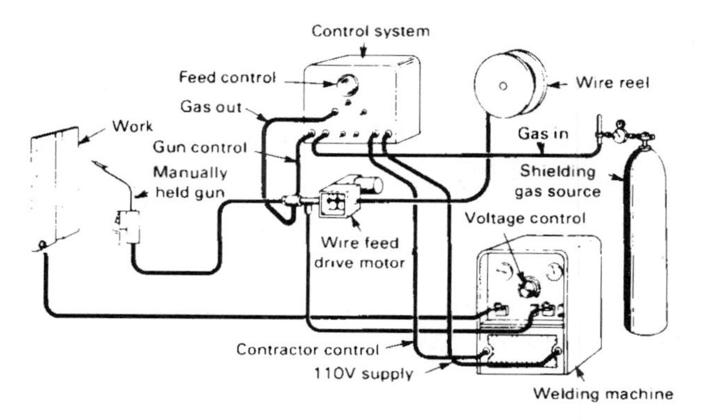

Fig. 192. Gas metal-arc welding equipment.

The welding machine or power source for consumable-electrode welding is called a constant voltage (CV) type of welder, meaning that its output voltage is essentially the same with different welding current levels. These CV power sources do not have a welding-current control and cannot be used for welding with electrodes. The welding-current output is determined by the load on the machine, which is dependent on the electrode wire-feed speed. The wire-feeder system must be matched to the constant-voltage power supply. At a given wire-feed speed rate, the welding machine will supply the proper amount of current to maintain a steady arc. Thus the electrode wire-feed rate determines the amount of welding current supplied to the arc.

The welding gun and cable assembly is used to carry the electrode wire, the welding current, and the shielding gas to the welding arc. The electrode wire is centered in the nozzle, with the shielding gas supplied concentric to it. The gun is held fairly close to the work to properly control the arc and provide an efficient gas shielding envelope. Guns for heavy-duty work at high currents, and guns using inert gas and medium currents, must be water cooled.

The shielding gas displaces the air around the arc to prevent contamination by the oxygen or nitrogen in the atmosphere. This gas shielding envelope must efficiently shield the area in order to obtain high-quality welds. The shielding gases normally used for gas metal-arc welding are argon, helium, or mixtures for nonferrous metals; CO_2 for steels; CO_2 with argon and sometimes helium for steel and stainless steel.

The electrode wire composition for gas metal-arc welding must be selected to match the metal being welded. The electrode wire size depends on the variation of the process and the welding position. All electrode wires are solid and bare except in the case of carbon steel wire, when a very thin protective coating (usually copper) is employed.

Gas Tungsten-Arc Welding

Gas tungsten-arc (TIG) welding is an arc welding process wherein coalescence is produced by heating with an arc between a single tungsten (nonconsumable) electrode and the work. It was invented by the aircraft industry and used extensively to weld hard-to-weld metals, primarily magnesium and aluminum, and also stainless steels. Shielding is obtained from an inert gas mixture. Filler metal may or may not be used. Fig. 193 shows the arc, the tungsten electrode, and the gas shield envelope all properly positioned above the work piece. The filler-metal rod is being fed manually into the arc and weld pool.

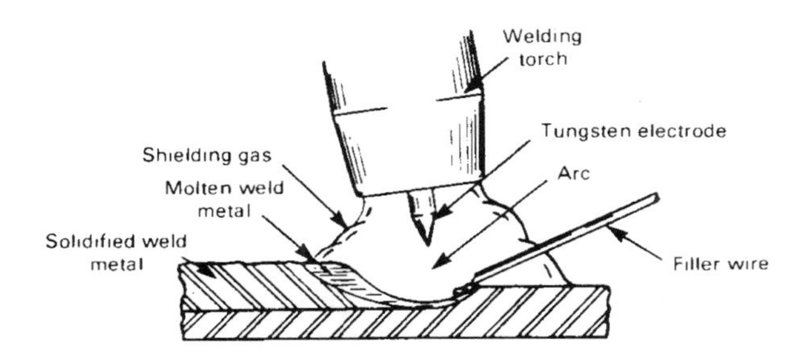

Fig. 193. Gas tungsten-arc welding.

Some of the outstanding features of gas tungsten-arc welding are: the top quality in hard-to-weld materials and alloys; practically no after-weld cleanup; no weld spatter; no slag production. The process can be used for welding aluminum, magnesium, stainless steel, cast iron, and mild steels. It will weld a wide range of metal thicknesses. The major components required for gas tungsten-arc welding are shown in Fig. 194.

A specially designed welding machine (power source) is used for tungsten-arc welding. Both AC and DC machines are built for the welding of specific materials: AC is usually used for welding aluminum and magnesium; DC for stainless steel, cast iron, mild steel, and several alloys. High-frequency current is used in starting the welding arc when using DC current, and continuously with AC current. A typical gas tungsten-arc welding machine operates with a range of 3 to 350 amperes, with 10 to 35 volts at a 60% duty cycle.

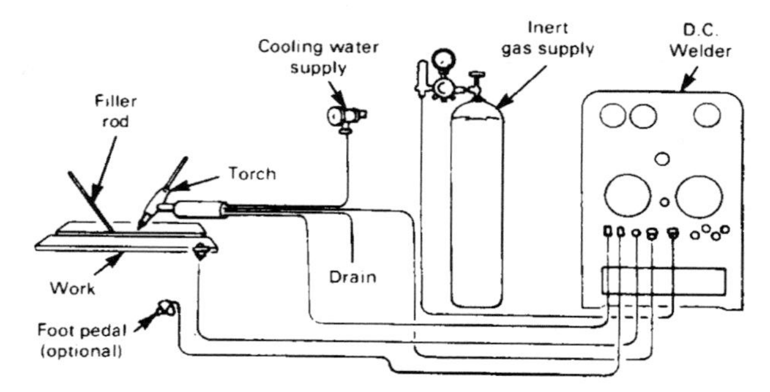

Fig. 194. Gas tungsten-arc welding equipment.

The torch holds the tungsten electrode and directs shielding gas and welding power to the arc. Most torches are water cooled; some air-cooled torches are in use. The electrodes are made of tungsten and tungsten alloys. They have a very high melting point (6170°F) and are practically nonconsumable. The electrode does not touch the molten weld puddle; properly positioned, it hangs over the work and the arc keeps the puddle liquid. Electrode tips contaminated by contact with the weld puddle must be cleaned or they cause a sputtering arc.

Filler metals are normally used except when very thin metal is welded. The composition of the filler metal should be matched to that of the base metal. The size of the filler-metal rod depends on the thickness of the base metal and the welding current. Filler metal is usually added to the puddle manually, but automatic feed may on occasion be used. An inert gas, either argon, helium, or a mixture of both, shields the arc from the atmosphere. Argon is more commonly used because it is easily obtainable and, being heavier than helium, provides better shielding at lower flow rates.

Welding Safety

Welding involves several potentially hazardous conditions: very high temperatures; use of explosive gases; possible exposure to harmful light, toxic fumes, etc.; molten metal spatter, flying particles, etc. Welding hazards, however, can be successfully controlled to ensure the safety of the welder.

Some of the basic actions required are as follows: protective clothing must be worn by the welder to shield skin from exposure to the brilliant light given off by the arc; a helmet is required to protect the face and eyes from the arc; fire-resistant protective clothing, shoes, leather gloves, jacket, apron, etc., are a necessity; a dark-colored filter glass in the helmet allows the welder to watch the arc while protecting the eyes.

Ventilation must be provided when welding in confined areas. The work area must be kept clean and the equipment properly maintained.

RIGGING

Weight Estimating

The first, and usually the most important, consideration when selecting tools and equipment for rigging work is the weight of the object to be moved. Reasonable accuracy in the determination of an object's weight is a requirement for safe rigging. When it is not known, and dependable information is not available, the weight must be estimated. This should be an approximate calculation, not a guess.

In most cases, a rigging weight estimate is made by roughly calculating the object's volume and multiplying this by the unit weight of the material of which it is made. As only an approximate figure is required, the calculation can be simplified by using approximate values, which will allow many of the calculations to be made mentally.

For example, most heavy objects are made of iron or steel which ranges in weight from 475 to 490 pounds per cubic foot. This value can be rounded off to 500. For cylindrical calculations the value of Pi (π) can be rounded off from 3.1416 to an even 3. The object's dimensions can be rounded off to the closest even numbers, preferably multiples of 10. When rounding off dimensions, alternately increasing and decreasing to get even numbers will help to cancel out errors.

In the example in Fig. 195, values and dimensions are rounded off to allow a quick and accurate weight estimate of the tank.

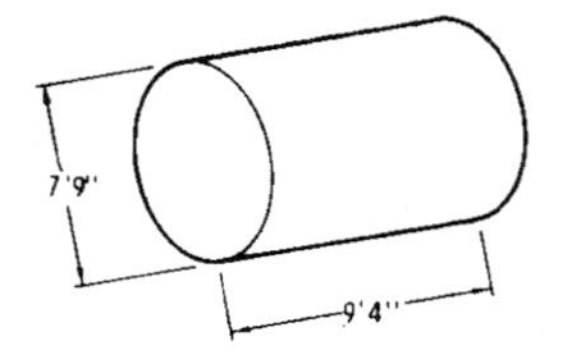

Fig. 195.

¾" Thick
Steel Plate

Round off 7'9" to 8' dia. or 4' rad.
Round off 9'4" to 9' length

Estimate Calculation
rad. × rad. × Pi. = end area
4 × 4 × 3 = 48 sq. ft. area
dia. × Pi. × length = shell area
8 × 3 × 9 = 216 sq. ft. shell area
312 sq. ft. *total area*

¾" *Thickness*

1 sq. ft. area has
1⁄16 cu. ft. volume
312 divided by 16 gives
20 cu. ft. *volume*
20 × 500 = 10,000 lbs. *total weight*

An alternate method is to obtain the weight per square foot of steel plate from a table of weights, and multiply this by the total area. The weight per square foot of ¾-inch steel plate is listed as 30.6 pounds.

Rounding off the values:

31 × 310 = 9,600 lbs. *total weight*

Estimating the weight of any regular-shaped object may be done in the same manner as the preceding example. In some cases several calculations may be required, as with the chambered roll illustrated in Fig. 196.

For calculating purposes, the roll is considered as made up of three parts, two shafts and a body. The total solid volume is the sum of the solid volumes of the three parts. To determine actual volume, the total chamber volume, which is the sum of the three chamber volumes, is subtracted from the total solid volume. Weight is then determined by multiplying the actual volume by the unit weight of the roll material.

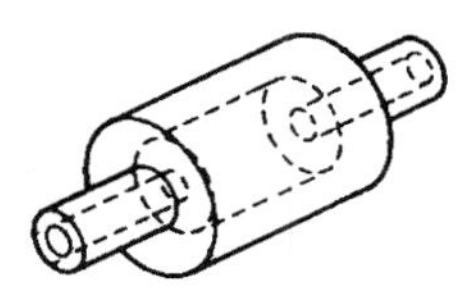

Fig. 196.

The weight of irregular-shaped objects may be estimated with a high degree of accuracy by visualizing the object as a regular shape, or made up of a group of regular shapes. For example, the irregular-shaped object in Fig. 197A may be visualized as a regular-shaped object of lesser dimensions as shown in Fig. 197B.

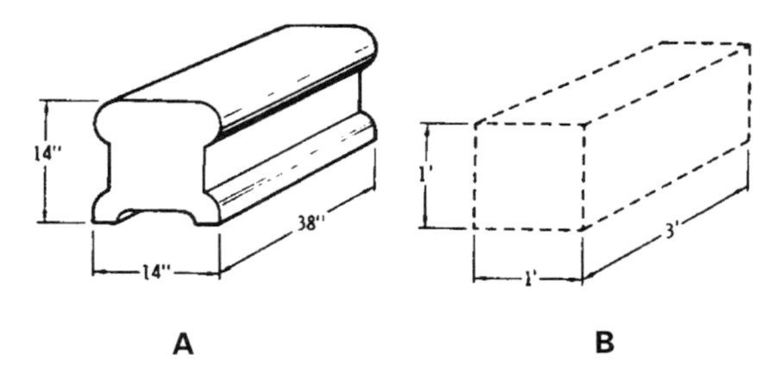

Fig. 197.

Machines are usually an assembly of components of varying shapes, sizes, and construction. They may be visualized as a group of regular-shaped solid units when weight estimating. Each unit must be reduced in size to approximate the actual volume of material it contains. The machine shown in Fig. 198A, for example, could be visualized as shown in Fig. 198B.

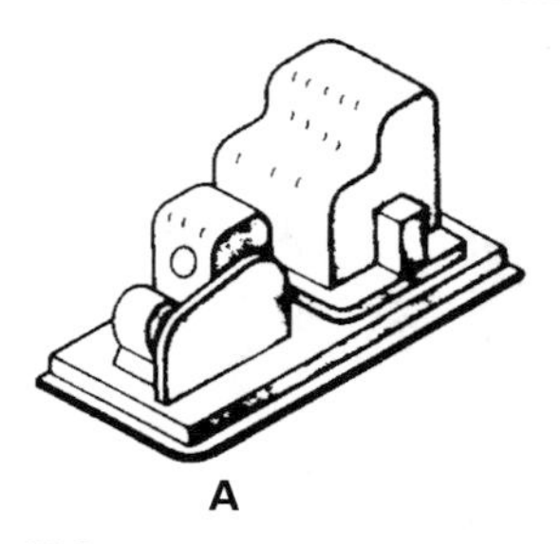

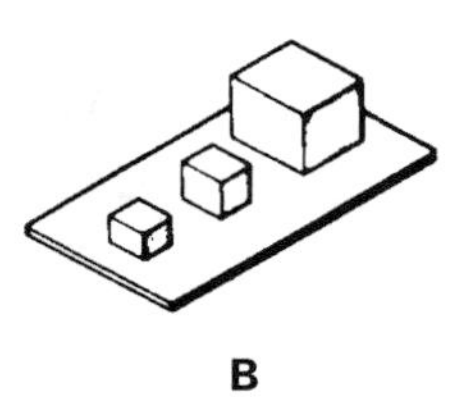

Fig. 198.

Weight of Steel Bar Stock in Pounds per Lin. Foot

Size	Square	Round
1	3.4	2.7
1 ½	7.7	6.0
2	13.6	13.6
3	30.6	24.0
4	54.4	42.7
5	85.0	66.8
6	122.4	96.1
7	166.7	130.8
8	217.6	171.0
9	283.1	222.3
10	340.0	267.0
11	411.4	323.1
12	489.6	384.5

Weight of Steel Plate in Pounds per Sq. Foot

Thickness	Weight
1/16	2.55
1/8	5.1
3/16	7.65
1/4	10.2
5/16	12.75
3/8	15.3
1/2	20.4
5/8	25.5
3/4	30.6
1	40.8
1 ¼	51.0
1 ½	61.2
2	81.6

Weights of Materials

Material	Weight per cu. in.	Weight per cu. ft.
Aluminum	.093	160
Brass	.303	524
Cast Iron	.260	450
Concrete	.083	144
Sand	.070	120
Steel	.281	490
Water	.036	62 ½
Wood	.020	36

Wire Rope

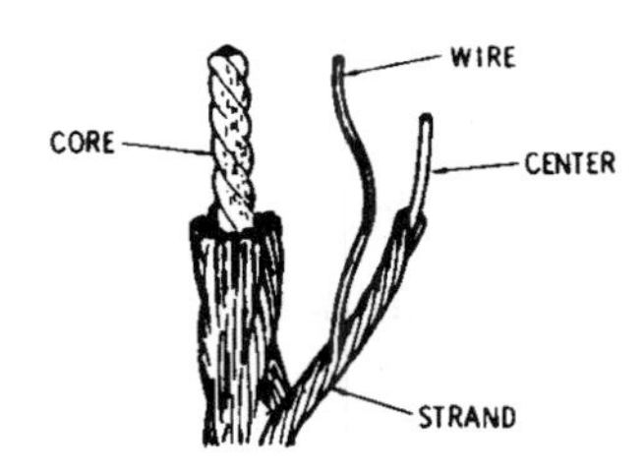

Fig. 199.

The basic element in the construction of wire rope is a single metallic *wire*. Several of these wires are laid helically around a *center* to form a *strand*. Finally, a number of strands are laid helically around a *core* to form the wire rope.

The primary function of the core is to serve as a foundation for the rope, to keep it round and to keep the strands correctly spaced and supported.

During construction, the wires that make up the strand may be laid around the center in either a clockwise or counterclockwise direction. The same is true of the strands when they are laid around the core. This direction of rotation is called the *lay* of the rope. In *right* lay rope the strands rotate around the core in a clockwise direction, as the threads do in a right-hand thread. In *left* lay the strands rotate counterclockwise, as do left-hand threads.

The terms *regular* and *lang* are used to designate direction of the wires around the center. Regular lay means that the wires rotate in a direction opposite to the direction of the strands around the core. This results in the wires being roughly parallel to the center line of the rope. Lang lay means the wires rotate in the same direction as the strands, resulting in the wires being at a diagonal to the rope center line.

A right regular lay rope is shown in Fig. 199. The strands rotate clockwise and the wires counterclockwise. This is the most widely used rope lay and is commonly referred to simply as "regular lay."

Wire rope is classified by number of strands and the approximate number of wires in each strand. For example, the 6 × 7 classification indicates the rope has 6 strands and that each strand contains 7 wires. The wires in a strand are placed in layers around a center wire, each layer containing six more wires than the preceding one. These arrangements are referred to by the number of wires in each layer.

The 7 wire strand is 6-1, the 19 wire strand is 12-6-1, the 37 wire strand is 18-12-6-1.

The designations for wire rope classifications are only nominal, as the actual number of wires in a strand varies with the style of construction. For example, the 6 × 19 classification is made up of wire rope having anywhere from 15 to 26 wires per strand.

Factor of Safety

The safe use of wire rope requires that loads be limited to a portion of the rope's ultimate or breaking strength. The safe load for a wire rope is determined by dividing its breaking strength by a *factor of safety*. Factors of safety for wire rope range from 5 for steady loads to 8 or more for uneven and shock loads.

For example, the breaking strength for a ½" diameter improved plow-steel rope is listed in Table 23 at 10.5 tons. If this rope were to be used with hoisting tackle at a factor of safety of 5, its maximum safe load would be one-fifth of the breaking strength, or 2.1 tons. If, however, it were to be used in a sling at a factor of safety of 8, its maximum safe load would be one-eighth of the breaking strength, or 1.4 tons.

Table 23. 6 × 19 Classification Ropes—Independent Wire Rope Core (IWRC)

Diameter in Inches	Breaking Strength in Tons of 2000 Pounds		
	Improved Plow Steel		Extra Improved Plow Steel
	Fiber Core	IWRC	IWRC
3/16	1.46	—	—
1/4	2.59	2.78	3.20
5/16	4.03	4.33	4.98
3/8	5.77	6.20	7.14
7/16	7.82	8.41	9.67
1/2	10.2	11.0	12.6
9/16	12.9	13.9	15.9
5/8	15.8	17.0	19.6
3/4	22.6	24.3	27.9
7/8	30.6	32.9	37.8
1	39.8	42.8	49.1

Efficiencies of End Attachments

Fitting	Nominal Efficiency, percent of catalog rated rope strength
Wire Rope Sockets	100
Spelter (Zinced) Attachments	100
Fittings (Swaged or Pressed)	100
*Torpedo Collar (with or without thimble)	100
Open Wedge Sockets	80-90
Clips (U-bolt Type)	80
Clips (Twin-based Type)	80
Spliced-in Thimbles: ¼ and smaller	90
5/16	89
3/8	88
½	86
5/8	84
¾	82
7/8 to 2½, incl.	80

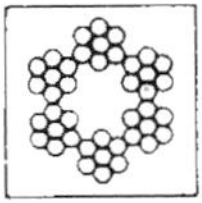

6X7 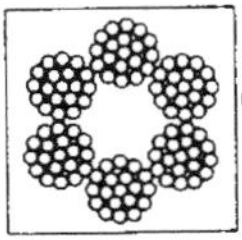 6X19 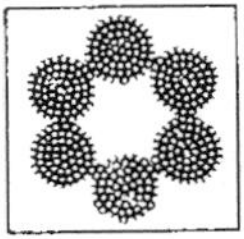 6X37

Fig. 200.

Wire Rope Attachment

The *U-bolt* or *Crosby* (Fig. 202A) wire rope clip is probably the most common method of attaching a wire rope to equipment. All U-bolt clips must be placed on the rope with U bolts bearing on the short or "dead" end of the rope. Illustrations of correct and wrong application of U-bolt clips are shown in Fig. 201.

An improved type wire rope clip called the *double-base safety* or *fist grip,* shown in Fig. 202B, has corrugated jaws to fit both parts of the rope, allowing it to be installed without regard to the live or dead part.

When making an eye attachment with clips, a thimble should be used, and the correct number of clips (listed in Table 24). All clips should be spaced not less than six rope diameters apart. Apply the clip farthest from the thimble first, at about 4 inches from the end of the rope, and screw up tightly. Next, put on the clip nearest the thimble and apply the nuts handtight. Then put on the one or more intermediate clips handtight. Take a strain on the rope, and while the rope is under this strain, tighten all the clips previously left loose. Tighten alternately on the two nuts so as to keep the clip square. After the rope has been in use a short time, retighten all clips.

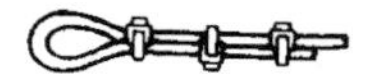

Fig. 201.

Table 24. Recommended Number of Clips

Rope Size	U-Bolt	Safety
¼ to ⅜	2 to 4	2
7/16 to ⅝	3 or 4	2
¾	4 or 5	3
⅞ to 1⅛	4 or 5	4
1¼ to 1½	5 to 8	5

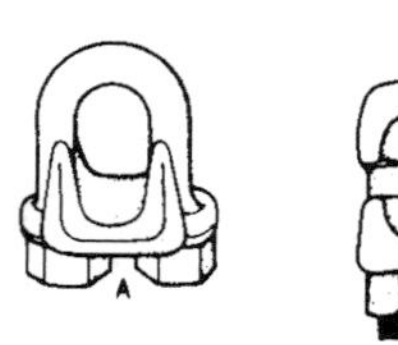

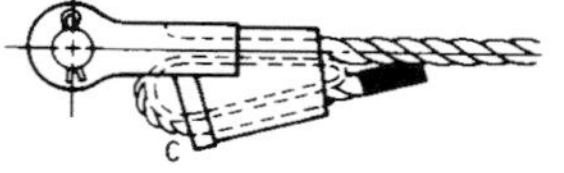

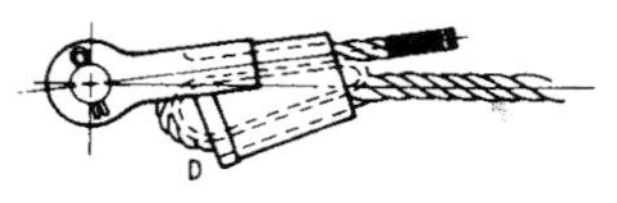

Fig. 202.

Wedge socket attachments are used on equipment where frequent changes are required. Care must be exercised to install the rope so that the pulling part is directly in line with the clevis pin as shown in Fig. 202C. If incorrectly installed, as shown in Fig. 202D, a sharp bend will be produced in the rope as it enters the socket.

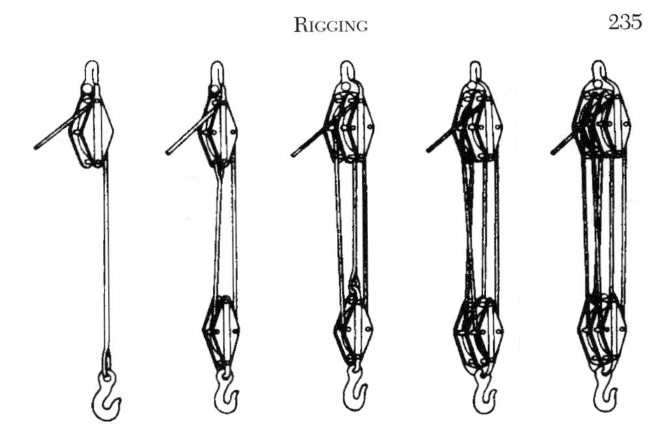

Fig. 203.

Multiple Reeving

A single rope supporting a load is referred to a single part line, and the tension in the rope is equal to the suspended weight. When a load is supported by a multipart wire rope tackle as shown in Fig. 203, and the rope is not moving, the load on each line, including the lead line, is equal to the weight of the load divided by the number of parts of rope supporting the load. When this load is raised, however, the loads on the individual supporting ropes change, increasing from the dead end to the lead line.

How to Measure Wire Rope

The measurement should be made carefully with calipers. Fig. 204 shows the correct and incorrect method of measuring the diameter of wire rope.

Wire Rope Slings

The single sling with loop ends, most widely used of all sling types, lends itself readily for use in a basket hitch, choker hitch, or as a straight rope. Blocking should always be used to protect the sling from sharp corners.

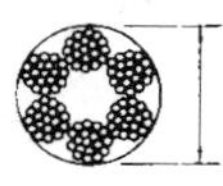

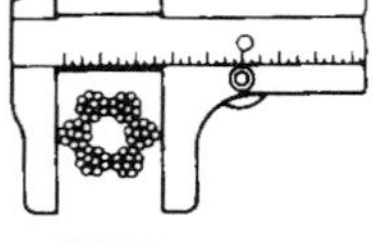

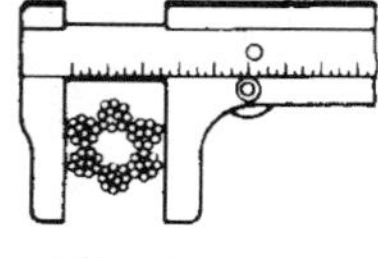

Fig. 204.

Table 25. Safe Loads in Tons

Nominal Size Inches	Single	Choker	U-sling	Basket	60-deg.	45-deg.	30-deg.
1/4	.5	.3	.7	.6	.57	.5	.3
5/16	.8	.6	1.1	1.0	.9	.7	.6
3/8	1.1	.8	1.5	1.4	1.3	1.1	.8
1/2	2.0	1.4	2.7	2.4	2.3	1.9	1.3
5/8	2.9	2.1	4.2	3.8	3.7	3.0	2.1
3/4	4.1	3.0	6.0	5.4	5.2	4.2	3.0
7/8	5.6	3.8	7.7	6.8	6.7	5.4	3.8
1	7.2	5.0	10.0	9.3	8.7	7.1	5.0
1 1/8	9.0	5.6	11.2	10.5	9.7	7.9	5.6

Sling end fittings of the more popular styles are shown in Fig. 205. Any combination of these, to suit the job requirements, is available.

The lengths of slings having loop or ring style end fittings are measured from the weight-bearing surface. Those with pin style end fittings are measured from the center of the pin.

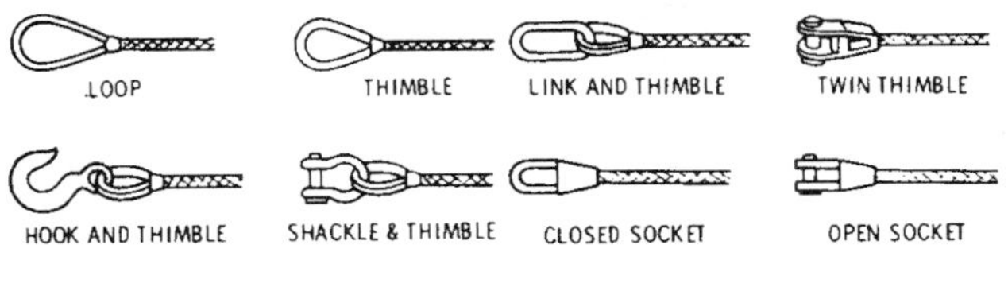

Fig. 205.

Eyebolts and Shackles

The strength of an eyebolt is influenced greatly by the direction of pull to which it is subjected. For loads involving angular forces, the shoulder type has several times the strength of the conventional eyebolt. Table 26 lists the safe loads by sizes and direction of load.

The shackle is the recommended fastener for attaching slings to eyebolts, chain, or wire rope. Table 27 lists the safe loads by nominal size and gives the dimensions of standard shackles.

Table 26. Ordinary Drop-Forged Steel Eyebolts

Size	SIZE	PULL	
½"	1,100 lb.	50 lb.	40 lb.
⅝"	1,800 lb.	90 lb.	65 lb.
¾"	2,800 lb.	135 lb.	100 lb.
⅞"	3,900 lb.	210 lb.	150 lb.
1"	5,100 lb.	280 lb.	210 lb.
1¼"	8,400 lb.	500 lb.	370 lb.
1½"	12,200 lb.	770 lb.	575 lb.
1¾"	16,500 lb.	1,080 lb.	800 lb.
2"	21,800 lb.	1,440 lb.	1,140 lb.

Table 27. Drop Forged Steel Shoulder-type Eyebolts

Size	SIZE	PULL	
¼"	300 lb.	30 lb.	40 lb.
½"	1,300 lb.	140 lb.	150 lb.
¾"	3,000 lb.	250 lb.	300 lb.
1 "	6,000 lb.	500 lb.	600 lb.
1 ¼"	9,000 lb.	800 lb.	900 lb.
1 ½"	13,000 lb.	1,200lb.	1,300 lb.
2 "	23,000 lb.	2,100 lb.	2,300 lb.
2 ½"	37,000 lb.	3,800 lb.	4,300 lb.

Fiber Rope

For many years the principal materials from which fiber rope was manufactured were *manila* (hemp) and *cotton*. Synthetic materials have in many cases replaced manila and cotton. While the reasons for this are numerous, probably the most important are increased resistance to deterioration, greater strength and greater pliability. Most rope is manufactured in three (3) or more strands. Three-strand construction has been the most widely used style for many years but since the eight (8) strand plaited construction has been introduced, it is widely used, particularly in synthetic materials. Increasing the number of strands reduces the size of each strand for a given rope size and increases the rope pliability.

Because of the variety of rope materials and constructions now in use, as well as the wide range of factors affecting rope behavior, it is impossible to cover the multitude of possible rope applications. Perhaps the single most important consideration in rope use is that of safety. While all aspects cannot be specifically detailed, several general safety considerations are important if the mechanic is to use rope properly and avoid possible rope failure. Some very important safety considerations are as follows:

1. Practically all rope failure accidents are caused by improper care and use rather than poor engineering or original product defect.
2. Rope must be adequate for the job. Choosing a rope of correct

size, material and strength must not be done haphazardly. Consult dealer, distributor or manufacturer for information and assistance if needed.

3. Do not overload rope. Sudden strains or shock loading can cause failure. Working load specifications may not be applicable when rope is subject to significant dynamic loading. Loads must be handled slowly and smoothly to minimize dynamic effects.
4. Avoid using rope that shows signs of aging and wear. If in doubt, destroy the used rope. If the fibers show wear in a given area, the rope should be respliced, downgraded or replaced.
5. Avoid chemical exposure. Rope can be severely damaged by contact with some chemicals. Special attention must be given for applications where exposure to either fumes or actual contact may occur.
6. Avoid overheating. Heat can seriously affect the strength of rope. The frictional heat from slippage on capstan or winch may cause localized heating which can melt synthetic fibers or burn natural fibers.
7. Never stand in line with rope under strain. If a rope or attachment fails it can recoil with sufficient force to cause physical injury. The snap-back action can propel fittings and rope with possible disastrous results.

Because of the wide range of rope use, rope conditions and exposure affecting the rope, it is impossible to make blanket recommendations as to the correct choice of rope to use. The following listing of rope characteristics can be an aid in making selections. Consult dealer or distributor for detail assistance.

Rope Characteristics

Manila—Made from fine Abaca (hemp) fiber. Excellent resistance to sunlight, low stretch and easy to knot. Good surface abrasion resistance.

Nylon—The strongest fiber rope manufactured. High elasticity allows absorption of shock loads. Resistant to rot, oils, gasoline, grease, marine growth and most chemicals. High abrasion resistance.

Polypropylene—A lightweight fiber with good strength. It floats, is resistant to rot, gasoline and most chemicals as well as being waterproof. Some manufacturers' products contain additives to reduce sunlight deterioration.

Polyester—Less strength than nylon fiber but better resistance to sunlight deterioration. Low stretch and excellent surface abrasion resistance. Other characteristics similar to nylon.

Table 28. New Rope Working Load in Pounds

Diameter	Safety Factor	Manila	Nylon	Polypro-pylene	Poly-ester
¼"	10	54	124	113	120
5/16"	10	90	192	171	180
⅜"	10	122	278	244	270
½"	9	264	525	420	520
⅝"	9	496	935	700	925
¾"	7	695	1420	1090	1400
1"	7	1160	2520	1800	2490
1⅛"	7	1540	3320	2360	3280
1¼"	7	1740	3760	2700	3700
1½"	7	2380	5320	3820	5260

Working loads are for rope in good condition with appropriate splices, in non-critical applications and under normal service conditions. Loads must be reduced for exceptional service conditions such as shock loads, sustained loads, etc.

Knots (Figs. 206 to 209)

Bight—Formed by simply bending the rope and keeping the sides parallel.

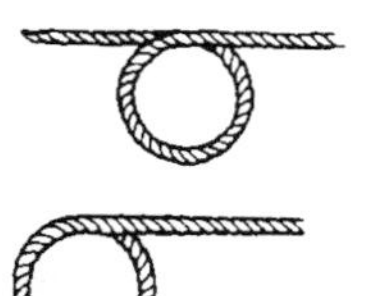

Loop or *Turn*—Formed by crossing the sides of a bight.

Round Turn—Further bending of one side of a loop.

Standing Part—That part of a rope that is not used in tying a knot; the long part which is not worked upon.

End—As the name implies, the very end of the rope.

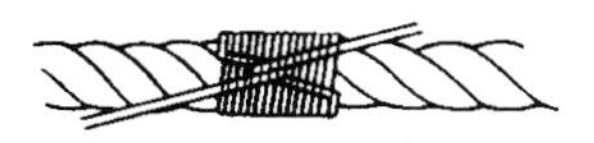

Whipping—Common, plain, or ordinary whipping is tied by laying a loop along the rope and then making a series of turns over it. The working end is finally stuck through this loop and the end hauled back out of sight. Both ends are then trimmed short. A whipping should be, in width, about equal to the diameter of the rope.

Scaffold Hitch—Lay the short end (a) of the rope over the top of the plank (Fig. 208A), leaving enough hanging down to the left to tie to the long rope, as shown in Fig. 208E. Wrap the long end (b) loosely twice around the plank, letting it hang down to the right as shown in Fig. 208A. Now, carry rope 1 over rope 2 and place it next to rope 3 as shown in Fig. 208B. Pick up rope 2 (Fig. 208C) and carry it over 1 and 3, and over the end of the plank. Take up the slack by pulling rope (a) to the left and rope (b) to the right. Draw ropes (a) and (b) above the plank as shown in Fig. 208D and join the short end (a) to the long rope (b) by an overhand bowline as shown in Fig. 208E. Pull the bowline tight, at the same time adjusting the lengths of the two ropes so that they hold the plank level. Attach a second rope to the other end of the plank in the same way and the scaffold is now ready for use.

Table 29. Anchor Shackles

Nominal Size	Tons Safe Load	Dimensions		
		A	B	C
⅜	.8	1 ½	1 1/16	7/16
½	1.4	2	⅞	⅝
⅝	2.2	2 ⅜	1 1/16	¾
¾	3.2	2 ⅞	1 ¼	⅞
⅞	4.3	3 ¼	1 ⅜	1
1	5.6	3 ⅝	1 11/16	1 ⅛
1 ⅛	6.7	4 ¼	1 ⅞	1 ¼
1 ¼	8.2	4 ¾	2	1 ⅜
1 ½	11.8	5 ½	2 ¼	1 ⅝
2	21.1	7 ¾	3 ¼	2 ¼

Screw Pin—Drop Forged Steel

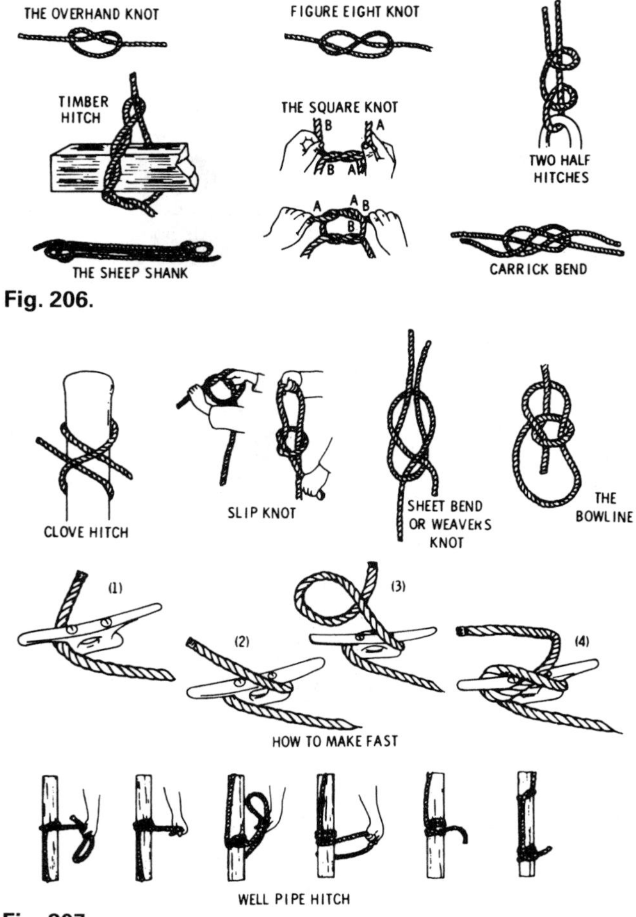

Fig. 206.

Fig. 207.

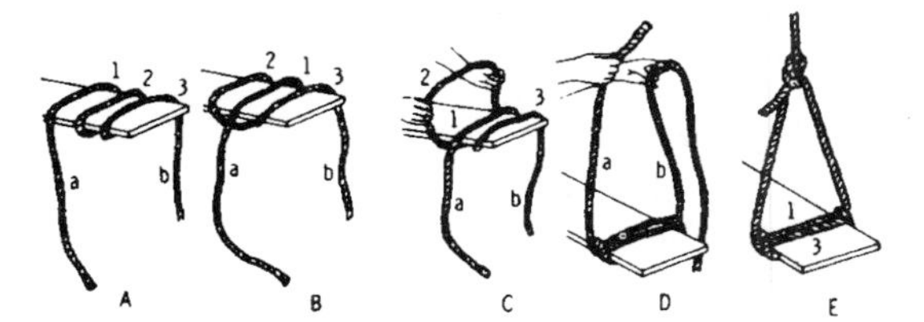

Fig. 208. Scaffold hitch.

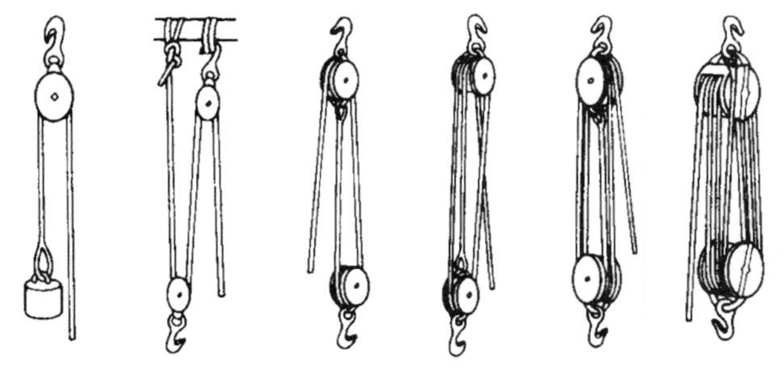

Fig. 209. Reeving rope falls.

HAND SIGNALS FOR CRANES AND HOISTS

With forearm vertical and forefinger upward, move hand in a horizontal circle.

With arm extended and palm downward, wave hand down and up.

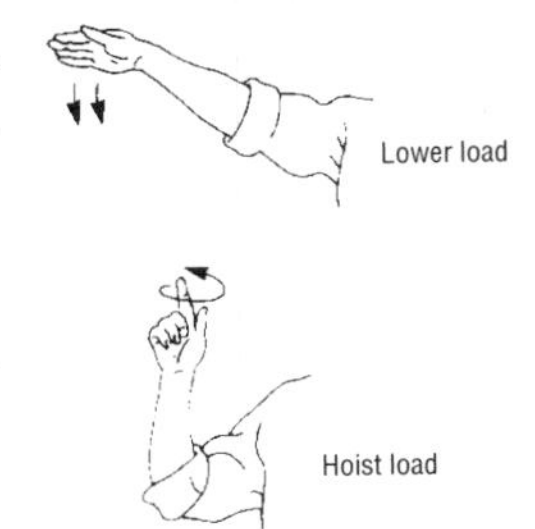

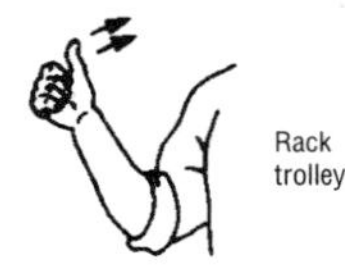

With arm extended and fingers clenched, jerk horizontally, pointing the direction with thumb.

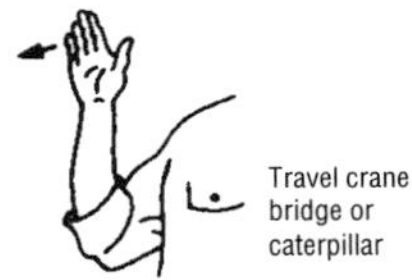

With arm extended, forearm nearly vertical, and hand open with fingers pointing upward, wave hand in direction of travel while facing in that direction.

With arm extended, fingers clenched, and thumb pointing upward, move hand up and down.

With arm extended, fingers clenched, and thumb pointing downward, move hand down and up.

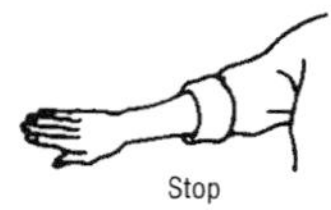

With arm extended and palm downward, hold position rigidly.

With arm extended and palm downward, move hand rapidly to right and left.

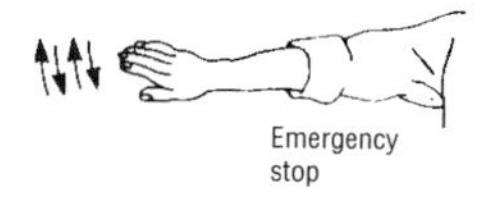

With arm extended, point forefinger in direction of travel.

Slew boom

PIPING

Thread Designations

American Standard

American Standard Pipe Threads are designated by specifying in sequence the nominal size, number of threads per inch, and the thread series symbols:

Nominal Size	*No. of Threads*	*Symbols*
3/8	18	NPT

Each of the letters in the symbols have the following significance:

N — American (Nat.) Standard
P — Pipe
T — Taper
C — Coupling
S — Straight
L — Locknut
R — Railing Fittings
M — Mechanical

Examples

3/8 — 18 NPT	American Standard Taper Pipe Thread
3/8 — 18 NPSC	Am. Std. Straight Coupling Pipe Thread
1/8 — 27 NPTR	Am. Std. Taper Railing Pipe Thread
1/2 — 14 NPSM	Am. Std. Straight Mechanical Pipe Thread
1 — 11½ NPSL	Am. Std. Straight Locknut Pipe Thread

Left-hand threads are designated by adding LH.

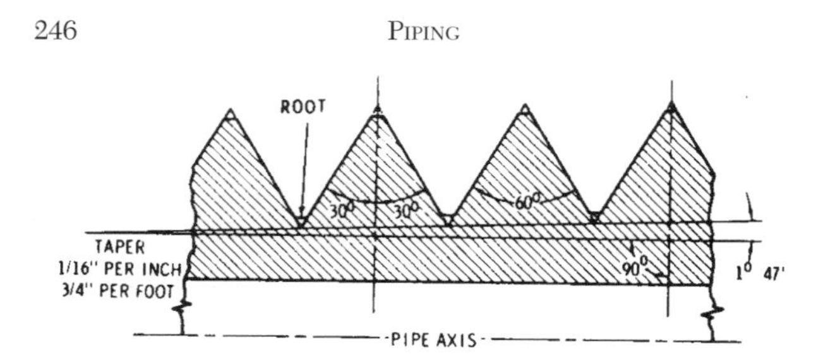

Fig. 210. American Standard Taper Pipe Thread Form

American Standard Taper Pipe Threads (NPT)

Basic Dimensions

Taper pipe threads are engaged or made up in two phases, *hand engagement* and *wrench makeup* (Fig. 211). Table 30 lists the basic hand engagement and wrench makeup for American Standard Taper Pipe Threads. Dimensions are rounded off to the closest 1/32 inch.

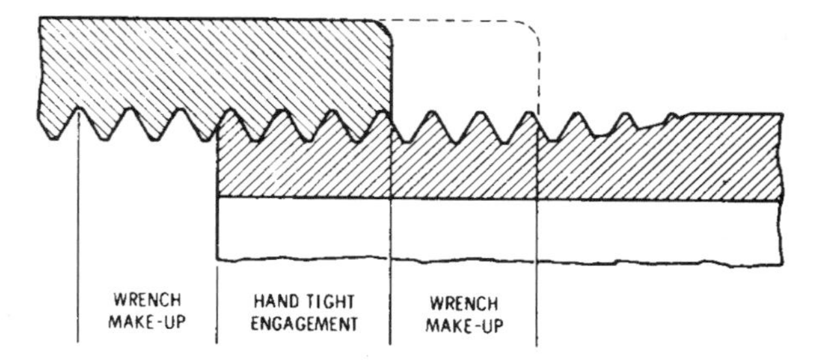

Fig. 211.

Note: Table 30 is a table of basic makeup dimensions. Commercial product may vary as much as one turn large or small and still be within standard tolerance. In actual shop practice, pipe threads are usually cut to give a connection which makes up less than the basic standard. Common practice is about 3 turns by hand and 3 to 4 turns by wrench.

Table 30. Taper Pipe Makeup Dimensions

Pipe	Threads	HAND TIGHT		WRENCH MAKE-UP		Dimension	Turns
Size	Per. In.	Dimension	Turns	Dimension	Turns	TOTAL MAKE-UP	
⅛	27	3/16	4 ½	3/32	2 ½	9/32	7
¼	18	7/32	4	3/16	3	13/16	7
⅜	18	¼	4 ½	3/16	3	7/16	7 ½
½	14	5/16	4 ½	7/32	3	17/32	7 ½
3/4	14	5/16	4 ½	7/32	3	17/32	7 ½
1	11 ½	⅜	4 ½	¼	3 ¼	⅝	7 ¾
1 ¼	11 ½	13/32	4 ½	9/32	3 ¼	11/16	8
1 ½	11 ½	13/32	4 ½	9/32	3 ¼	11/16	8
2	11 ½	7/16	5	¼	3	11/16	8
2 ½	8	11/16	5 ½	⅜	3	1 1/16	8 ½
3	8	¾	6	⅜	3	1 ⅛	9

Pipe Measurement

Dimensions on pipe drawings specify the location of center lines and/or points on center lines; they do not specify pipe lengths. This system of distance dimensioning and measurement is also followed in the fabrication and installation of pipe assemblies (Fig. 212).

To determine actual pipe lengths, allowances must be made for the length of the fittings and the distance threaded pipe is made up into the fittings. The method of doing this is to subtract an amount called *take-out* from the *center-to-center* dimension. The relationships of takeout to other threaded pipe connection distances, termed *make-up*, *center-to-center*, and *end-to-end*, are illustrated in Fig. 213.

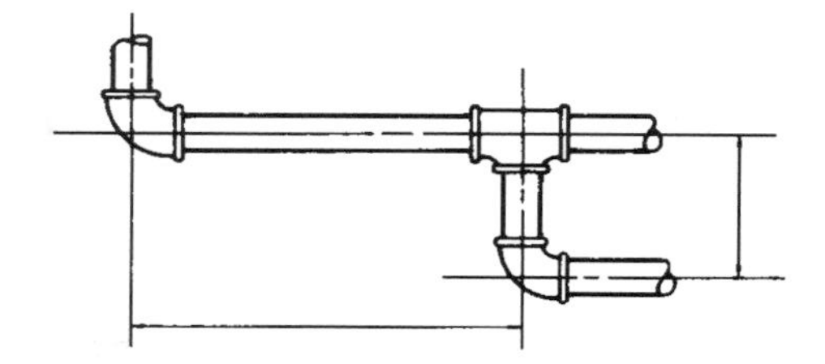

Fig. 212.

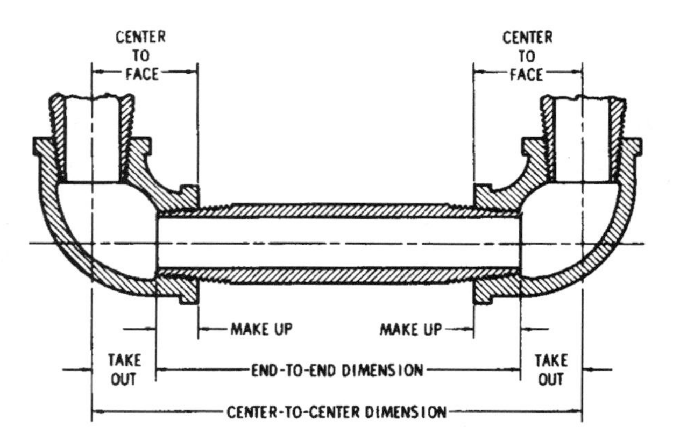

Fig. 213.

To determine end-to-end pipe length, the take-out is subtracted from the center-to-center dimension. Standard tables may be used for this purpose (Table 31). These tables should be used with judgment, however, since commercial product tolerance is one turn plus or minus. On critical connections, materials should be checked and compensation made for variances.

Table 31. Takeout Allowances

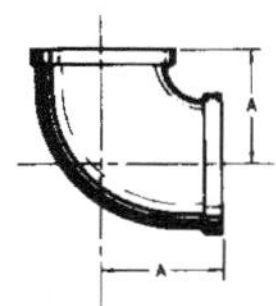

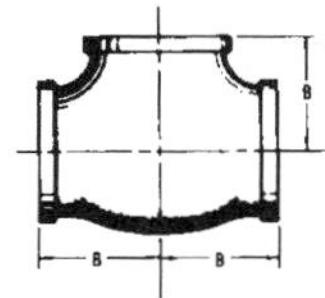

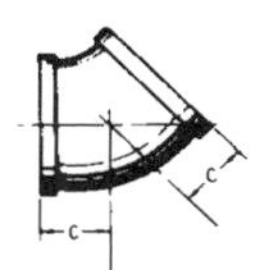

	90^{8} Elbow		Tee		45^{8} Elbow	
Pipe Size	A	Take Out	B	Take Out	C	Take Out
1/8	11/16	7/16	11/16	7/16	9/16	1/4
1/4	13/16	7/16	13/16	7/16	3/4	3/8
3/8	15/16	9/16	15/16	9/16	13/16	7/16
1/2	1 1/8	5/8	1 1/8	5/8	7/8	3/8
3/4	1 5/16	3/4	1 5/16	3/4	1	7/16
1	1 1/2	7/8	1 1/2	7/8	1 1/3	9/16
1 1/4	1 3/4	1 1/8	1 3/4	1 1/8	1 5/16	11/16
1 1/2	1 15/16	1 1/4	1 15/16	1 1/4	1 7/16	3/4
2	2 1/4	1 5/8	2 1/4	1 5/8	1 11/16	1

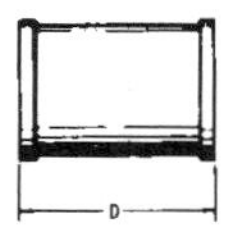

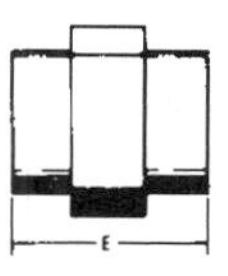

		Coupling		Union	
Pipe Size	Thread Make-Up	D	Take Out	E	Take Out
1/8	1/4	1	1/4	1 1/2	3/4
1/4	3/8	1 1/8	3/8	1 5/8	7/8
3/8	3/8	1 1/4	3/8	1 3/4	1
1/2	1/2	1 3/8	3/8	1 7/8	1
3/4	9/16	1 1/2	3/8	2 1/8	1 1/16
1	9/16	1 3/4	1/2	2 3/8	1 1/4
1 1/4	5/8	2	3/4	2 5/8	1 3/8
1 1/2	5/8	2 1/2	7/8	3	1 1/2
2	11/16	2 1/2	1 1/4	3 1/4	1 3/4

As the threads are made up in the fitting, high forces and pressures are developed by the wedging action of the taper. Also, frictional heat is developed as the surfaces are deformed to match the variations in the thread form. It is important that the threads be clean and well lubricated, and that the connection is not screwed up fast enough to generate excessive heat. Use of a lubricant called "dope" allows thread surfaces to deform and mate without galling and seizing. The dope also helps to plug openings resulting from improper threads, and acts as a cement.

Table 32. Commercial Pipe Sizes and Wall Thicknesses

Nominal Pipe Size	Outside Dia.	Nominal Wall Thickness					
		Sched. 5	Sched. 10	Sched. 40 Std.	Sched. 80 Ex. St.	Sched. 160	Ex. Ex. Strong
⅛	.405	—	.049	.068	.095	—	—
¼	.540	—	.065	.088	.119	—	—
⅜	.675	—	.065	.091	.126	—	—
½	.840	—	.083	.109	.147	.187	.294
¾	1.050	.065	.083	.113	.154	.218	.308
1	1.315	.065	.109	.133	.179	.250	.358
1 ¼	1.660	.065	.109	.140	.191	.250	.382
1 ½	1.900	.065	.109	.145	.200	.281	.400
2	2.375	.065	.109	.154	.218	.343	.436
2 ½	2.875	.083	.120	.203	.276	.375	.552
3	3.500	.083	.120	.216	.300	.438	.600
3 ½	4.000	.083	.120	.226	.318	—	—
4	4.500	.083	.120	.237	.337	.531	.674
5	5.563	.109	.134	.258	.375	.625	.750
6	6.625	.109	.134	.280	.432	.718	.864
8	8.625	.109	.148	.322	.500	.906	.875

The requirements for tight makeup of threaded pipe connections are: good quality threads; clean threads; proper dope for the application; slow final makeup to avoid heat generation.

Pipe schedule numbers shown in Table 32 indicate pipe strength. The higher the number in a given size, the greater the strength. The schedule number indicates the approximate values of the following expression:

$$\text{Schedule number} = \frac{1000 \times \text{internal pressure}}{\text{allowable stress in pipe}}$$

Copper Water Tube

Seamless copper water tube, popularly called *copper tubing*, is widely used for plumbing, water lines, heater coils, fuel oil lines, gas lines, etc. Standard copper tubing is commercially available in three (3) types, designated as type "K," "L," and "M." The type "K" has the thickest wall and is generally used for underground installations. The type "L" has a thinner wall and is most widely used for general plumbing, heating, etc. The type "M" is the thinwall tubing used for low-pressure and drainage applications.

The outside diameter of copper tubing is uniformly $^1/_8$" larger than the nominal size; for example, the outside diameter of 1" tubing is $1^1/_8$". All three types of copper tubing in given size have the same outside diameter. The inside diameter of each type will vary as the wall thickness varies (Table 33).

The three types of tubing are made in either "hard" or "soft" grade. Hard tubing is used for applications where lines must be straight without kinks or pockets. Soft tubing is used for bending around obstructions, inaccessible places, etc.

Table 33. Standard Copper Tubing Dimensions

Nominal Size	Outside Dia.	Inside Diameter		
		Type "K"	Type "L"	Type "M"
⅜	½	.402	.430	.450
½	⅝	.528	.545	.569
⅝	¾	.652	.668	.690
¾	⅞	.745	.785	.811
1	1 ⅛	.995	1.025	1.055
1 ¼	1 ⅜	1.245	1.265	1.291
1 ½	1 ⅝	1.481	1.505	1.527
2	2 ⅛	1.959	1.985	2.009
2 ½	2 ⅝	2.435	2.465	2.495
3	3 ⅛	2.907	2.945	2.981
3 ½	3 ⅝	3.385	3.425	3.459
4	4 ⅛	3.857	3.905	3.935

Plastic Pipe

Plastic pipe, available in a great variety of types and sizes, is used in a multiplicity of applications. This is in part because it can be readily cut, fitted and assembled before joining. Also because the joining is accomplished with suitable cement that is quick setting, it does not require heating, thus resulting in an efficient and relatively low cost pipe system. Another reason, and in many instances the principal reason, for the use of plastic pipe is that it is the most effective and economical piping system available to handle corrosive materials. While cement jointing of plastic pipe is commonly practiced, the heavier wall thicknesses, schedules 80 and 120, may be threaded and joined in the same manner as regular threaded pipe. The outside diameter of plastic pipe is the same as that of commercial metal pipe, and while the inside diameter is slightly smaller, the difference is negligible. Table 34 lists the outside and inside diameters of plastic pipe in the schedule 40, 80, and 120 strength groups. Pipe made from thermoplastic (becoming soft when heated), is limited to applications where exposure temperatures are well below its softening temperature. One of the principal uses for plastic pipe is that of waste water handling. Pipe for this purpose is available, with suitable cleaners and cement, at pipe and plumbing supply outlets. Because this is a relatively low tempera-

Table 34. Plastic Pipe Dimensions

Nominal Size	Outside Dia.	Inside Diameter		
		Schedule 40	Schedule 80	Schedule 120
1/8	.405	.261	.203	—
1/4	.540	.354	.288	—
3/8	.675	.483	.408	—
1/2	.840	.608	.528	.480
3/4	1.050	.810	.725	.690
1	1.315	1.033	.935	.891
1 1/4	1.660	1.364	1.256	1.204
1 1/2	1.900	1.592	1.476	1.423
2	2.375	2.049	1.913	1.845
2 1/2	2.875	2.445	2.289	2.239
3	3.500	3.042	2.864	2.758
4	4.500	3.998	3.786	3.572
6	6.625	6.031	5.709	5.434

ture, low hazard application, there is little reason to be concerned about material formulation. Other applications, particularly if hazardous materials are involved, should be referred to suppliers or others qualified to make formulation selections to match application exposures.

Plastic pipe systems employ a variety of fittings, valves, pumps, etc., to complement the pipe. Cement joined pipe systems measurements are made in the same manner as copper water type, i.e., the pipe bottoms in the fitting socket and take-out dimensions are calculated by subtracting the socket depth from center-to-face dimensions. Threaded plastic pipe measurements are handled in the same manner as those of regular metal threaded pipe.

45-Degree Offset

The *travel* distance of a 45-degree offset is calculated in the same manner as the diagonal of a square. Multiply the distance across flats by 1.414. The *run* and *offset* represent the two equal sides of a square and the *travel* the diagonal, is shown in Fig. 214.

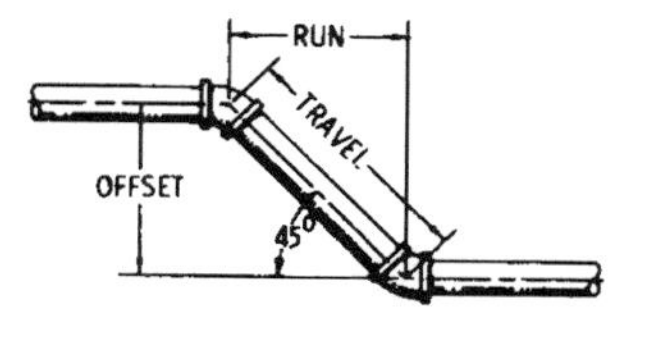

Fig. 214.

There are also occasions when the travel is known and the offset and run dimensions are wanted. This may be done in the same manner as finding the distance across the flats of a square when the across-corners dimension is known. Multiply the travel dimension by 0.707.

Examples

What is the travel of a 16-inch 45-degree offset?

$$16 \times 1.414 = 22.625 \text{ or } 22\frac{5}{8} \text{ inches.}$$

What is the offset and run of a 45-degree offset having a travel of 26 inches?

$$26 \times 0.707 = 18.382 \text{ or } 18\tfrac{3}{8} \text{ inches.}$$

Other Offsets

The dimensions of piping offsets of several other common angles may be calculated by multiplying the known values by the appropriate constants listed in the following table:

Angle	To find Travel Offset Known	To Find Travel Run Known	To Find Run Travel Known	To Find Offset Travel Known
60°	1.155	2.000	0.500	0.866
30°	2.000	1.155	0.866	0.500
22½°	2.613	1.082	0.924	0.383
11¼°	5.126	1.000	0.980	0.195

45-Degree Rolling Offset

The 45-degree offset is often used to offset a pipe line in a plane other than the horizontal or vertical. This is done by rotating the offset out of the horizontal or vertical plane, and is known as a *rolling offset.* The rolling offset can best be visualized as contained in an imaginary isometric box as shown in Fig. 215.

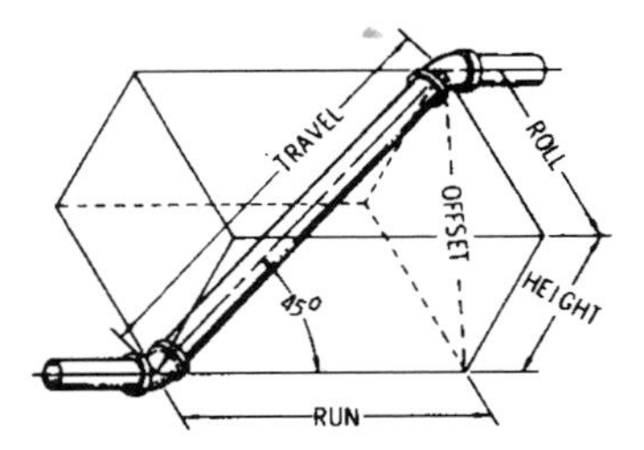

Fig. 215.

The run and offset distances are equal, as they are in the plain 45-degree offset; however, there are two additional dimensions, *roll* and *height*. Two right-angle triangles must now be considered. The original one remains the same, with the offset and run as equal sides and the travel as the hypotenuse. The new triangle has the roll and height as sides and the offset as the hypotenuse.

The method of finding distances for plain 45-degree offsets is also used for calculating rolling offset distances. In addition, the sum-of-the-squares equations are used to find the values of the second triangle.

Sum of the Squares—The sum-of-the-squares equation states that the hypotenuse of a right angle triangle squared, is equal to the side opposite squared plus the side adjacent squared. This equation is commonly written as follows:

$$c^2 = a^2 + b^2$$

Substitution of pipe offset terms in this equation and its rearrangements gives the following equations:

$$\begin{aligned} offset^2 &= roll^2 + height^2 \\ run^2 &= roll^2 + height^2 \\ roll^2 &= offset^2 - height^2 \\ roll^2 &= run^2 - height^2 \\ height^2 &= offset^2 - roll^2 \\ height^2 &= run^2 - roll^2 \end{aligned}$$

Depending on what the known values are, it may sometimes be necessary to solve two equations to find the distance wanted.

Examples

What is the *travel* for a 6" *roll* with a 7" *height*?

Using the formula:

$$offset^2 = roll^2 + height^2$$
$$offset^2 = (6 \times 6) + (7 \times 7) = 36 + 49 = 85$$
$$offset^2 = \sqrt{85} = 9.22 = 9\tfrac{7}{32} \text{ inches}$$

Followed by:

$$travel = offset \times 1.414$$
$$travel = 9\tfrac{7}{32} \times 1.414 = 13.035 = 13\tfrac{1}{32} \text{ inches}$$

What is the *roll* for a 11" *offset* with 8" *height*?
Using the formula:

$$roll^2 = offset^2 - height^2$$
$$roll^2 = (11 \times 11) - (8 \times 8) = 121 - 64 = 57$$
$$roll^2 = \sqrt{57} = 7.55 = 7\tfrac{9}{16} \text{ inches}$$

What is the *height* for a 16" *offset* with 12" *roll*?
Using the formula:

$$height^2 = offset^2 - roll^2$$
$$height^2 = (16 \times 16) - (12 \times 12) = 256 - 144 = 112$$
$$height^2 = \sqrt{112} = 10.583 = 10\tfrac{9}{32} \text{ inches}$$

Flanged Pipe Connections

Flanged pipe connections are widely used, particularly on larger size pipe, as they provide a practical and economical piping connection system. Flanges are commonly connected to the pipe by screw threads or by welding. Several types of flange facings are in use, the simplest of which are the plain *flat face*, and the *raised face*.

The plain flat-faced flange is usually used for cast iron flanges where pressures are under 125 pounds. Higher pressure cast iron flanges and steel flanges are made with a raised face. Generally, full-face gaskets are used with flat-face flanges and ring gaskets with raised-face flanges. The function of the gasket is to provide a loose, compressible substance between the faces with sufficient body resiliency and strength to make the flange connection leak-proof.

The assembly and tightening of a pipe-flange connection is a relatively simple operation; however, certain practices must be followed to obtain a leakproof connection. The gasket must line up evenly with the inside bore of the flange face, with no portion of it extending into the bore. When tightening the bolts, the flange faces must be kept parallel and the bolts tightened uniformly.

The tightening sequence for round flanges is shown in Fig. 216A. The sequence is to lightly tighten the first bolt, then move directly across the circle for the second bolt, then move a quarter way around the circle for the third, and directly across for the fourth, continuing the sequence until all are tightened.

When tightening an oval flange, the bolts are tightened across the short center line as shown in Fig. 216B.

A four-bolt flange, either round or square, is tightened with a simple criss-cross sequence as shown in Fig. 216C.

Do not snug up the bolts on the first go-around. This can tilt the flanges out of parallel. If using an impact wrench, set the wrench at about one-half final torque for the first go-around. Pay particular attention to the hard-to-reach bolts.

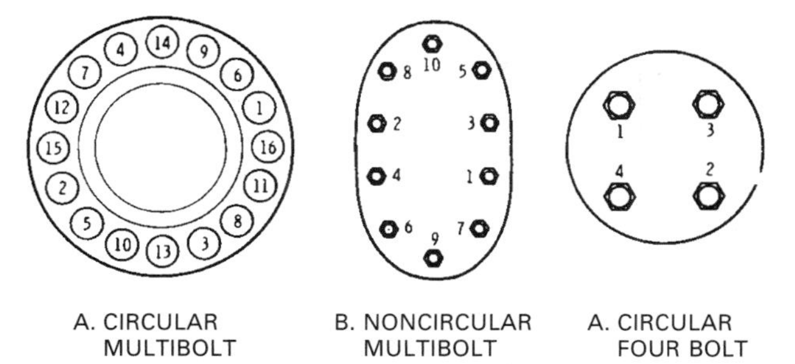

Fig. 216.

Pipe Flange Bolt-Hole Layout

The mating of pipe flanges to other flanges or circular parts requires correct layout of bolt holes. In addition, the holes must be located around the circle to line up when the flanges are mated. The usual practice is to specify the location of the holes as either "On" or "Off" the vertical center line. The shop term commonly applied to "On" the center line layout is "One Hole Up" and to "Off" the center line is "Two Holes Up." An "Off" the center line or "Two Holes Up" layout is illustrated inn Fig. 217.

While bolt holes may be laid out with a protractor using angular measurements to obtain uniform spacing, this method is most satisfactory when there are six (6) or fewer holes. Also, layout by stepping off spacing around the circle with dividers, by trial and error, is a time-consuming operation. To eliminate the trials and errors, a system of multipliers or constants may be used to calculate the chordal distance between bolt hole centers. Simply multiply the constant for the appropriate number of bolt holes by the bolt circle diameter to determine the chordal distance between holes.

Layout Procedure

1. Lay out horizontal and vertical center lines.
2. Lay out bolt circle.
3. Find value of "B" (multiply bolt circle dia. by constant).
4. For two-holes-up layout, divide "B" by 2 for value of "C."
5. Measure distance "C" off the center line and locate the center of the first bolt hole.
6. Set dividers to dimension "B" and layout center points by swinging arcs, starting from first center point.

Table 35. Flange Hole Constants

Number Of Bolt Holes	Constant
4	.707
6	.500
8	.383
10	.309
12	.259
16	.195
20	.156
24	.131
28	.112
32	.098
36	.087
40	.079

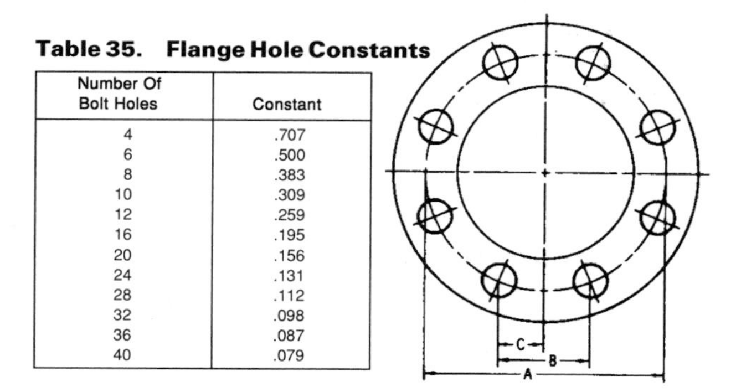

Fig. 217.

Valves

The principal function of two-way pipe valves is to open and close a line to flow. Commonly used for this purpose are *gate, globe, needle,* and *plug* valves.

Gate—Fluid flows through a gate valve in a straight line. Its construction offers little resistance to flow and causes a minimum of pressure drop. A gate-like disc—actuated by a stem screw and handwheel—moves up and down at right angles to the path of flow, and seats against two seat faces to shut off flow.

Gate valves are best for services that require infrequent valve operation and where disc is kept either fully opened or closed.

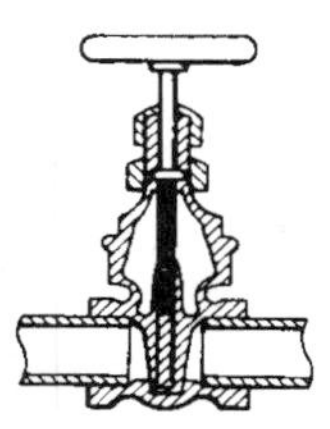

Fig. 218.

Globe—Fluid changes direction when flowing through a globe valve. The construction increases resistance to—and permits close regulation of—fluid flow. Disc and seat can be quickly and conveniently replaced or reseated. Angle valves are similar in design to globe valves. They are used when making a 90-degree turn in a line, as they reduce the number of joints and give less resistance to flow than the elbow and globe valve.

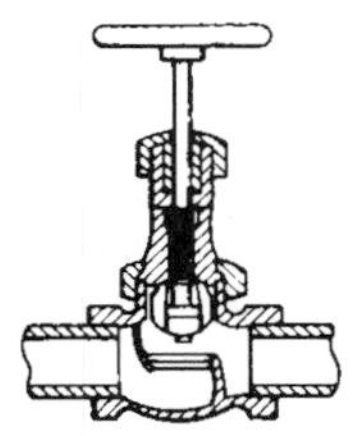

Fig. 219.

Needle—The needle valve can be used to shut off flow; however, it is designed primarily as a throttling valve. The pointed disc can be adjusted in the mating seat to give small increments of flow change. As the port diameter is smaller than the connection size, resistance to flow is high, making the needle valve unsuitable for high-volume flow.

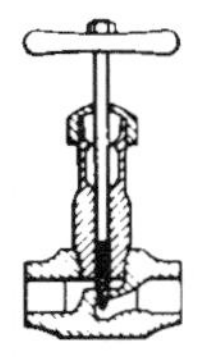

Fig. 220.

Plug—Fluid flows through a plug valve in a straight line. Its advantages are low cost, small pressure loss because of its straight-through construction, and fast operation. Only a quarter turn is needed to fully open or close it.

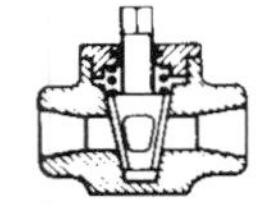

Fig. 221.

STEAM TRAPS

To efficiently drain condensate from steam lines, steam traps must be correctly located and the piping properly arranged. The following basic rules will go a long way toward providing satisfactory operation of steam traps (Fig. 222).

1. Provide a separate trap for each piece of equipment or apparatus. Short circuiting (steam follows path of least resistance to trap) may occur if more than one piece of apparatus, coil, etc. is connected to a single trap.
2. Tap steam supply off the top of the steam main to obtain dry steam and avoid steam line condensate.
3. Install a supply valve close to the steam main to allow maintenance and/or revisions without steam main shutdown.
4. Install a steam supply valve close to the equipment entrance to allow equipment maintenance work without supply line shutdown.
5. Connect condensate discharge line to lowest point in equipment to avoid water pockets and water hammer.
6. Install shutoff valve upstream of condensate-removal piping to cut off discharge of condensate from equipment and allow service work to be performed.
7. Install strainer and strainer flush valve ahead of trap to keep rust, dirt, and scale out of working parts and to allow blowdown removal of foreign material from strainer basket.
8. Provide unions on both sides of trap for its removal and/or replacement.
9. Install test valve downstream of trap to allow observance of discharge when testing.
10. Install check valve downstream of trap to prevent condensate flow-back during shutdown or in the event of unusual conditions.
11. Install downstream shutoff valve to cut off equipment condensate piping from main condensate system for maintenance or service work.
12. Do not install a bypass unless there is some urgent need for it. Bypasses are an additional expense to install and are frequently left open, resulting in loss of steam and inefficient operation of equipment.

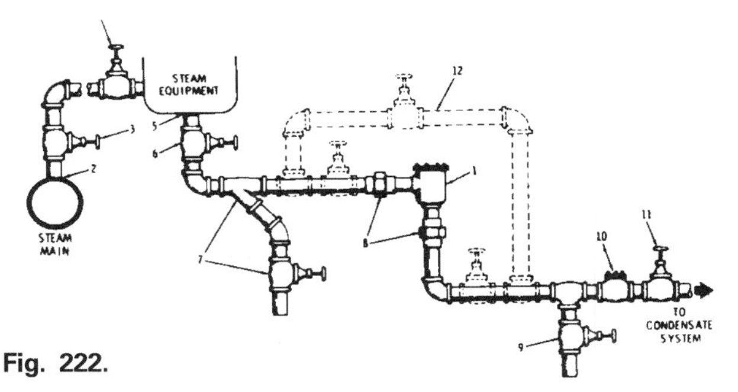

Fig. 222.

Steam Trap Troubleshooting

In cases of improper functioning of steam equipment, a few simple checks of the steam system should be made before looking for trap malfunction. The following preliminary checks should precede checking the operation of a stream trap.

1. Check the steam supply pressure—it should be at or above the minimum required.
2. Check to be sure all valves required to be open are in the full-open position (supply, upstream shutoff, downstream shut-off).
3. Check to be sure all valves required to be closed are in the tight closed position (bypass, strainer, test).

The initial step in checking the operation of a steam trap is to check its temperature. Since a properly functioning steam trap is an automatic valve which allows condensate to be discharged but closes to prevent the escape of steam, it should operate very close to the steam temperature. For exact checks, a surface pyrometer should be used. A convenient and dependable operating test is simply to sprinkle water on the trap. If the water spatters, rapidly boils, and vaporizes, the trap is hot and probably is very close to steam temperature.

If it is found that the trap is very close to steam temperature, the next step is to determine if condensate or steam is being discharged.

When a test valve is provided for this purpose, the check is made by closing the downstream shutoff valve and opening the test valve. The discharge from the test valve should be carefully observed to determine if condensate or live steam is escaping. If the trap being tested is the type that has an opening and closing cycle, condensate should flow from the test valve and then it should stop as the trap shuts off. The flow should resume when the trap opens again, etc. Steam should not discharge from the test valve if the trap is operating properly. If the trap is a continuous discharge type, there should be a continuous discharge of condensate but no steam.

In the event the installation does not have a test valve, the trap may be checked by listening to its operation. The ideal instrument to do this is an industrial stethoscope. If not available, a suitable device for this purpose is a screw driver or metal rod. By holding one end against the trap and the other end against the ear, the sound of the trap's operation may be heard. If the trap is operating properly, the flow of condensate should be heard for a few seconds, a click as the valve closes, and then silence, indicating the valve has closed tight. This cycle of sounds should repeat in a regular pattern. The listening procedure, however, is not suitable for checking continuous discharge traps, as these automatically regulate to an open position in balance with the condensate flow.

Another check on trap operation is to open the strainer valve and observe the discharge at this point. There should be an initial gush of condensate and steam as the valve is opened, followed by a continuous flow of live steam. If condensate flows for a prolonged period before steam is observed, the condensate is not being properly discharged from the system.

Unsatisfactory performance of a stream unit may not be due to improper steam trap operation. When testing steam traps, conditions other than trap malfunction must also be considered. Some of the common faults that cause troubles are the following:

1. Inadequate steam supply
2. Incorrectly sized trap
3. Improperly connected piping
4. Improper pitch of condensate lines
5. Inadequate condensate lines

STEAM LINES

Opening Steam Supply Valve

The opening of valves controlling steam flow in steam supply lines, called steam *mains*, requires care and correct procedure. The expansion or growth of the piping system as the temperature increases when steam is introduced, must be carefully controlled. Also the air in the lines, and the large volume of condensate formed as the line heats up, must be removed. To facilitate removal of condensate during normal operation, as well as at start-up, steam lines incorporate *drip pockets, drip legs, drip valves* as shown in Fig. 223.

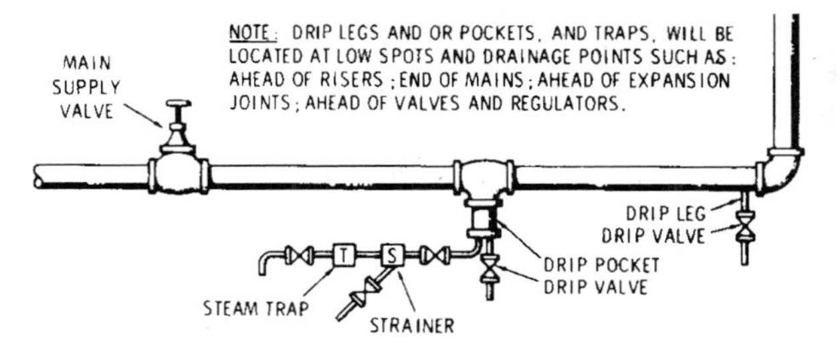

Fig. 223.

The following procedure should be followed when opening a steam main supply valve:

1. Open all drip valves full open to act as air vents and condensate discharge openings. Check setting of distribution valves to be sure steam goes only to those branch lines ready to receive it.
2. Open main supply valve slowly and in stages to control steam flow volume and provide gradual heatup of the line.
3. Watch discharge at drip valves. Do not close drip valves until warmup condensate has been discharged (except for next item).

Table 36. Steam Systems Troubleshooting Guide

CONDITION	REASON	CORRECTIVE ACTION
TRAP BLOWS LIVE STEAM	1. NO PRIME (bucket traps) a. Trap not primed when originally installed b. Trap not primed after cleanout c. Open or leaking by-pass valve d. Sudden pressure drops	1. a. Prime the trap b. Prime the trap c. Remove or repair by-pass valve d. Install check valve ahead of trap
	2. VALVE MECHANISM DOES NOT CLOSE a. Scale or dirt lodged in orifice b. Worn or defective valve or disc mechanism	2. a. Clean out the trap b. Repair or replace defective parts
	3. RUPTURED BELLOWS (thermostatic traps)	3. Replace bellows
	4. BACK PRESSURE TOO HIGH (thermodynamic trap) a. Worn or defective parts b. Trap stuck open c. Condensate return line or pig tank undersized	4. a. Repair or replace defective parts b. Clean out the trap c. Increase line or pig tank size
	5. BLOWING FLASH STEAM Forms when condensate released to lower or atmospheric pressure	5. Normal condition No corrective action necessary

Table 36. Steam Systems Troubleshooting Guide (Cont.)

CONDITION	REASON	CORRECTIVE ACTION
TRAP DOES NOT DISCHARGE	1. PRESSURE TOO HIGH a. Trap pressure rating too low b. Orifice enlarged by normal wear c. Pressure reducing valve set too high or broken d. System pressure raised	1. a. Install correct trap b. Replace worn orifice c. Readjust or replace pressure reducing valve d. Install correct pressure change assembly
	2. CONDENSATE NOT REACHING TRAP a. Strainer clogged b. Obstruction in line to trap inlet c. By-pass opening or leaking d. Steam supply shut off	2. a. Blow out screen or replace b. Remove obstruction c. Remove or repair by-pass valve d. Open steam supply valve
	3. TRAP CLOGGED WITH FOREIGN MATTER	3. Clean out and install strainer
	4. TRAP HELD CLOSED BY DEFECTIVE MECHANISM	4. Repair or replace mechanism
	5. HIGH VACUUM IN CONDENSATE RETURN LINE	5. Install correct pressure change assembly
	6. NO PRESSURE DIFFERENTIAL ACROSS TRAP a. Blocked or restricted condensate return line b. Incorrect pressure change assembly	6. a. Remove restriction b. Install correct pressure change assembly

Table 36. Steam Systems Troubleshooting Guide (Cont.)

CONDITION	REASON	CORRECTIVE ACTION
CONTINUOUS DISCHARGE FROM TRAP	1. TRAP TOO SMALL a. Capacity undersized b. Pressure rating of trap too high	1. a. Install properly sized larger trap b. Install correct pressure change assembly
	2. TRAP CLOGGED WITH FOREIGN MATTER a. Dirt or foreign matter in trap internals b. Strainer plugged	2. a. Clean out and install strainer b. Clean out strainer
	3. BELLOWS OVERSTRESSED (Thermostatic traps)	3. Replace bellows
	4. LOSS OF PRIME	4. Install check valve on inlet side
	5. FAILURE OF VALVE TO SEAT a. Worn valve and seat b. Scale or dirt under valve and in orifice c. Worn guide pins and lever	5. a. Replace worn parts b. Clean out the trap c. Replace worn parts

Table 36. Steam Systems Troubleshooting Guide (Cont.)

CONDITION	REASON	CORRECTIVE ACTION
SLUGGISH OR UNEVEN HEATING	1. TRAP HAS NO CAPACITY MARGIN FOR HEAVY STARTING LOADS	1. Install properly sized larger trap
	2. INSUFFICIENT AIR HANDLING CAPACITY (bucket traps)	2. Use thermic buckets or increase vent size
	3. SHORT CIRCUITING (group traps)	3. Trap each unit individually
	4. INADEQUATE STEAM SUPPLY a. Steam supply pressure valve has changed b. Pressure reducing valve setting off	4. a. Restore normal steam pressure b. Readjust or replace reducing valve
BACK PRESSURE TROUBLES	1. CONDENSATE RETURN LINE TOO SMALL	1. Install larger condensate return line
	2. OTHER TRAPS BLOWING STEAM INTO HEADER	2. Locate and repair other faulty traps
	3. PIG TANK VENT LINE PLUGGED	3. Clean out pig tank vent line
	4. OBSTRUCTION IN CONDENSATE RETURN LINE	4. Remove obstruction
	5. EXCESS VACUUM IN CONDENSATE RETURN LINE	5. Install correct pressure change assembly

4. Condensate should not be drained from drip pockets. An accumulation of condensate is necessary in drip pockets as they are in the line to do the following:
   A. To let condensate escape by gravity from the fast moving steam,
   B. To store condensate until the pressure differential is great enough for the steam trap to discharge,
   C. Provide storage of condensate until there is positive pressure in the line,
   D. Provide static head, enabling trap to discharge before a positive pressure exists in the line.
5. Check to see that line pressure comes up to the required operating pressure.
6. Check operation of all steam traps draining condensate from line to be sure they are operating properly (check temperature, discharge, etc.).

AUTOMATIC SPRINKLER SYSTEMS

Wet Sprinkler Systems

A *wet sprinkler system* (Fig. 224) is described by the National Board of Fire Underwriters as "a system employing automatic sprinklers attached to a piping system containing water and connected to a water supply so that water discharges immediately from sprinklers opened by a fire." A vital component of a wet sprinkler system is the *alarm check valve,* also called *wet sprinkler valve.* Its function is to direct water to alarm devices and sound the sprinkler alarm. *It does NOT control the flow of water into the system.*

The alarm check valve is located in the pipe riser at the point the water line enters the building. An underground valve with an indicator post is usually located a safe distance outside the building. In design and operation the alarm check valve is a globe type check valve. A groove cut in the seat is connected by passage to a threaded outlet on the side of the valve body. When the valve lifts, the water can flow to the outlet. There is also a large drain port above the seat which is connected to the *drain valve.* Two additional ports, one above and one below the seat, allow the attachment of pressure gauges.

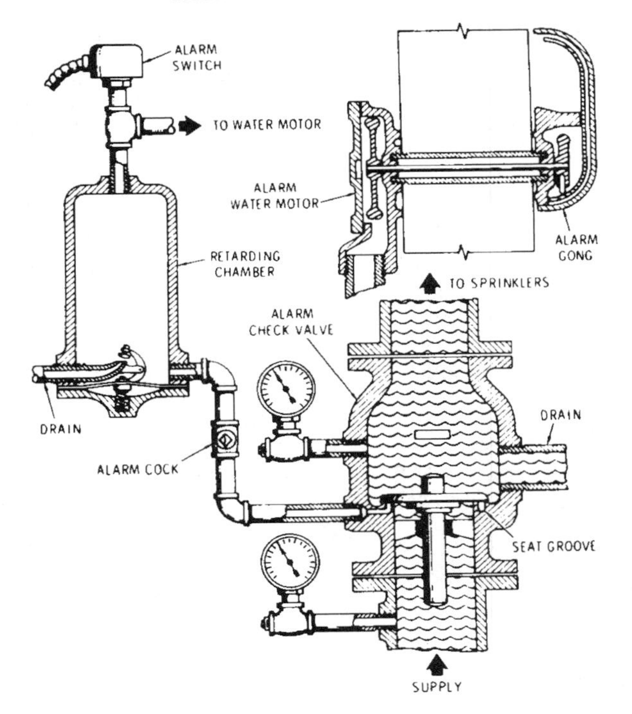

Fig. 224.

A plug or stop-cock type valve called the *alarm control cock* is connected to the outlet from the seat groove. This cock controls the water flow to the alarm devices, allowing their silencing. This cock *MUST* be in the alarm position when the system is "set," or the alarm check valve will be unable to perform its function of sounding an alarm.

Operation of Wet Sprinkler Alarm System

When a sprinkler head opens, or for any reason water escapes from a wet sprinkler system, the flow through the alarm check valve causes the check valve disc to lift. Water entering the seat groove flows through the alarm cock to the retarding chamber. The function of the retarding chamber is to avoid unnecessary alarms which might be caused by slight leakage. It will allow a small volume water flow without actuating the alarm. When there is a large flow, as occurs when a sprinkler opens, the chamber is quickly filled and pressure closes the diaphragm-actuated drain valve in the bottom of the retarding chamber. The electrical alarm is then actuated by the pressure of the water. The water also flows to the water motor, causing it to be rotated and whirl hammers inside the alarm gong, thus mechanically sounding the bell alarm.

Placing a Wet Sprinkler System in Service

1. Check system to be sure it is ready to be filled with water. If the system has been shut down because a head has opened, be sure the head has been replaced with one of proper rating.
2. Open "vent valves"—located at high points.
3. Place "alarm cock" in CLOSED position. This will prevent sounding of alarm while flowing water fills systems.
4. Place "drain valve" in nearly closed position. A trickle of water should flow from drain valve during filling.
5. Open "indicator post valve" slowly. When the system has filled, there will be a quieting of the sound of rushing water. Open valve full open, then back off a quarter turn.
6. Observe the water flow at the vent valves. When a steady flow of water occurs (no air), close vent valves.
7. Check the water flow by quickly opening the "drain valve" and closing it. The water pressure should drop about 10 pounds when the valve is opened and immediately return to full pressure when valve is closed. Excessive pressure drop indicates insufficient water flow.
8. Open "alarm cock"—system is now in SET condition.
9. Test—Open "drain valve" several turns. Water should flow, sounding the mechanical gong alarm and the electrical alarm.

10. Close "drain valve"—If the alarms have functioned properly, the system is operational.

Dry-Pipe Sprinkler Systems

A *dry-pipe sprinkler system* is described by the National Board of Fire Underwriters as: "a system employing automatic sprinklers attached to a piping system containing air under pressure, the release of which as from the opening of a sprinkler permits the water pressure to open a valve known as a *dry-pipe valve*. The water then flows into the piping system and out the open sprinkler."

The dry-pipe valve is located in the pipe riser at the point the water line enters the building. An underground valve with indicator post is usually located a safe distance outside the building.

The dry-pipe valve is a dual style valve, as both an air valve and a water valve are contained inside its body. These two internal valves may be separate units, one positioned above the other, or they may be combined in a single unit, one within the other. The function of the air valve is to retain the air in the piping system and to hold the water valve closed, thus restraining the flow of water. As long as sufficient air pressure is maintained in the system the air valve can do this, since it has a larger surface than the water valve. The air in the system acting on this large surface provides enough force to hold the water valve closed against the pressure of the water (Fig. 225).

Operation of Dry-Pipe Sprinkler Systems

When a sprinkler head opens, or for any reason air escapes from a dry pipe system, the air pressure above the internal air valve is reduced. When the air pressure falls to the point where its force is exceeded by the force of the water below the internal valve, both valves are thrown open. This allows an unobstructed flow of water through the dry pipe valve into the system piping and to the open sprinklers. As the valves are thrown open, water fills the intermediate chamber. To avoid unnecessary alarms the chamber is equipped with a "velocity drip valve." This drip valve is normally open to the atmosphere and allows drainage of any slight water leakage past the internal water valve seat.

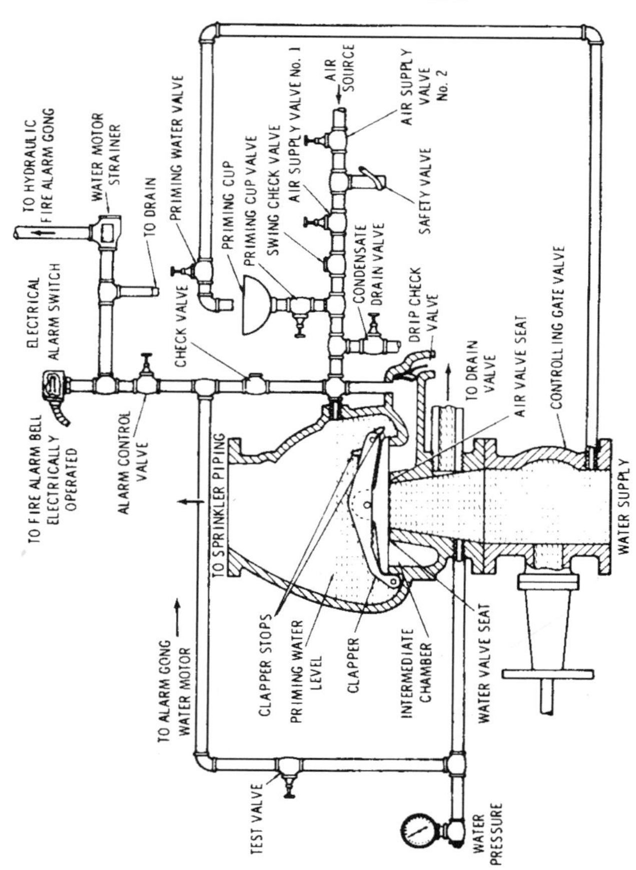

TO HYDRAULIC FIRE ALARM GONG
WATER MOTOR STRAINER
TO DRAIN
PRIMING WATER VALVE
PRIMING CUP
PRIMING CUP VALVE
SWING CHECK VALVE
AIR SUPPLY VALVE No. 1
AIR SOURCE
AIR SUPPLY VALVE No. 2
SAFETY VALVE
ELECTRICAL ALARM SWITCH
CHECK VALVE
CONDENSATE DRAIN VALVE
DRIP CHECK VALVE
TO DRAIN VALVE
AIR VALVE SEAT
CONTROLLING GATE VALVE
TO FIRE ALARM BELL ELECTRICALLY OPERATED
ALARM CONTROL VALVE
TO SPRINKLER PIPING
WATER SUPPLY
TO ALARM GONG WATER MOTOR
CLAPPER STOPS
PRIMING WATER LEVEL
CLAPPER
INTERMEDIATE CHAMBER
WATER VALVE SEAT
TEST VALVE
WATER PRESSURE

Fig. 225.

When a sprinkler head opens and falling air pressure allows the internal valves to be thrown open, the intermediate chamber instantly fills with water and the velocity drip valve is forced closed. The water now under pressure in the chamber flows to the electrical alarm switch and to the alarm gong.

Placing a Dry-Pipe Sprinkler System in Service

1. Close the valve controlling water flow to the system. This may be located in the riser or it may be an underground valve with an indicator post. If a fire has occurred and water is flowing from opened sprinklers, approval of the person in authority must be obtained before closing the valve.
2. Open the "drain valve" and allow the water to drain from the sprinkler system piping.
3. Open all "vent" and "drain" valves throughout the system. Vents will be located at the high points and drains at all trapped and low portions of the piping system.
4. Manually push open the "velocity drip valve." Also open the drain valve for the "dry pipe valve body," if one is provided.
5. Remove the cover plate from the dry pipe valve and carefully clean the rubber facings and seat surfaces of the internal air and water valves. DO NOT use rags or abrasive wiping materials. Wipe the seats clean with the bare fingers.
6. Unlatch the "clapper" and carefully close the internal air and water valves.
7. Replace the dry pipe valve cover and close the drain valve, if one is provided.
8. Open the "priming cup valve" and "priming water valve" to admit priming seal water into the dry pipe valve to the level of the pipe connection. The priming water provides a more positive seal to prevent air from escaping past the air valve seat into the intermediate chamber.
9. Drain excess water by opening "condensate drain valve." Close tightly when water no longer drains from valve.
10. Open "air supply valve" and admit air to build up a few pounds of pressure in the system.
11. Check all open vents and drains throughout the system to be sure all water has been forced from the low points. As soon

as dry air exhausts at the various open points, the openings should be closed. Close air supply valve.

12. Replace any open sprinklers with new sprinkler heads of the proper rating.
13. Open air supply valve and allow system air pressure to build up to the required pressure. The air pressure required to keep the internal valves closed varies directly with water supply pressure. Consult pressure setting tables.
14. Open the system water supply valve slightly to obtain a small flow of water to the dry pipe valve.
15. When water is flowing clear at the drain valve, slowly close it, allowing water pressure to gradually build up below the internal water valve as observed on the water pressure gauge.
16. When water pressure has reached the maximum below the internal water valve, open the supply valve to full open position. Back off valve about a quarter turn from full open.
17. Test the alarms—open the "test valve," or if system has a three-position test cock, place cock in TEST position. Water should flow to the electrical alarm switch and also to the alarm gong water motor.
18. If alarms have functioned properly, close the test valve or place the three-position test cock in ALARM position.

The alarm test is a test of the functioning of the alarm system only and does not indicate the condition of the dry pipe valve. The dry pipe valve operation is tested by opening a vent valve to allow air in the piping system to escape, causing the dry pipe valve to trip. It must then be reset, going through the procedure listed above.

CARPENTRY

Commercial Lumber Sizes

Two words, *timber* and *lumber*, are commonly used to describe the principal material used by carpenters. In the early stages of lumber production the word timber is usually applied to wood while in its natural state. Wood after cutting and sawing into standard commercial pieces is called *lumber*.

Stock lumber may be *green*, meaning the wood contains a large percentage of its natural moisture, or it may be *seasoned*. Seasoning is the process, either naturally or by exposure to heat, of removing about 85% of the moisture contained in freshly cut timber. Lumber is usually classified according to the three types into which it is rough sawed. These are: *dimension stock*, which is 2 inches thick and from 4 to 12 inches wide; *timbers*, which are 4 to 8 inches thick and 6 to 10 inches wide; and *common boards*, which are 1 inch thick and 4 to 12 inches wide.

Rough lumber is *dressed* or *surfaced* by removing about ⅛ inch from each side and about ⅜ inch from the edges. This planing operation is called *dressing* or *surfacing*, and the letter "D" for dressed or "S" for surfaced is used to indicate how many sides or edges are planed. For example, D1S or S1S indicates one side has been planed. Lumber planed on both sides and edges is designated S4S.

Traditional practice in sawmill operation was to saw logs to nominal dimensions, i.e., dimension stock was sawed to 2-inch thickness, boards to 1 inch, etc. Seasoning or drying further reduced the thickness roughly ⅛ inch, and the width a lesser amount. Table

Table 37. New and Old Standard Lumber Sizes

Lumber Classification	Nominal Size		Actual S4S Size		Old S4S Size	
	Thickness	Width	Thickness	Width	Thickness	Width
Dimension	2"	4"	1½"	3½"	1⅝"	3⅝"
	2"	6"	1½"	5½"	1⅝"	5⅝"
	2"	8"	1½"	7¼"	1⅝"	7½"
	2"	10"	1½"	9¼"	1⅝"	9½"
	2"	12"	1½"	11¼"	1⅝"	11½"
Timbers	4"	6"	3½"	5½"	3⅝"	5⅝"
	4"	8"	3½"	7¼"	3⅝"	7½"
	4"	10"	3½"	9¼"	3⅝"	9½"
	6"	6"	5½"	5½"	5⅝"	5⅝"
	6"	8"	5½"	7¼"	5⅝"	7½"
	6"	10"	5½"	9¼"	5⅝"	9½"
	8"	8"	7¼"	7¼"	7½"	7½"
	10"	10"	9¼"	9¼"	9½"	9½"
Common Boards	1"	4"	¾"	3½"	25/32"	3⅝"
	1"	6"	¾"	5½"	25/32"	5⅝"
	1"	8"	¾"	7½"	25/32"	7⅝"
	1"	10"	¾"	9¼"	25/32"	9½"
	1"	12"	¾"	11¼"	25/32"	11½"

37 lists the dimensions of lumber produced as standard during the many years this practice was followed. The old standard sizes are shown in the Old S4S column.

The practice of sawing to nominal dimensions is no longer followed. Instead, finished standard dimensions have been established and are used to determine sawmill dimensions. This allows more efficient use of logs, as lumber can be rough sawed to lesser dimensions, providing only enough extra material to surface plane to the new, reduced standard finished dimensions. The present standard lumber sizes are listed in the Actual S4S column in Table 37.

Building Layout

The first step in building layout is to establish by measurement from boundary markers or other reliable positions the wall line for one side of the building, and one building corner location point on the line. A stake is driven into the ground at this point and a nail driven into the top of the stake to accurately mark the corner point. The other corner point stakes are then located by measurement and by squaring the corners.

The corners are squared by use of the 3-4-5 triangle measurement system. Measurements are made along the sides in multiples of 3 and 4, and along the diagonal in multiples of 5. The 6-8-10 and 9-12-15 combination measurements are often used. To further assure square corners, the diagonals are checked by measurement to see if they are the same length.

After the corner stakes are accurately located, *batter board* stakes are driven in at each corner about 4 feet beyond the building lines, and the batter boards are attached. Batter boards are the usual method used to retain the outline of a building layout. The height of the boards may also be positioned to conform to the height of above-grade foundation walls. A line is held across the top of opposite boards at the corners and adjusted using a plumb bob so that it is exactly over the nails in the stakes. Saw kerfs are cut where the lines touch the boards so that accurate line locations are assured after the corner stakes are removed.

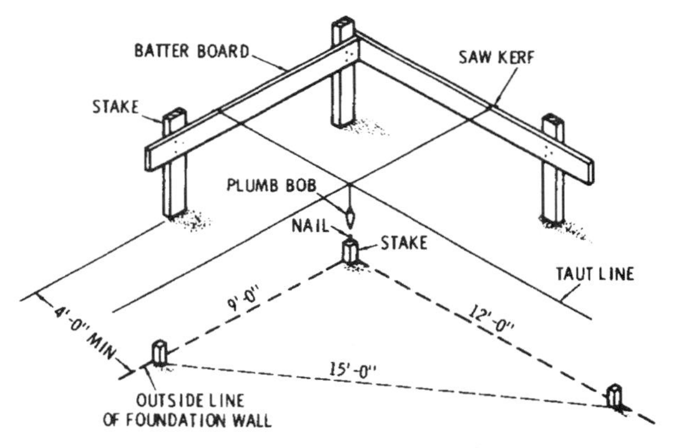

Fig. 226. Layout stakes and batter boards.

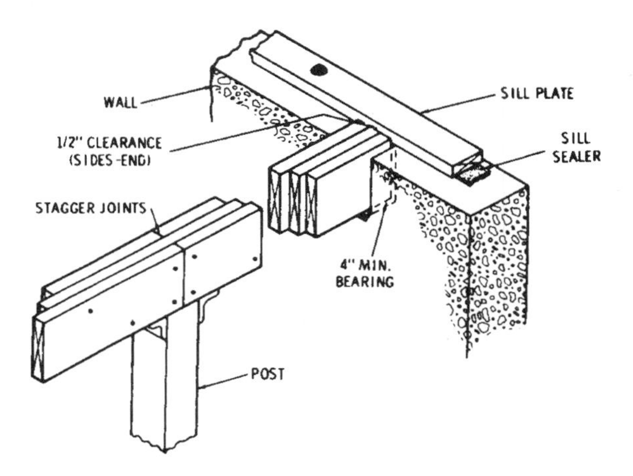

Fig. 227. Built-up wooden girders.

Wood-Frame Building Construction

While the details of wood-frame building construction may vary in different localities, the fundamental principles are the same. The following figures, charts, and tables show established methods of construction and accepted practices used in wood-frame building construction.

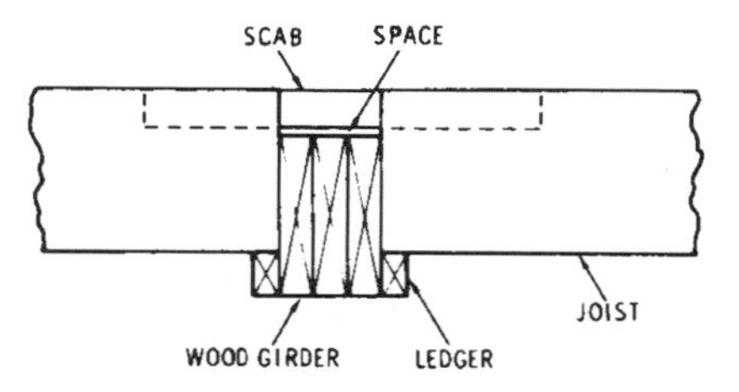

Fig. 228. Connecting scab used to tie joists together.

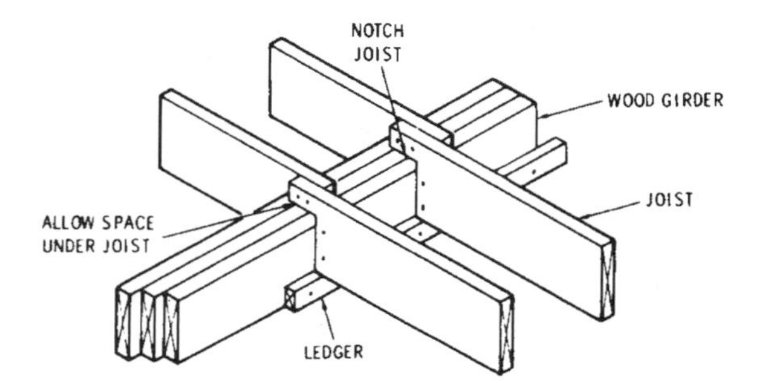

Fig. 229. Floor joists notched to fit over girder.

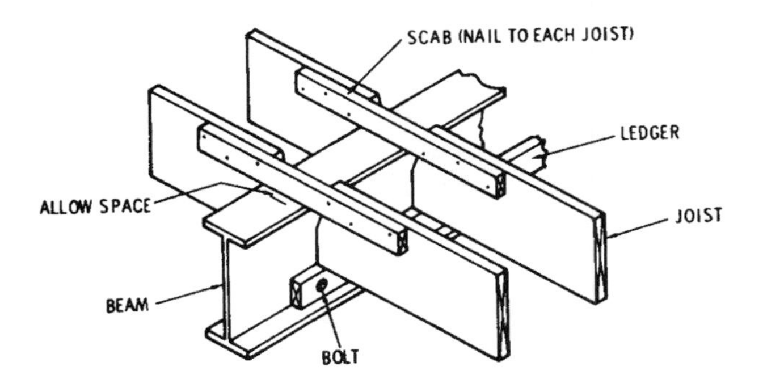

Fig. 230. Floor joists resting on wooden ledger fastened to "I" beam girder.

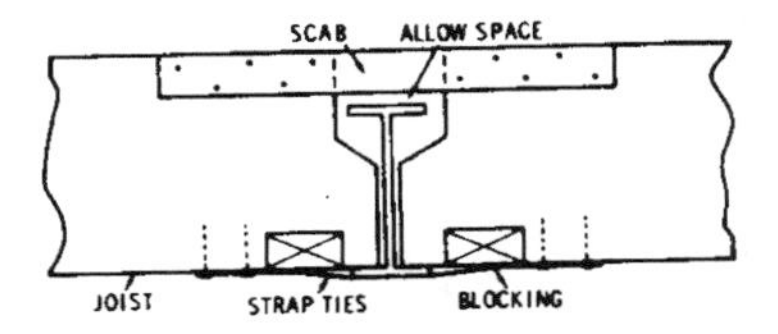

Fig. 231. Floor joists resting directly on "I" beam girder and connected at top with scab board.

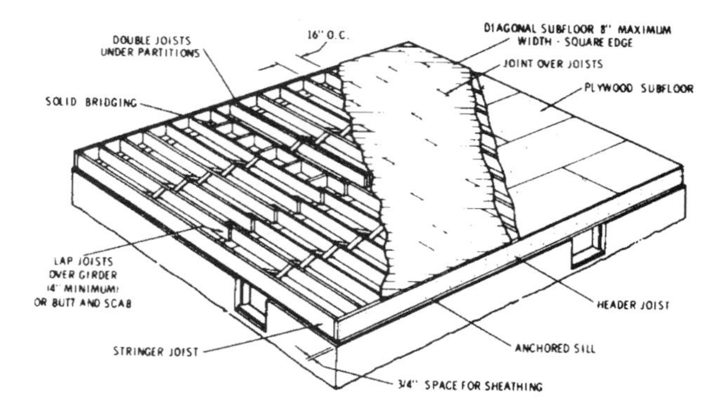

Fig. 232. Platform construction—details of floor joists and sub-flooring

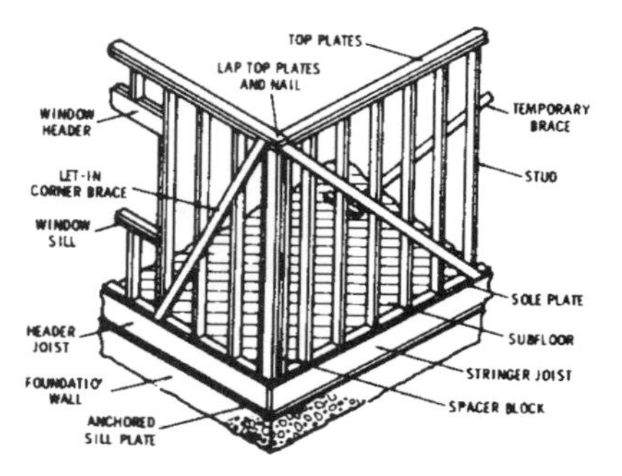

Fig. 233. Platform construction—wall studs with let-in bracing and double top plates.

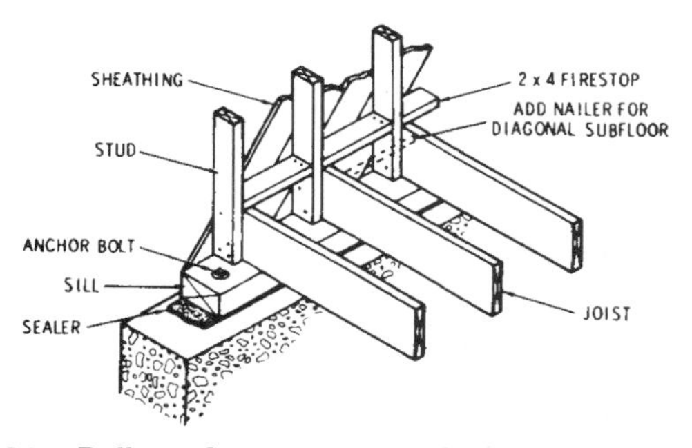

Fig. 234. Balloon-frame construction—wall studs and floor joists rest on anchored sill.

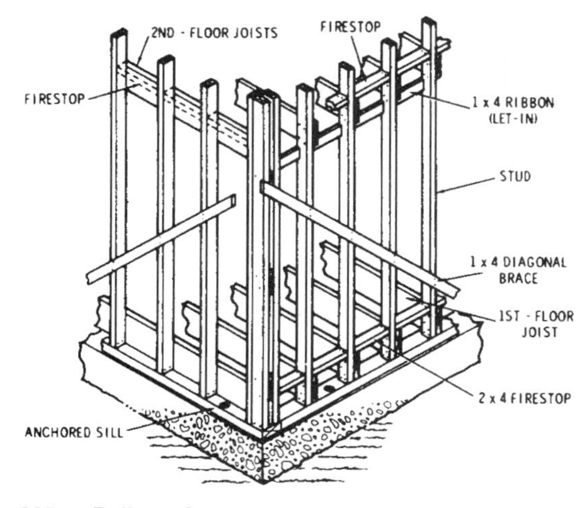

Fig. 235. Balloon-frame construction—second-floor joists rest on 1" x 4" ribbons that have been let into the wall studs. Fire stops prevent spread of fire through open wall passages.

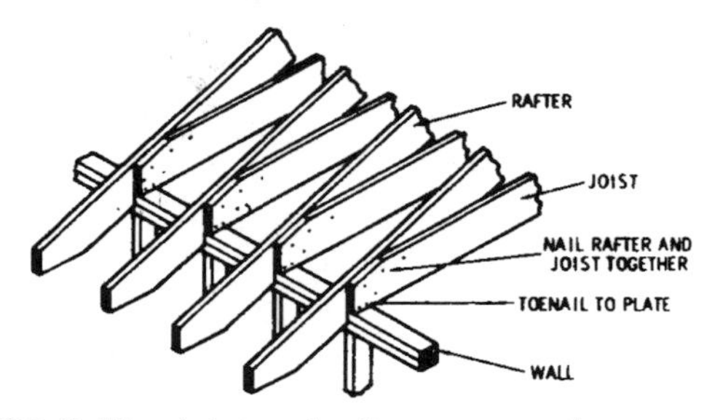

Fig. 236. Ceiling joist and rafter construction.

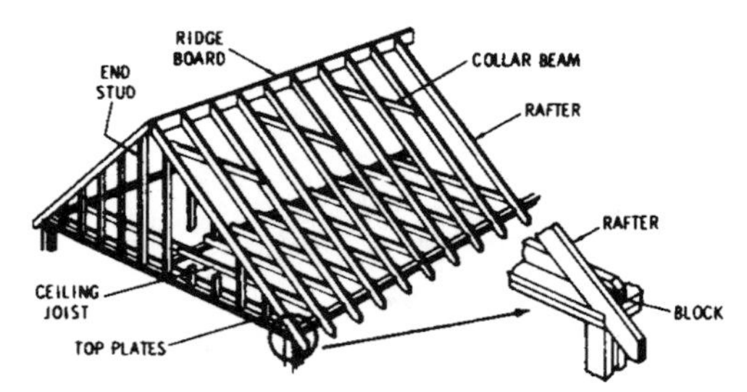

Fig. 237. Roof frame construction.

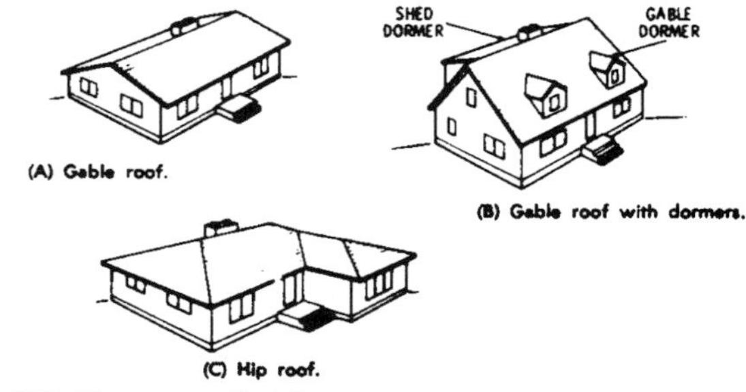

Fig. 238. Frame roof styles.

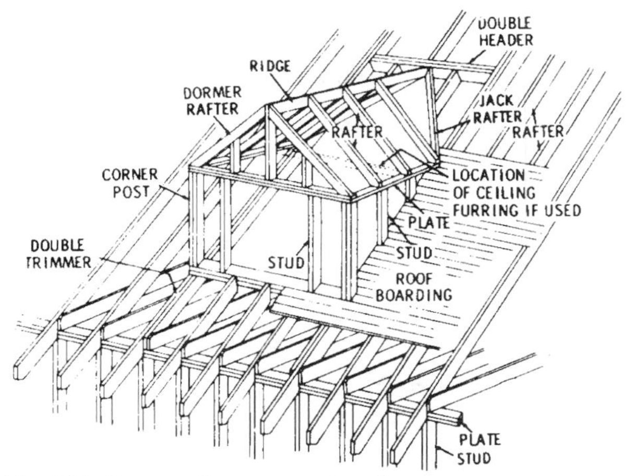

Fig. 239. Dormer-frame construction.

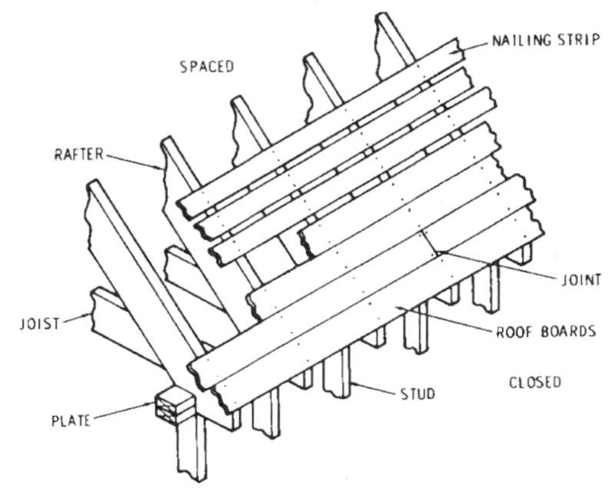

Fig. 240. Board roof sheathing—spaced for wood shingles or closed for asphalt shingles.

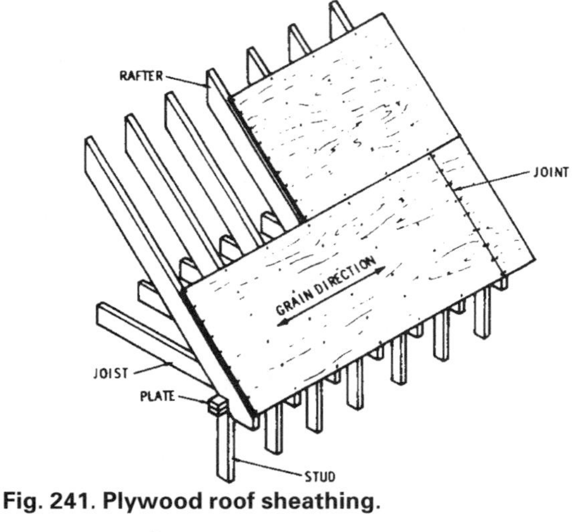

Fig. 241. Plywood roof sheathing.

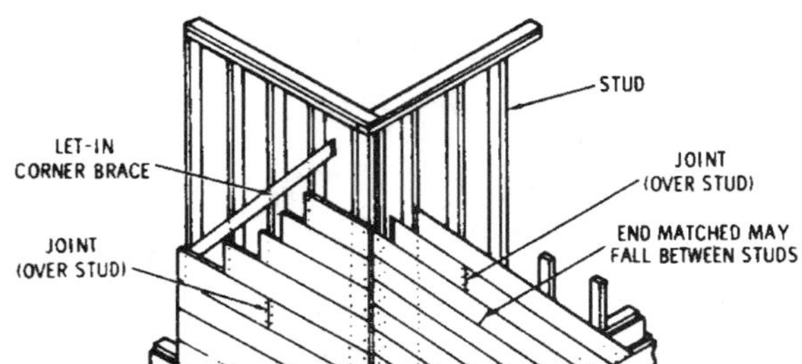

Fig. 242. Horizontal or diagonal applied board wall sheathing.

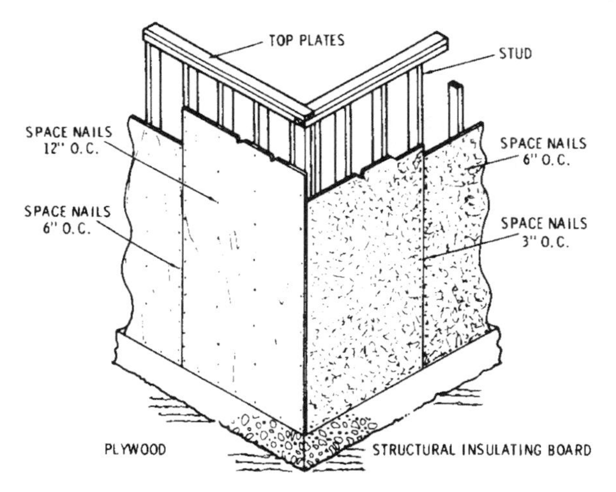

Fig. 243. Plywood or insulating board wall sheathing.

Manufactured Wood Trusses

A fundamental change is taking place in the way wooden buildings are constructed. Factory-assembled components are being used with ever-increasing frequency. *Manufactured wood trusses*, made to standard dimensions in a controlled factory environment using selected materials, are replacing conventional "stick-built" structures. Two types of wood trusses most widely used are *floor trusses* and *roof trusses*. The most common roof truss configurations are shown in Fig. 244. Each truss is an assembly of precision cut selected wood members securely fastened together with *metal connector plates*. In addition to reduced cost and improved quality, manufactured trusses make possible longer clear spans, reducing the requirement for interior bearing partitions.

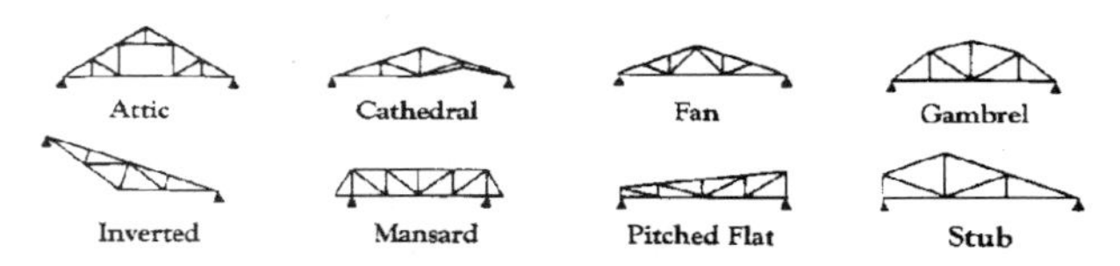

Fig. 244. Common Roof Truss Configurations.

Table 38. Standard Wood Screw Dimensions

Screw Gauge	Screw Diameter	Head Diameter		
		Flat	Round	Oval
0	0.060	0.112	0.106	0.112
1	0.073	0.138	0.130	0.138
2	0.086	0.164	0.154	0.164
3	0.099	0.190	0.178	0.190
4	0.112	0.216	0.202	0.216
5	0.125	0.242	0.228	0.242
6	0.138	0.268	0.250	0.268
7	0.151	0.294	0.274	0.294
8	0.164	0.320	0.298	0.320
9	0.177	0.346	0.322	0.346
10	0.190	0.371	0.346	0.371
11	0.203	0.398	0.370	0.398
12	0.216	0.424	0.395	0.424
13	0.229	0.450	0.414	0.450
14	0.242	0.476	0.443	0.476
15	0.255	0.502	0.467	0.502
16	0.268	0.528	0.491	0.528
17	0.282	0.554	0.515	0.554
18	0.294	0.580	0.524	0.580
20	0.321	0.636	0.569	0.636
22	0.347	0.689	0.611	0.689
24	0.374	0.742	0.652	0.742
26	0.400	0.795	0.694	0.795
28	0.426	0.847	0.735	0.847
30	0.453	0.900	0.777	0.900

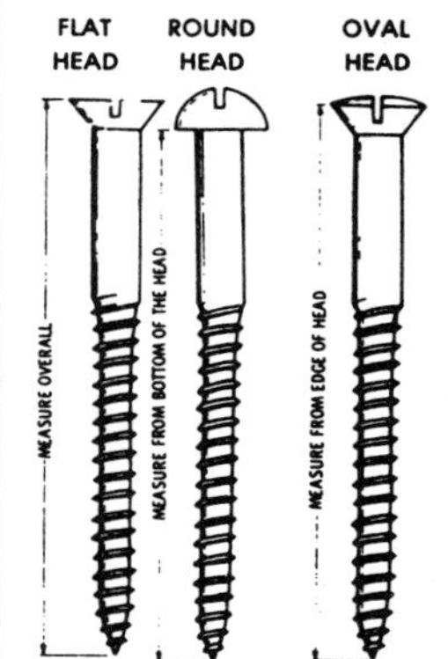

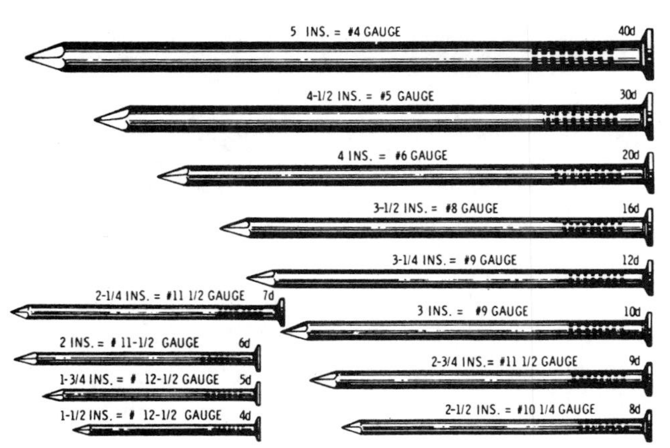

Fig. 245. Common Wire Nails.

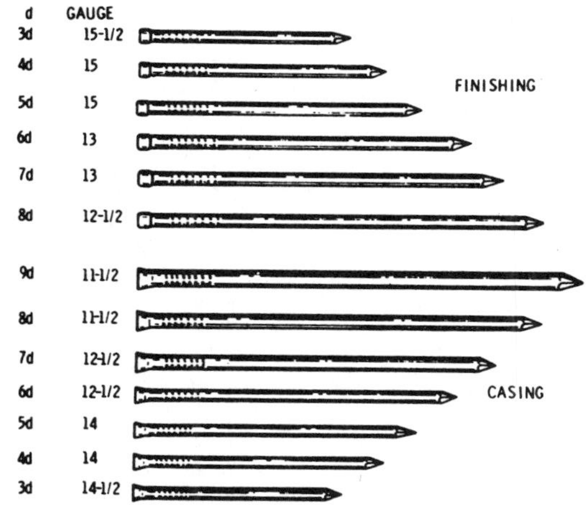

Fig. 246. Finishing and Casing Nails.

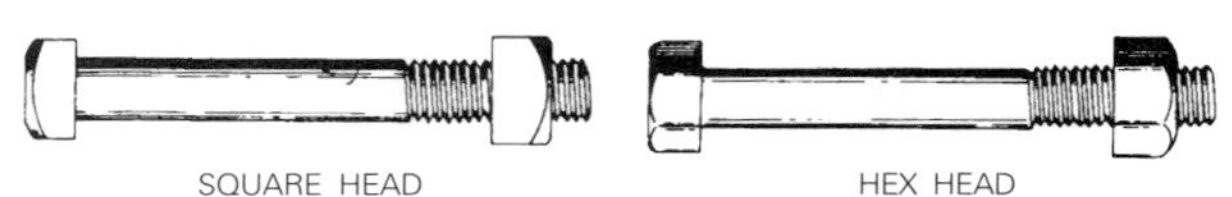

Table 39. Nominal Dimensions of Standard Machine Bolts

Diameter	UNC Threads Per Inch	Square Head			Hex Head		
		Width Across Flats	Width Across Corners	Head Height	Width Across Flats	Width Across Corners	Head Height
1/4	20	3/8	17/32	11/64	7/16	1/2	11/64
5/16	18	1/2	45/64	13/64	1/2	9/16	7/32
3/8	16	9/16	51/64	1/4	9/16	2 1/32	1/4
7/16	14	5/8	57/64	19/64	5/8	4 7/64	19/64
1/2	13	3/4	1 1/16	21/64	3/4	5 5/64	11/32
5/8	11	15/16	1 21/64	27/64	1 5/16	1 3/32	27/64
3/4	10	1 1/8	1 19/64	1/2	1 1/8	1 19/64	1/2
7/8	9	1 5/16	1 55/64	19/32	1 5/16	1 33/64	37/64
1	8	1 1/2	2 1/8	21/32	1 1/2	1 47/64	43/64
1 1/8	7	1 11/16	2 25/64	3/4	1 11/16	1 61/64	3/4
1 1/4	7	1 7/8	2 21/32	27/32	1 7/8	2 11/64	27/32
1 3/8	6	2 1/16	2 59/64	29/32	2 1/16	2 3/8	29/64
1 1/2	6	2 1/4	3 3/16	1	2 1/4	2 19/32	1

Power Nailers and Staplers

Power nailers and *staplers*, first used in production operations, are now commonly used for carpentry projects to promote efficiency and provide dependable fastenings. The nail and staple have certain features that determine which is best used for specific applications. For example, a staple has more holding power in wood than a nail because it has two legs holding it. However, when conditions of lateral stress exist, or in an application where longer fasteners are needed, nails are the correct fastener to use. The types of nails available for use in power nailers varies from the small brad nail for fine finish work to the 5" long spike for extra heavy fastening. There are three principal shank types: *smooth* for general applications; *ring* for superior holding in softwood applications; *screw* for hardwood applications. The diamond point nail is the most commonly used for general softwood applications.

Shown in Fig. 247 is a compressed air-powered nailer incorporating a magazine that holds a clip of nails which feed automatically for rapid controlled nailing. Other models have larger capacity maga-

Table 40. Schedule of Common Wire Nail Use in Wood Frame Building Construction

Joining	Nailing method	Nails		
		Number	Size	Placement
Header to joint	End-nail	3	16d	
Joist to sill or girder	Toenail	2	10d or	
		3	8d	
Header and stringer joist to sill	Toenail		10d	16 in. on center
Bridging to joist	Toenail each end	2	8d	
Ledger strip to beam, 2 in. thick		3	16d	At each joist
Subfloor, boards:				
1 by 6 in. and smaller		2	8d	To each joist
1 by 8 in.		3	8d	To each joist
Subfloor, plywood:				
At edges			8d	6 in. on center
At intermediate joists			8d	8 in. on center
Subfloor (2 by 6 in., T&G) to joist or girder	Blind-nail (casing) and face-nail	2	16d	
Soleplate to stud, horizontal assembly	End-nail	2	16d	At each stud
Top plate to stud	End-nail	2	16d	
Stud to soleplate	Toenail	4	8d	
Soleplate to joist or blocking	Face-nail		16d	16 in. on center
Doubled studs	Face-nail, stagger		10d	16 in. on center
End stud of intersecting wall to exterior wall stud	Face-nail		16d	16 in. on center
Upper top plate to lower top plate	Face-nail		16d	16 in. on center
Upper top plate, laps and intersections	Face-nail	2	16d	
Continuous header, two pieces, each edge			12d	12 in. on center
Ceiling joist to top wall plates	Toenail	3	8d	

Table 40. Schedule of Common Wire Nail Use in Wood Frame Building Construction (Cont'd)

Ceiling joist laps at partition	Face-nail	4	16d	
Rafter to top plate	Toenail	2	8d	
Rafter to ceiling joist	Face-nail	5	10d	
Rafter to valley or hip rafter	Toenail	3	10d	
Ridge board to rafter	End-nail	3	10d	
Rafter to rafter through ridge board	Toenail	4	8d	
	Edge-nail	1	10d	
Collar beam to rafter:				
2 in. member	Face-nail	2	12d	
1 in. member	Face-nail	3	8d	
1-in. diagonal let-in brace to each stud and plate				
(4 nails at top)		2	8d	
Built-up corner studs:				
Studs to blocking	Face-nail	2	10d	Each side
Intersecting stud to corner studs	Face-nail		16d	12 in. on center
Built-up girders and beams, three or more members				
Wall sheathing:	Face-nail		20d	32 in. on center, each side
1 by 8 in. or less, horizontal				
1 by 6 in. or greater, diagonal	Face-nail	2	8d	At each stud
Wall sheathing, vertically applied plywood:	Face-nail	3	8d	At each stud
⅜ in. and less thick				
½ in. and over thick	Face-nail		6d	6 in. edge
Wall sheathing, vertically applied fiberboard:	Face-nail		8d	12 in. intermediate
½ in. thick				
25/32 in. thick	Face-nail			
Roof sheathing, boards, 4-, 6-, 8-in. width	Face-nail			1½ in. roofing nail } 3 in. edge and
Roof sheathing, plywood:	Face-nail	2	8d	1¾ in. roofing nail } 6 in. intermediate
⅜ in. and less thick				At each rafter
½ in. and over thick	Face-nail		6d	
	Face-nail		8d	6 in. edge and 12 in. intermediate

DUO-FAST CORP.

Fig. 247. Compressed Air Powered Nailer.

The *power stapler* is similar in appearance and operation to the power nailer shown in Fig. 247. The fasteners it drives (staples) are available in a great variety of wire gauges, crown widths and leg lengths. Staplers are in common use for carpentry operations involving application of sheathing, decking, roofing, etc., as well as in many finishing operations.

Nail Selection Guide for Power Nailer

Shank Types

Smooth—Smooth-shank nails are available in a wide range of lengths and shank diameters for light, standard, and heavy-duty applications. Smooth-shank nails provide excellent holding power in most wood applications and are available with round heads, modified D-heads, or duplex heads for applications that require easy removal of the nail.

Ring—Ring-shank nails offer superior holding power in softwood applications and are excellent for crating and container construction where fastened joints are subject to lateral stress.

Screw—Excellent for hardwood applications, such as pallets. Screw-shank nails provide greater maximum withdrawal loads than smooth-shank nails.

Staple Selection Guide

Crown Widths/Wire Sizes

Fine-Wire Wide Crown—Recommended for applications where the staple crown should not cut into the material being fastened.

Heavy-Wire Wide Crown—Used for construction-related applications such as asphalt roofing and sheathing where broad holding power is required.

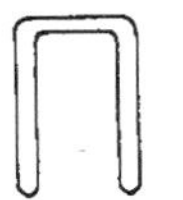

Fine-Wire Intermediate Crown—Holds well when driven into many types of soft material, such as cloth and vinyl. Maintains a neat, clean appearance when driven.

Heavy-Wire Intermediate Crown—Recommended for heavier wood applications, such as decking and wall sheathing, where superior holding power is required.

Fine-Wire Narrow Crown—Used for applications where appearance is important and the staple must be nearly invisible.

Heavy-Wire Narrow Crown—Recommended for furniture, cabinet, and construction-related applications.

STAIR LAYOUT

Knowledge and understanding of certain terms and practices help considerably in the layout and installation of simple stairs. This is true for stairs as simple as a few steps to a full floor level of stairs, both straight and platform type.

Straight Stairs *Platform Stairs*

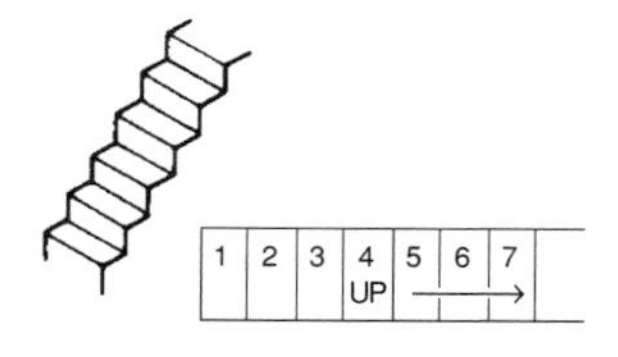

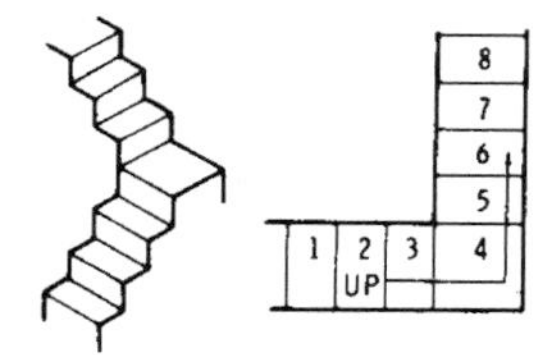

Fig. 248.

Terms

Total Rise—The vertical distance from floor surface to floor surface.

Total Run—The straight length from first rise to final rise.

Tread Rise—Vertical height of one rise.

Tread Run—Distance from one rise to next rise.

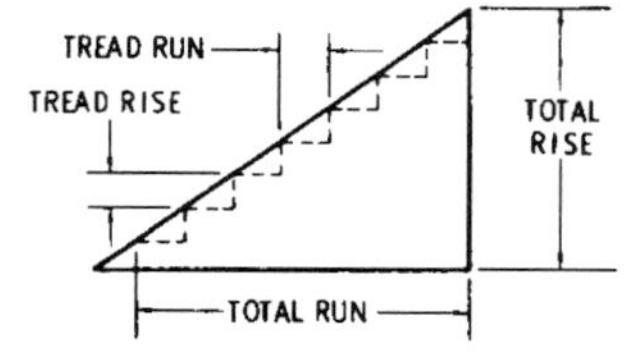

Conditions

A stair layout usually must comply with certain fixed conditions and/or specific dimensions, such as height, size of opening, available space, direction of run, etc. In addition, there are certain general conditions which the layout must meet, such as tread-rise dimension, tread-run dimension, etc.

While there is no standard tread-rise dimension as such, experience has shown that it should be in the area of 7 inches. Too large a rise results in a steep stairway, sometimes referred to as a "hard" stairway, because it is difficult both to ascend and descend. Also, the tread rise must be proportional to the tread run; that is, the tread run narrows as the tread rise increases and vice versa. There are numerous methods or rules to determine the dimensions of the tread rise and run. A very simple one, easy to remember and quite satisfactory for general use, is that the sum of the tread rise and run dimensions should total 17 inches. The tread rise dimension should be held as close as possible to the optimum 7 inches. Another rule in common use is that the run dimension plus twice the rise dimension should equal 24.

Dimensions

Specific layout dimensions for stairs are determined from the given or known dimensions and/or conditions, such as size of opening, height from floor to floor, available space, direction of run, etc. The manner or method of layout, therefore, is dictated by the given or known information. For instance, the most frequently encountered situation is one where the total rise (distance from level to level) is known, and a set of straight stairs is to be installed. If space is not limited, all necessary dimensions for layout of a comfortable set of stairs may be calculated from this one dimension.

Example

A straight set of stairs is to be constructed and installed from one level to the level above. The vertical distance from floor level to floor level is 12 feet 8 inches.

Total Rise—Vertical distance between levels 12' 8", or 104".

Tread Rise—The calculation of the tread rise is done in three steps. First, determine the approximate number of rises by dividing the total rise in inches by 7. Second, select a whole number that is close to the number calculated. This will be the number of rises or steps in the stairs. Third, calculate the tread rise dimension by dividing the total rise measurement by the number of rises or steps selected.

Using the example stated above, 7 will divided into 104 almost 15 times; therefore the selection is between 14 and 15. If 15 rises are selected, the tread rise will be 6$\frac{15}{16}$". If 14 rises are selected, the treads rise will be 7$\frac{7}{16}$".

Tread Run—The tread run is calculated by subtracting the tread rise dimension from 17. As this is an approximate figure, the actual tread run dimension is rounded off to the closest convenient fraction; for instance, for a $6\frac{15}{16}$" tread rise a 10" run may be used, while for a $7\frac{7}{16}$" rise a $9\frac{1}{2}$" run would be proper.

Total Run—The total run is calculated by multiplying the tread run dimension by one less than the number of rises. Using the figures from the above example, the total run for 14 rises would be $123\frac{1}{2}$" ($13 \times 9\frac{1}{2}$), and the total run for 15 rises would be 140" (14×10).

The two examples stated above are illustrated in Fig. 249.

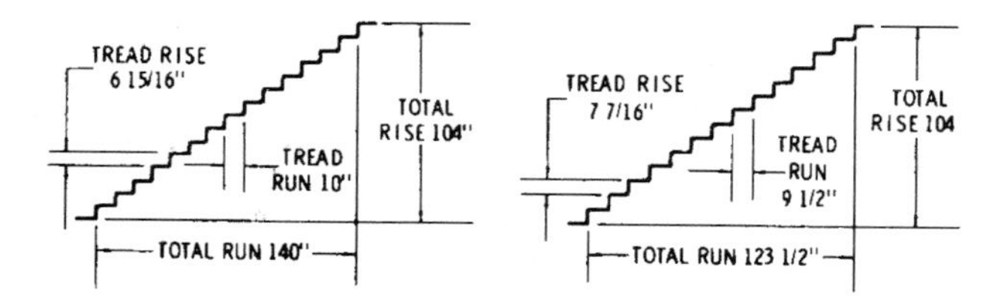

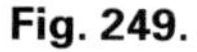

Fig. 249.

Stairs are usually supported by either *stair horses* or by *stair stringers*. Stair horses are an underneath style of support member, with the stair tread resting on the steps of the horses. Stair stringers are side support members; the treads are located between the stringers and are supported by end attachment.

When constructing simple stairs, the stair horse style of construction is usually used with wood materials and the stair stringer style when structural steel materials are used (Fig. 250).

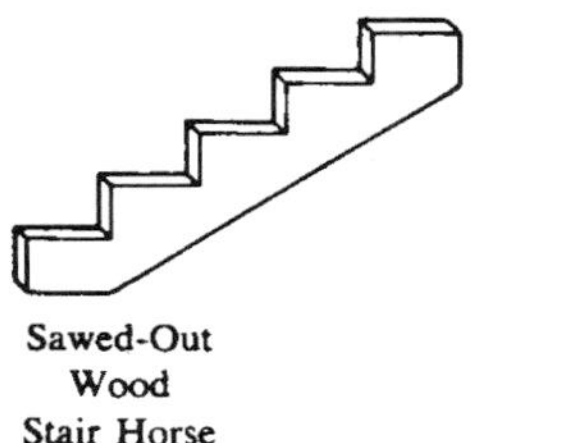

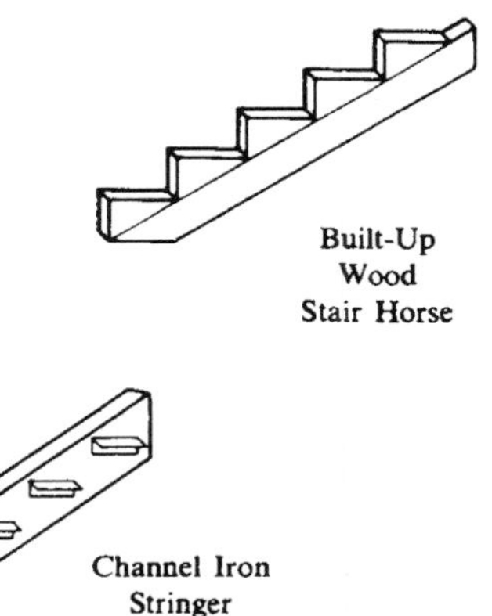

Fig. 250.

Stair Length

The length of stair horses or stringers may be greater or less than the stair length, depending on the construction that is used. The actual length of the horse or stringer is determined during the layout. A few construction styles are shown in Fig. 251.

The usual practice followed to determine stair length is to lay out to scale the total rise and run on a steel square as shown in Fig. 252.

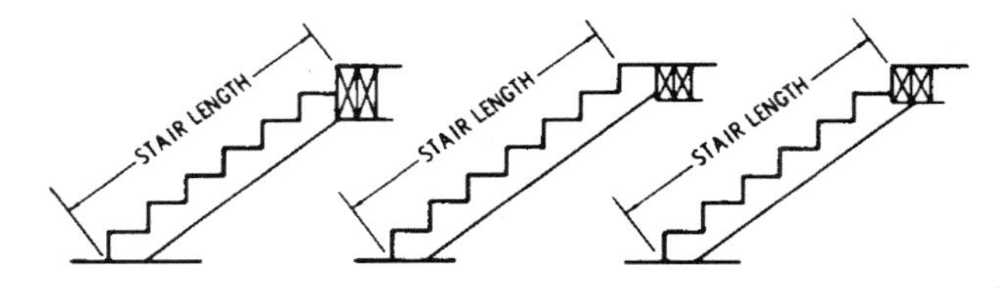

Fig. 251.

Measuring between the two points on the legs of the square will give the stair length. While the stair length is not precisely the hypotenuse of a right triangle whose sides are the total rise and the total run, in practice it is considered to be. The error, as shown below, is slight. Stair length may also be calculated using either trigonometry or the sum-of-the-squares equation.

The length of the rough stock needed for layout will depend on the style of construction. In many cases it must be greater than the stair length.

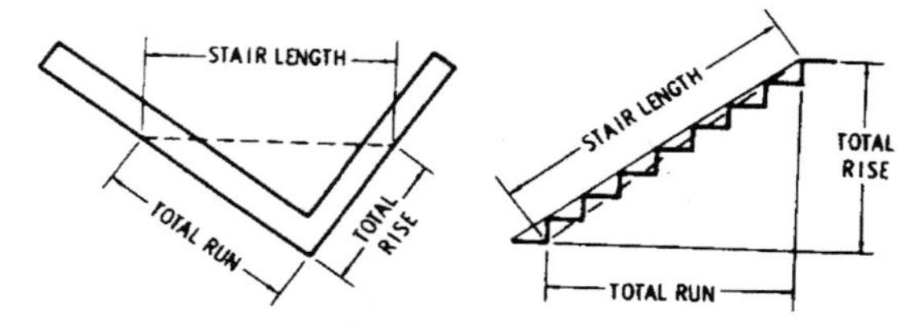

Fig. 252.

Stair Horse Layout

The stair horse should be laid out with the top edge of the stock as the layout *top line*. The edge of the steps are laid out touching this top line or layout line. This will result in a minimum amount of cutout and will leave supporting material below. The layout is started from the left end, holding the body of the square in the left hand. The tongue is held in the right hand with the outside point of the square down or facing away from the top line, as shown in Fig. 253.

As shown in Fig. 252, the tread rise measurement point on the tongue of the square, and the tread run measurement point on the body of the square, are positioned on the top line. The mark for the tread run cut is made against the body and the tread rise cut mark is made against the tongue of the square.

The square is then moved to the next position, again with the run and rise measurement points on the body and tongue matching the

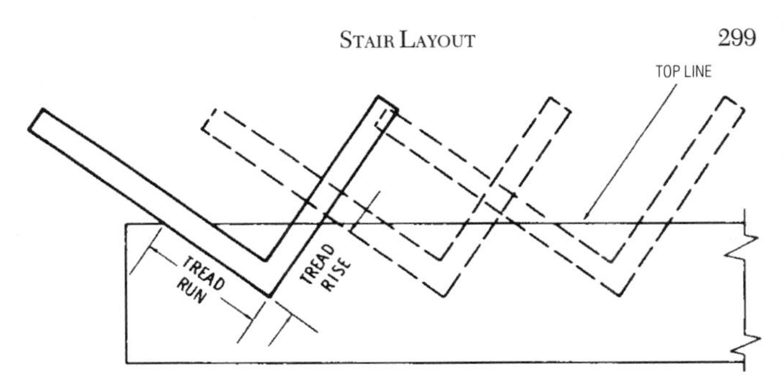

Fig. 253.

top line. The intersecting points of adjacent layouts become the points or edges of the steps. The square is moved along in this manner and layouts made until the required number of steps have been laid out.

Stair Stringer Layout

Because stair stringers usually act as enclosures on the sides, it is desirable to have stringer material extend above the treads. To provide this material above the treads, a stair stringer is usually laid out from the bottom or *lower line*. This layout is also made starting from the left, but the square is reversed as shown in Fig. 254.

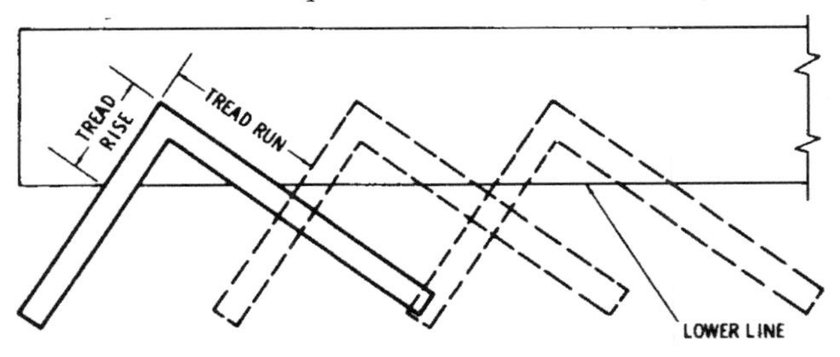

Fig. 254.

Stair Drop

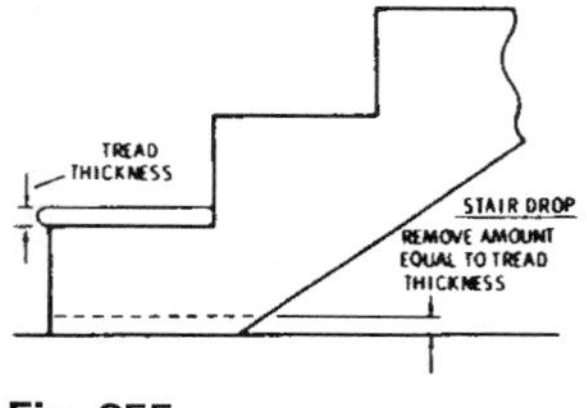

Fig. 255.

Layout procedure up to this point has made no provision for the thickness of the stair tread. To compensate for the tread thickness, the stair horse or stringer must be "dropped." If this is not done, the first step will be the thickness of the tread greater than the proper tread rise, and the top step will be the same amount less than the proper rise. Removing the correct amount from the bottom of the horse or stringer automatically corrects the height of both the bottom and top rises (Fig. 255).

If finish flooring is to be laid after stairs are installed, the amount that should be removed to allow for "drop" will be the difference between tread thickness and the finish flooring.

ELECTRICITY

Because of the potential danger ever present with electrical energy, a basic requirement when working with electricity is that there be no guesswork or chance-taking. Therefore, activities in the area of electrical work should be restricted to those things with which one has experience or about which one has specific knowledge or understanding.

Electrical Terms

Electromotive Force—The force which causes electricity to flow when there is a difference of potential between two points. The unit of measurement is the volt.

Direct Current—The flow of electricity in one direction. This is commonly associated with continuous direct current which is nonpulsating, as from a storage battery.

Alternating Current—The flow of electricity that is continuously reversing or alternating in direction, resulting in a regularly pulsating flow.

Voltage—The value of the electromotive force in an electrical system. It may be compared to pressure in a hydraulic system.

Amperage—The quantity and rate of flow in an electrical system. It may be compared to the volume of flow in a hydraulic system.

Resistance—The resistance offered by materials to the movement of electrons, commonly referred to as the flow of electricity. The unit of measurement commonly used is the *ohm*.

Cycle—The interval or period during which alternating current (using zero as a starting point) increases to maximum force in a positive direction, reverses and decreases to zero, then increases to maximum force in a negative direction, then reverses again and decreases to zero value. One cycle of such a flow is visually represented in Fig. 256.

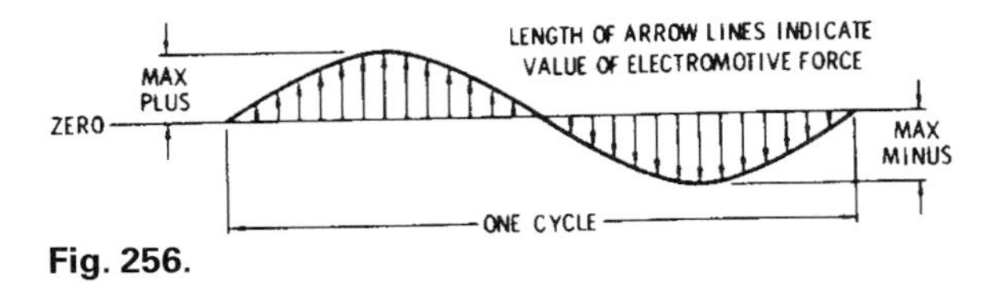

Fig. 256.

Frequency—The number of complete cycles per second of the alternating current flow. The most widely used alternating current frequency is 60 cycles per second. This is the number of complete cycles per second; thus the pulsation rate is twice this, or 120 pulses per second. Frequency is now specified as so many hertz. The term *hertz* is defined as cycles per second and is abbreviated Hz.

Phase—The word "phase" applies to the number of current surges that flow simultaneously in an electrical circuit. Fig. 256 is a graphic representation of single-phase alternating current. The single line represents a current flow that is continuously increasing or decreasing in value.

Three-phase current has three separate surges of current flowing together. In any given instant, however, their values differ, as the peaks and valleys of the pulsations are spaced equally apart (Fig. 257). The waveforms are lettered A, B, and C to represent the

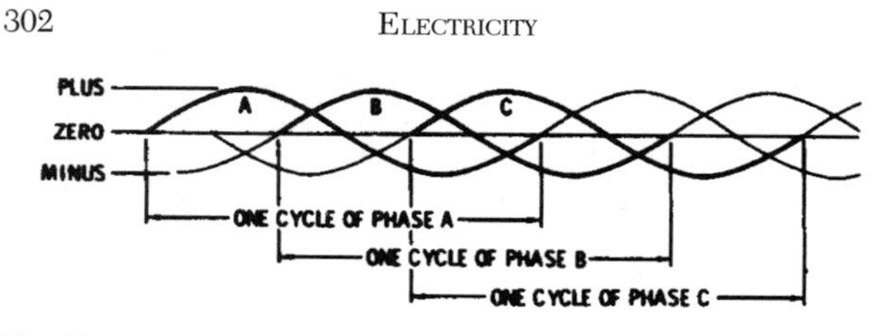

Fig. 257.

alternating current flow for each phase during a complete cycle. In three-phase current flow, any one current pulse is always one-third of a cycle out of matching with another.

Watt—The *watt* is the electrical unit of power, or the rate of doing work. One watt represents the power that is used when one ampere of current flows in an electrical circuit with a voltage or pressure of one volt.

Watt Hour—The *watt hour* expresses watts in time measurement of hours. For example, if a 100-watt lamp is in operation for a 2-hour period, it will consume 200 watt hours of electrical energy.

Kilowatt Hour—One *kilowatt* is equal to 1000 watts. One kilowatt hour is the electrical energy expended at the rate of one kilowatt (1000 watts) over a period of one hour.

Electrical Calculations

Most simple electrical calculations associated with common electrical power circuits involve the use of two basic formulas. These are the *Ohm's Law* formula and the basic *electrical power* formula. By substitution of known values into these formulas, and their rearrangements, unknown values may be easily determined.

Ohm's Law

This is the universally used electrical law stating the relationship of current, voltage, and resistance. This is done mathematically by the formula shown below. Current is stated in *amperes* and abbreviated *I*. Resistance is stated in *ohms* and abbreviated *R*, and voltage in *volts* abbreviated *E*.

$$Current = \frac{voltage}{resistance} \text{ or } I = \frac{E}{R}$$

The arrangement of values gives two other forms of the same equation as follows:

$$R = \frac{E}{I} \text{ and } E = I \times R$$

Example

An ammeter placed in a 110-volt circuit indicates a current flow of 5 amperes; what is the resistance of the circuit?

$$R = \frac{E}{I} \text{ or } R = \frac{110}{5} \text{ or } R = 22 \text{ ohms}$$

Power Formula

This formula indicates the rate at any given instant at which work is being done by current moving through a circuit. Voltage and amperes are abbreviated *E* and *I* as in Ohm's Law, and watts are abbreviated *W*.

$$Watts = voltage \times amperes \text{ or } W = E \times I$$

The two other forms of the formula, obtained by rearrangement of the values, are:

$$E = \frac{W}{I} \text{ or } I = \frac{W}{E}$$

Example

Using the same values as used in the example above, 5 amperes flowing in a 110-volt circuit, how much power is consumed?

$$W = E \times I \text{ or } W = 110 \times 5 \text{ or } W = 550 \text{ watts}$$

Example

A 110-volt appliance is rated at 2000 watts; can this appliance be plugged into a circuit fused at 15 amps?

$$I = \frac{W}{E} \text{ or } I = \frac{2000}{110} \text{ or } I = 18.18 \text{ amperes}$$

Obviously the fuse would blow if this appliance were plugged into the circuit.

Electrical Circuits

An electrical circuit is composed of elements such as lamps, switches, motors, resistors, wires, cables, batteries or other voltage sources, etc. All are conductors or conducting devices which form an electrical path called a circuit. To represent the various elements of a circuit on paper, circuit diagrams composed of lines and symbols are used. These are called *schematic diagrams*, and symbols used to represent the circuit elements, including the voltage

Table 41. Symbols Used in Industrial Applications

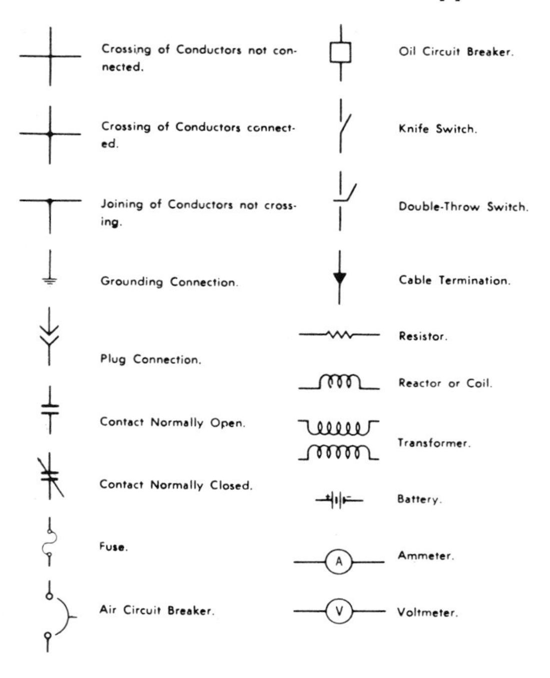

source, are standardized. Some of the symbols commonly used in industrial applications are shown in Table 41.

Electrical circuits may be classified as *series* circuits, *parallel* circuits, or a combination of series and parallel circuits. A series circuit is one where all parts of the circuit are electrically connected end to end. The current flows from one terminal of the power-source through each element and to the other power-source terminal. The same amount of current flows in each part of the circuit. An example of a series circuit is illustrated in Fig. 258.

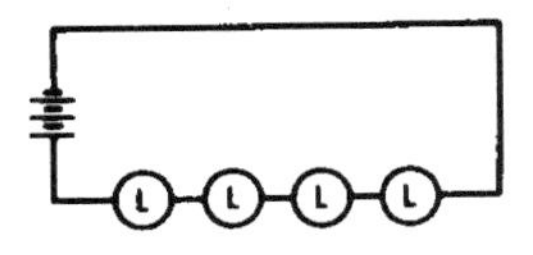

Fig. 258.

A parallel circuit is one where each element is so connected that it has direct flow to both terminals of the power source. The voltage across any element in a parallel circuit is equal to the voltage of the source, or power supply. An example of a parallel circuit is illustrated in Fig. 259.

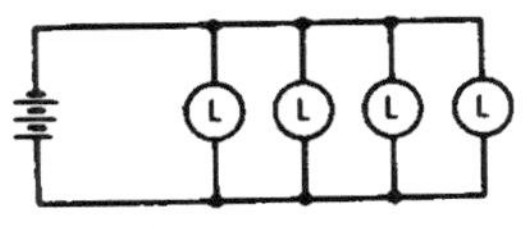

Fig. 259.

The relationship of values in series and parallel circuits using Ohm's Law and the power formula are illustrated and compared in the following examples.

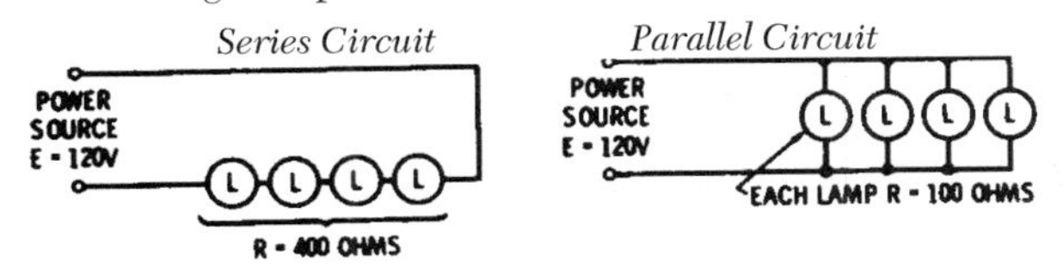

Fig. 260.

Current Flow Thru Series Circuit

$$I = \frac{E}{R} = \frac{120}{400} = .3 \text{ amp}$$

Voltage Across One Lamp

$E = IR = .3 \times 100 = 30$ *volts*

Current Flow Thru One Lamp

$$I = \frac{E}{R} = \frac{30}{100} = .3 \text{ amps}$$

Power Used By One Lamp

$W = EI = 30 \times .3 = 9$ *watts*

Power Used By Circuit

$W = EI = 120 \times .3 = 36$ *watts*

Current Flow Thru Parallel Circuit

$$I = \frac{E}{R} = \frac{120}{25} = 4.8 \text{ amps}$$

Current Flow Thru One Lamp

$$I = \frac{E}{R} = \frac{120}{100} = 1.2 \text{ amps}$$

Voltage Cross One Lamp

$E = IR = 1.2 \times 100 = 120$ *volts*

Power Used By One Lamp

$W = EI = 120 \times 1.2 = 144$ *watts*

Power Used By Circuit

$W = EI = 120 \times 4.8 = 576$ *watts*

Electrical Wiring

The term *electrical wiring* is applied to the installation and assembly of electrical conductors. The size of the wire used for electrical conductors is specified by gauge number according to the American Wire Gauge System. The usual manner of designation is by abbreviation AWG. Table 42 lists the AWG numbers and the corresponding specifications using the *mil* unit to designate a 0.001-inch measurement.

The *circular mil* unit used in the table is a measurement of cross-sectional area based on a circle one mil in diameter.

Switches are the most widely used of all electric wiring devices. They are connected in series with the devices they control, and allow current to flow when closed and interrupt current flow when open. One of the most common of switch applications is the control of one or more lamps from a single location. The schematic diagram for such a circuit is illustrated in Fig. 261 and a sketch of actual wiring connections is shown in Fig. 262.

Table 42. Size, Resistance, and Weight of Standard Annealed Copper Wire.

Size of Wire, AWG	Diameter of Wire Mils	Cross Section Circular Mils	Resistance, Ohms per 1000 Ft. at 68 °F or 20 °C	Weight, Pounds per 1000 Ft.
0000	460	212,000	0.0500	641
000	410	168,000	0.062	508
00	365	133,000	0.078	403
0	325	106,000	0.098	319
1	289	83,700	0.124	253
2	258	66,400	0.156	201
3	229	52,600	0.197	159
4	204	41,700	0.248	126
5	182	33,100	0.313	100
6	162	26,300	0.395	79.5
7	144	20,800	0.498	63.0
8	128	16,500	0.628	50.0
9	114	13,100	0.792	39.6
10	102	10,400	0.998	31.4
11	91	8,230	1.26	24.9
12	81	6,530	1.59	19.8
13	72	5,180	2.00	15.7
14	64	4,110	2.53	12.4
15	57	3,260	3.18	9.86
16	51	2,580	4.02	7.82
17	45	2,050	5.06	6.20
18	40	1,620	6.39	4.92
19	36	1,290	8.05	3.90
20	32	1,020	10.15	3.09
21	28.5	810	12.80	2.45
22	25.3	642	16.14	1.94
23	22.6	509	20.36	1.54
24	20.1	404	25.67	1.22
25	17.9	320	32.37	0.970
26	15.9	254	40.81	0.769
27	14.2	202	51.47	0.610
28	12.6	160	64.90	0.484
29	11.3	127	81.83	0.384
30	10.0	101	103.2	0.304
31	8.9	79.7	130.1	0.241
32	8.0	63.2	164.1	0.191

Table 43. Current Capacities of Copper Wires (Amperes)

Wire Size	In Conduit or Cable		In Free Air		Weather-proof Wire
	Type RHW*	Type TW, R*	Type RHW*	Type TW, R*	
14	15	15	20	20	30
12	20	20	25	25	40
10	30	30	40	40	55
8	45	40	65	55	70
6	65	55	95	80	100
4	85	70	125	105	130
3	100	80	145	120	150
2	115	95	170	140	175
1	130	110	195	165	205
0	150	125	230	195	235
00	175	145	265	225	275
000	200	165	310	260	320

*Types "RHW", "TW", or "R" are identified by markings on outer cover

Table 44. Adequate Wire Sizes — Weatherproof Copper Wire

Load in Building Amperes	Distance in Feet from Pole to Building	*Recommended Size of Feeder Wire for Job
Up to 25 amperes 120 volts	Up to 50 feet	No. 10
	50 to 80 feet	No. 8
	80 to 125 feet	No. 6
20 to 30 amperes 240 volts	Up to 80 feet	No. 10
	80 to 125 feet	No. 8
	125 to 200 feet	No. 6
	200 to 350 feet	No. 4
30 to 50 amperes 240 volts	Up to 80 feet	No. 8
	80 to 125 feet	No. 6
	125 to 200 feet	No. 4
	200 to 300 feet	No. 2
	300 to 400 feet	No. 1

*These sizes are recommended to reduce "voltage drop" to a minimum

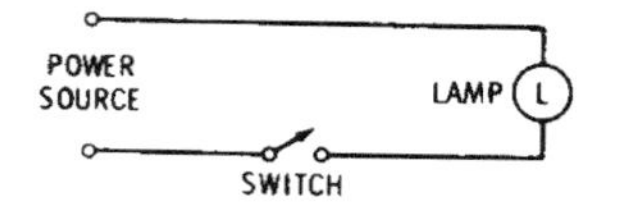

Fig. 261.

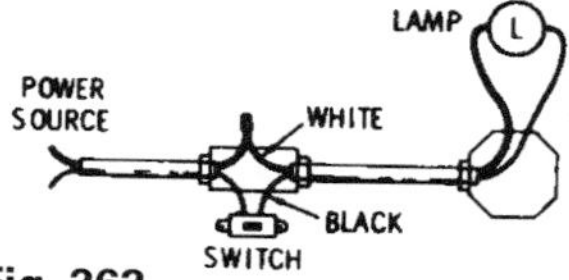

Fig. 262.

Table 45. Circuit Wire Sizes for Individual Single-phase Motors

Horsepower of Motor	Volts	Approximate Starting Current Amperes	Approximate Full Load Current Amperes	LENGTH OF RUN IN FEET from Main Switch to Motor								
				Feet	25	50	75	160	150	200	300	400
¼	120	20	5	Wire Size	14	14	14	12	10	10	8	6
⅓	120	20	5.5	Wire Size	14	14	14	12	10	8	6	6
½	120	22	7	Wire Size	14	14	12	12	10	8	6	6
¾	120	28	9.5	Wire Size	14	12	12	10	8	6	4	4
¼	240	10	2.5	Wire Size	14	14	14	14	14	14	12	12
⅓	240	10	3	Wire Size	14	14	14	14	14	14	12	10
½	240	11	3.5	Wire Size	14	14	14	14	14	12	12	10
34/4	240	14	4.7	Wire Size	14	14	14	14	14	12	10	10
1	240	16	5.5	Wire Size	14	14	14	14	14	12	10	10
1 ½	240	22	7.6	Wire Size	14	14	14	14	12	10	8	8
2	240	30	10	Wire Size	14	14	14	12	10	10	8	6
3	240	42	14	Wire Size	14	12	12	12	10	8	6	6
5	240	69	23	Wire Size	10	10	10	8	8	6	4	4
7 ½	240	100	34	Wire Size	8	8	8	8	6	4	2	2
10	240	130	43	Wire Size	6	6	6	6	4	4	2	1

Table 46. Types and Usage of Extension Cords

	Type	Wire Size	Use
Ordinary Lamp Cord	POSJ SPT	No. 16 or 18	In residences for lamps or small appliances.
Heavy-duty—with thicker covering	S, SJ or SJT	No. 10, 12, 14 or 16	In shops, and outdoors for larger motors, lawn mowers, outdoor lighting, etc.

Table 47. Ability of Cord to Carry Current (2- or 3-Wire Cord)

Wire Size	Type	Normal Load	Capacity Load
No. 18	S, SJ, SJT or POSJ	5.0 Amp. (600W)	7 Amp. (840W)
No. 16	S, SJ, SJT or POSJ	8.3 Amp. (1000W)	10 Amp. (1200W)
No. 14	S	12.5 Amp. (1500W)	15 Amp. (1800W)
No. 12	S	16.6 Amp. (1900W)	20 Amp. (2400W)

Table 48. Selecting the Length of Cord Set

Light Load (to 7 amps.)	Medium Load (7-10 amps.)	Heavy Load (10-15 amps.)
To 25 Ft.—Use No. 18	To 25 Ft.—Use No. 16	To 25 Ft.—Use No. 14
To 50 Ft.—Use No. 16	To 50 Ft.—Use No. 14	To 50 Ft.—Use No. 12
To 100 Ft.—Use No. 14	To 100 Ft.—Use No. 12	To 100 Ft.—Use No. 10

NOTE: As a safety precaution be sure to use only cords which are listed by Underwriters' Laboratories. Look for the Underwriters' seal when you buy.

To control lamps from two points requires a switch called a three-way switch. It has three terminals, one of which is so arranged that current is carried through it to either of the other two. Its function is to connect one wire to either of two other wires. The diagram in Fig. 263 shows a lamp circuit controlled from two points, using three-way switches. The actual connection boxes and the wires in the cable between the boxes for the three-way switch circuit are shown in Fig. 264.

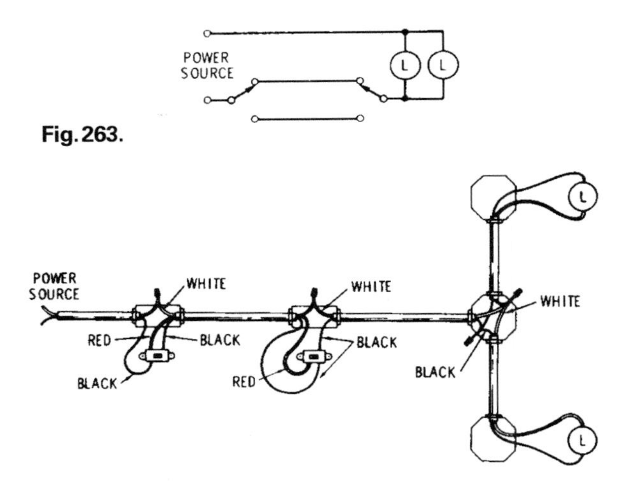

Fig. 263.

Fig. 264.

SHOP GEOMETRY

Lines

When lines touch at one point they are said to be *tangent* to one another. Two lines that cross are said to *intersect*. Two lines that are always the same distance apart are said to be *parallel*.

LINE TANGENT TO ARC

TANGENT ARCS

INTERSECTING LINES

PARALLEL LINES

Angles

There are four (4) common types of angles: *straight*, *right*, *acute*, and *obtuse*.

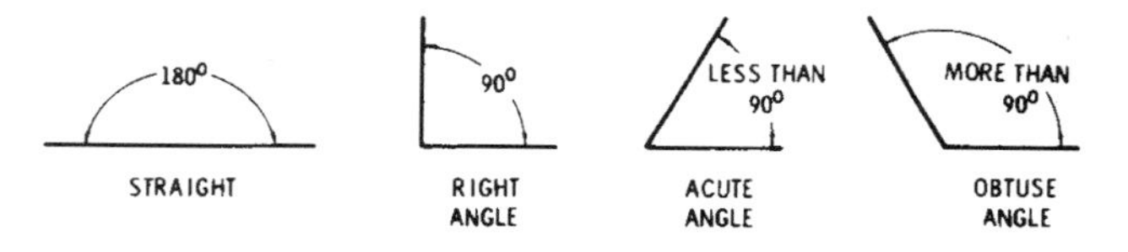

Circles

A *circle* is a closed curve on which all points are the same distance from a point (center) inside the circle. A line extending from the center to the closed curve line is the *radius*. The straight line that passes through the center of the circle and ends at the opposite side of the closed curved line is the *diameter*. An *arc* is a portion of the curved line. A *chord* refers to the straight line segment between the end points of an arc.

The distance around the circle is called the *circumference*. The circumference is measured in standard linear units or in angular measure. The total angular distance around the circumference of a circle is 360 degrees.

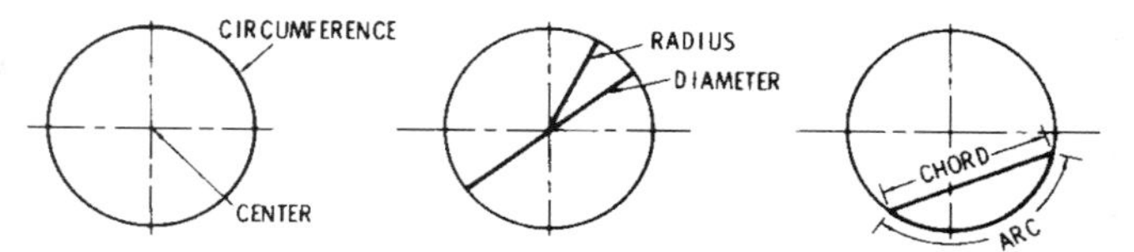

Triangles

A *triangle* consists of three straight lines joined at the end points to form a closed flat shape. The lines are called *sides* and the angles formed are called *inside angles*. Triangles are described by their included angles. The most commonly used is the *right triangle* in which one angle is a right (90°) angle. An *acute triangle* has three acute angles and an *obtuse triangle* has one angle greater than 90°.

The sum of the three inside angles of any triangle is always 180 degrees.

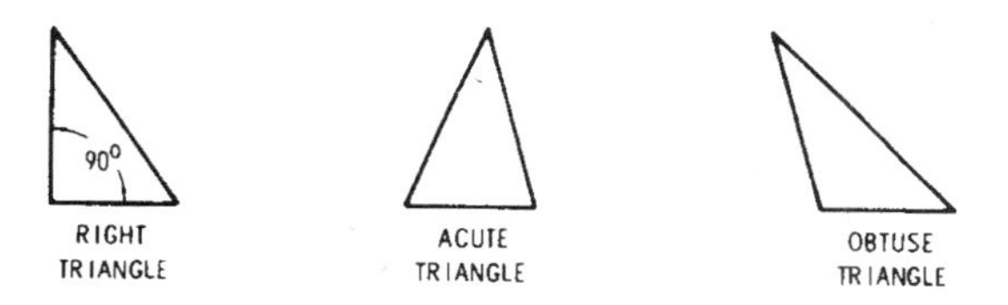

Quadrilaterals

Four-sided figures are called *quadrilaterals*. When all four angles are right angles, and two pair of sides are of equal length, the shape is a *rectangle*. When all four angles are right angles, and the four sides are equal, it is called a *square*.

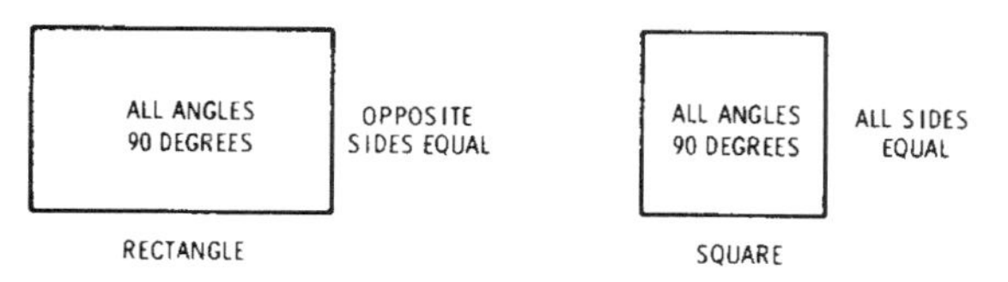

Regular Polygons

All plane figures with three or more sides are called *polygons*. Polygons having equal sides and equal angles are called *regular polygons*. The names given regular polygons imply the number of equal sides.

Number of Sides	Name	Number of Sides	Name
3	Triangle	7	Heptagon
4	Square	8	Octagon
5	Pentagon	9	Nonagon
6	Hexagon	10	Decagon

Geometrical Construction

Practical applications of geometric principles are in the construction of parallel, perpendicular, and tangent lines, dividing of straight and curved lines, and the bisecting of angles. These principles have further broad application in the layout of geometric shapes.

Dividing a Line

A line may be divided into equal parts by first drawing a line at any angle and length to the given line from one end point; second, stepping off on the angular line the same number of equal spaces as the line is to be divided into; third, connecting the last point with the other end point on the given line; fourth, drawing lines parallel to this connecting line. Dividing a line into three equal parts is illustrated below:

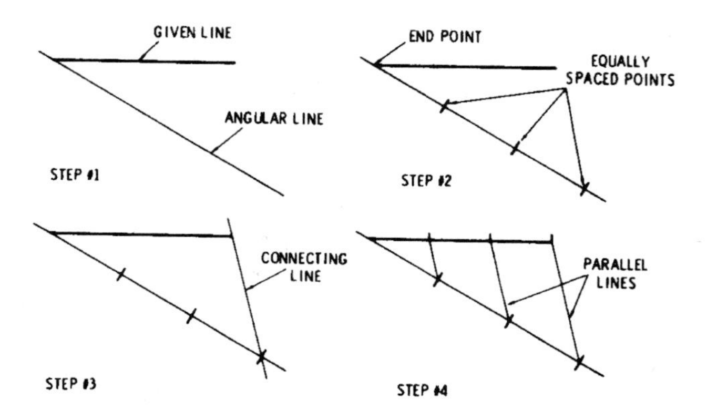

Erecting a Perpendicular Line

To erect a perpendicular or right-angle line at a given point on a line, use the following three steps. First, swing equal arcs from the given point to intersect the given line at points (a) and (b). Second, increase the radius by about one-half and swing arcs from points (a) and (b) to locate points (c) and (d). Third, draw a line through point (c) and (d) and the given point.

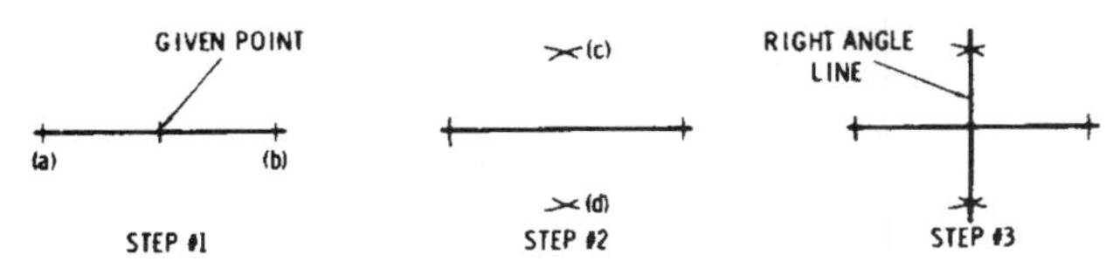

Polygon Construction

Square Inside a Circle

1. Draw a diameter line across circle.
2. Construct a perpendicular through center.
3. Connect the points of the diameter lines.

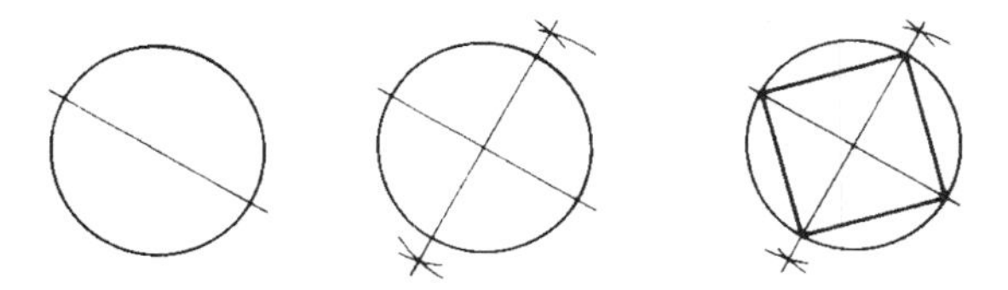

Square Outside a Circle

1. Draw a diameter line across circle.
2. Construct a perpendicular.
3. Construct tangent lines at the points of the diameter lines.

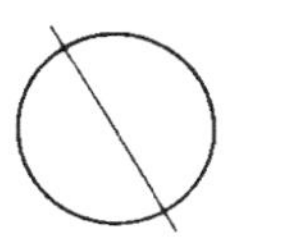
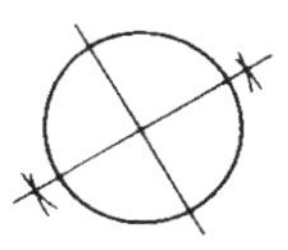
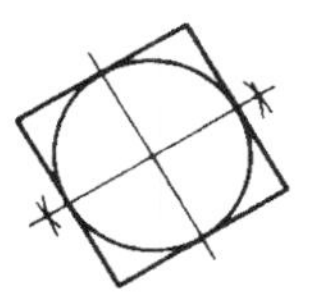

Hexagon Inside a Circle

1. Using the radius of the circle, step off arcs around the circle.
2. Connect the points where the arcs intersect the circle.

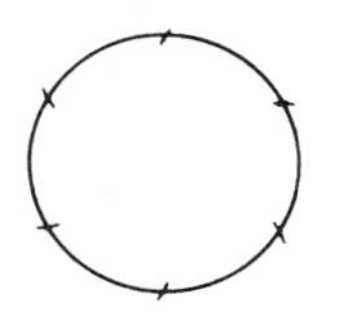

Hexagon Outside a Circle

1. Using the radius of the circle, step off arcs around the circle.
2. Construct a tangent line at each intersecting point.

Steps similar to these may be followed for constructing other polygons inside and outside of circles.

Dimensions of Polygons and Circles

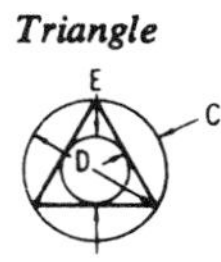

$$E = \text{side} \times .57735$$
$$D = \text{side} \times 1.1547 = 2E$$
$$\text{Side} = D \times .866$$
$$C = E \times .5 = D \times .25$$

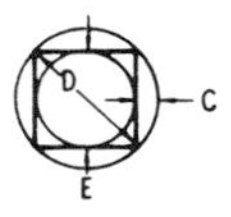

$$E = \text{side} = D \times .7071$$
$$D = \text{side} \times 1.4142 = \text{Diagonal}$$
$$\text{Side} = .7071$$
$$C = D \times .14645$$

$$E = \text{side} \times 1.3764 = D \times .809$$
$$D = \text{side} \times 1.7013 = E \times 1.2361$$
$$\text{Side} = D \times .5878$$
$$C = D \times .0955$$

Pentagon

$$E = \text{side} \times 1.7321 = D \times .866$$
$$D = \text{side} \times 2 = E \times 1.1547$$
$$\text{Side} = D \times .5$$
$$C = D \times .067$$

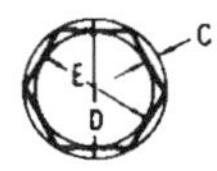

$$E = \text{side} \times 2.4142 = D \times .9239$$
$$D = \text{side} \times 2.6131 = E \times 1.0824$$
$$\text{Side} = D \times .3827$$
$$C = D \times .038$$

Octagon

SHOP TRIGONOMETRY

Right-Angle Triangles

All triangles are made up of six parts—three sides and three angles. A right-angle triangle is one having one angle of 90 degrees. A 90-degree angle is termed a *right angle*. The three sides of a right-angle triangle are called the *side opposite, side adjacent,* and *hypotenuse*. The hypotenuse is always the side directly across from, or opposite, the 90-degree angle. The other two sides are opposite one angle and adjacent to the other. Therefore, they may be called either opposite or adjacent sides in reference to the two angles as illustrated below.

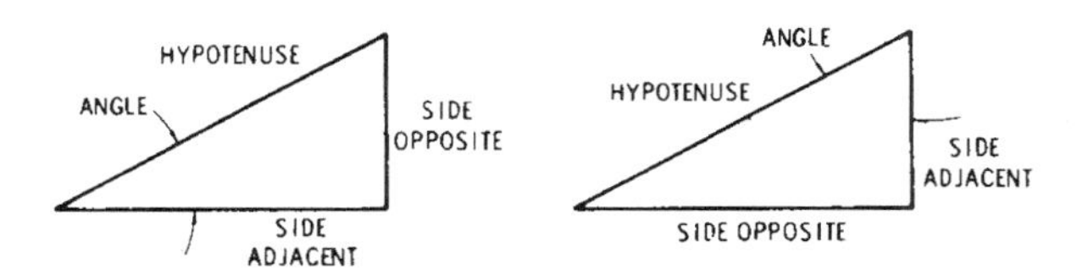

Trigonometric calculations employ numerical values called *trigonometric functions.* These values represent the ratios between the sides of triangles and are identified by the names *sine, coisine, tangent, cotangent, secant,* and *cosecant*. Each angle has a specific numerical value for each of its functions. These values are given in tables of trigonometric functions.

When one of the angles of a right-angle triangle (other than the 90° angle) and the length of one of the sides are known, the length of the other sides may be determined by use of the appropriate formula below:

$$\text{Length of side opposite} = \begin{cases} \text{hypotenuse} \times \text{sine} \\ \text{hypotenuse} \div \text{cosecant} \\ \text{side adjacent} \times \text{tangent} \\ \text{side adjacent} \div \text{cotangent} \end{cases}$$

$$\text{Length of side adjacent} = \begin{cases} \text{hypotenuse} \times \text{cosine} \\ \text{hypotenuse} \div \text{secant} \\ \text{side opposite} \times \text{cotangent} \\ \text{side opposite} \div \text{tangent} \end{cases}$$

$$\text{Length of hypotenuse} = \begin{cases} \text{side opposite} \times \text{cosecant} \\ \text{side opposite} \div \text{sine} \\ \text{side adjacent} \times \text{secant} \\ \text{side adjacent} \div \text{cosine} \end{cases}$$

When the lengths of two sides of a right-angle triangle are known, the angles may be determined in two steps using the trigonometric functions of the angles.

Step #1

Using the appropriate formula, calculate the numerical value of the function of the angle.

Step #2

Using a table of trigonometric functions, find the angle which corresponds to the function calculated by formula.

$$\text{Sine} = \frac{\text{SideOpposite}}{\text{Hypotenuse}} \qquad \text{Cotangent} = \frac{\text{Side Adjacent}}{\text{Side Opposite}}$$

$$\text{Cosine} = \frac{\text{Side Adjacent}}{\text{Hypotenuse}} \qquad \text{Secant} = \frac{\text{Hypotenuse}}{\text{Side Adjacent}}$$

$$\text{Tangent} = \frac{\text{Side Opposite}}{\text{Side Adjacent}} \qquad \text{Cosecant} = \frac{\text{Hypotenuse}}{\text{Side Opposite}}$$

Example:

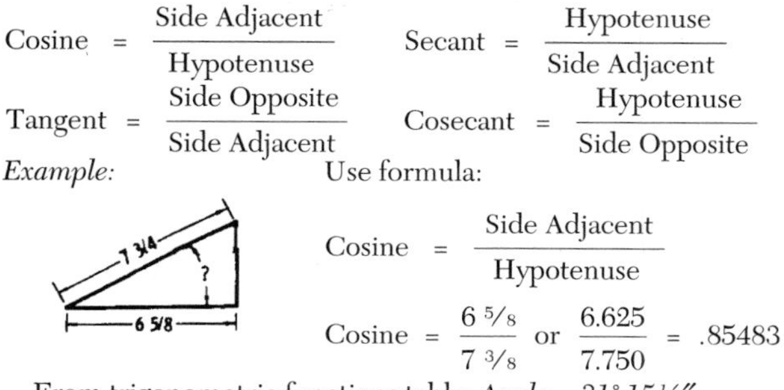

Use formula:

$$\text{Cosine} = \frac{\text{Side Adjacent}}{\text{Hypotenuse}}$$

$$\text{Cosine} = \frac{6\ 5/8}{7\ 3/8} \text{ or } \frac{6.625}{7.750} = .85483$$

From trigonometric functions table: *Angle = 31° 15½″.*

Sum of the Squares

When the lengths of two sides of a right-angle triangle are known, the length of the third side may be determined by the use of the sum-of-the-squares formula. It states that the square of the hypotenuse is equal to the sum of the squares of the other two sides. The basic formula is commonly written:

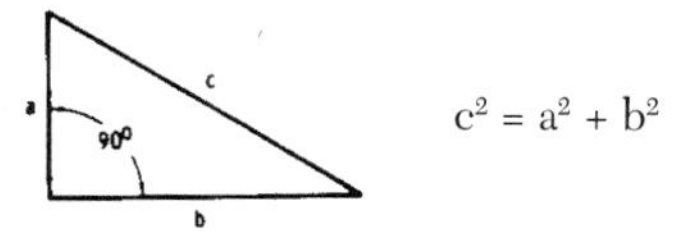

$$c^2 = a^2 + b^2$$

To find the length of the third side of a right-angle triangle when the lengths of two sides are known, the known values are substituted in the appropriate equation.

$$c = \sqrt{a^2 + b^2} \qquad a = \sqrt{c^2 + b^2} \qquad b = \sqrt{c^2 + a^2}$$

Example:

Find the length of c when a is 9 and b is 11. Use formula:

$$c = \sqrt{a^2 + b^2}$$

$$c = \sqrt{(9 \times 9) + (11 \times 11)} \text{ or } \sqrt{81 + 121} \text{ or } \sqrt{202}$$

$$c = \sqrt{202} \text{ or approximately } 14\,7/12$$

Note

Any triangle having sides with a 3-4-5 length ratio is a right-angle triangle.

Proof

$$c^2 = a^2 + b^2$$
$$5^2 = 3^2 + 4^2$$
$$25 = 9 + 16$$
$$25 = 25$$

Examples

6—8—10
12—16—20
15—20—25
18—24—30
24—32—40
30—40—50

Natural Trigonometric Functions

Degree	Sine	Cosine	Tangent	Secant	Degree	Sine	Cosine	Tangent	Secant
0	.00000	1.0000	.00000	1.0999	46	.7193	.6947	1.0355	1.4395
1	.01745	.9998	.01745	1.0001	47	.7314	.6820	1.0724	1.4663
2	.03490	.9994	.03492	1.0006	48	.7431	.6691	1.1106	1.4945
3	.05234	.9986	.05241	1.0014	49	.7547	.6561	1.1504	1.5242
4	.06976	.9976	.06993	1.0024	50	.7660	.6428	1.1918	1.5557
5	.08716	.9962	.08749	1.0038	51	.7771	.6293	1.2349	1.5890
6	.10453	.9945	.10510	1.0055	52	.7880	.6157	1.2799	1.6243
7	.12187	.9925	.12278	1.0075	53	.7986	.6018	1.3270	1.6616
8	.1392	.9903	.1405	1.0098	54	.8090	.5878	1.3764	1.7013
9	.1564	.9877	.1584	1.0125	55	.8192	.5736	1.4281	1.7434
10	.1736	.9848	.1763	1.0154	56	.8290	.5592	1.4826	1.7883
11	.1908	.9816	.1944	1.0187	57	.8387	.5446	1.5399	1.8361
12	.2079	.9781	.2126	1.0223	58	.8480	.5299	1.6003	1.8871
13	.2250	.9744	.2309	1.0263	59	.8572	.5150	1.6643	1.9416
14	.2419	.9703	.2493	1.0306	60	.8660	.5000	1.7321	2.0000
15	.2588	.9659	.2679	1.0353	61	.8746	.4848	1.8040	2.0627
16	.2756	.9613	.2867	1.0403	62	.8829	.4695	1.8807	2.1300
17	.2924	.9563	.3057	1.0457	63	.8910	.4540	1.9626	2.2027
18	.3090	.9511	.3249	1.0515	64	.8988	.4384	2.0503	2.2812
19	.3256	.9455	.3443	1.0576	65	.9063	.4226	2.1445	2.3662
20	.3420	.9397	.3640	1.0642	66	.9135	.4067	2.2460	2.4586
21	.3584	.9336	.3839	1.0711	67	.9205	.3907	2.3559	2.5598
22	.3746	.9272	.4040	1.0785	68	.9272	.3746	2.4751	2.6695
23	.3907	.9205	.4245	1.0864	69	.9336	.3584	2.6051	2.7904
24	.4067	.9135	.4452	1.0946	70	.9397	.3420	2.7475	2.9238
25	.4226	.9063	.4663	1.1034	71	.9455	.3256	2.6042	3.0715
26	.4384	.8988	.4877	1.1126	72	.9511	.3090	3.0777	3.2361
27	.4540	.8910	.5095	1.1223	73	.9563	.2924	3.2709	3.4203
28	.4695	.8829	.5317	1.1326	74	.9613	.2756	3.4874	3.6279
29	.4848	.8746	.5543	1.1433	75	.9659	.2588	3.7321	3.8637
30	.5000	.8660	.5774	1.1547	76	.9703	.2419	4.0108	4.1336
31	.5150	.8572	.6009	1.1663	77	.9744	.2250	4.3315	4.4454
32	.5299	.8480	.6249	1.1792	78	.9781	.2079	4.7046	4.8097
33	.5446	.8387	.6494	1.1924	79	.9816	.1908	5.1446	5.2408
34	.5592	.8290	.6745	1.2062	80	.9848	.1736	5.6713	5.7588
35	.5736	.8192	.7002	1.2208	81	.9877	.1564	6.6138	6.3924
36	.5878	.8090	.7265	1.2361	82	.9903	.1392	7.1154	7.1853
37	.6018	.7986	.7536	1.2521	83	.9925	.12187	8.1443	8.2055
38	.6157	.7880	.7813	1.2690	84	.9945	.10453	9.5144	9.5668
39	.6293	.7771	.8098	1.2867	85	.9962	.08716	11.4301	11.474
40	.6428	.7660	.8391	1.3054	86	.9976	.06976	14.3007	14.335
41	.6561	.7547	.8693	1.3250	87	.9986	.05234	19.0811	19.107
42	.6691	.7431	.9004	1.3456	88	.9994	.03490	28.6363	28.654
43	.6820	.7314	.9325	1.3673	89	.9998	.01745	57.2900	27.299
44	.6947	.7193	.9657	1.3902	90	1.0000	Inf.	Inf.	Inf.
45	.7071	.7071	1.0000	1.4142					

Geometric Formulas

Triangle

$$\text{area (A)} = \frac{bh}{2}$$

Square

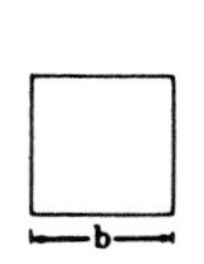

$$\text{area (A)} = b^2$$

Rectangle

$$\text{area (A)} = ab$$

Cone

$$\text{area (A)} = \pi RS$$
$$= \pi R\sqrt{R^2 + h^2}$$
$$\text{volume (V)} = \frac{\pi R^2 h}{3}$$
$$= 1.047R^2h$$
$$= 0.2618D^2h$$

Cylinder

$$\text{cylindrical surface} = \pi Dh$$
$$\text{total surface} = 2\pi R(R + h)$$
$$\text{volume (V)} = \pi R^2 h$$
$$= \frac{c^2 h}{4\pi}$$

Sphere

$$\text{area (A)} = 4\pi R^2$$
$$= \pi D^2$$
$$\text{volume (V)} = \frac{4}{3}\pi R^3$$
$$= 1/6\pi D^3$$

Cube

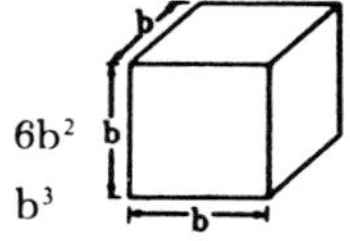

$$\text{area (A)} = 6b^2$$
$$\text{volume (V)} = b^3$$

Rectangular Solid

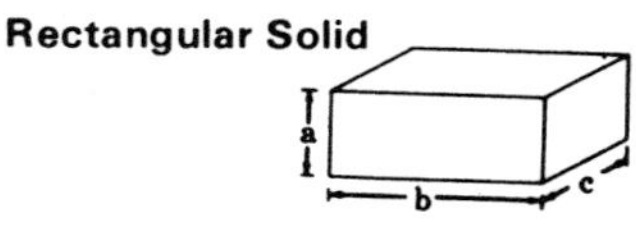

$$\text{area (A)} = 2\,(ab + bc + ac)$$
$$\text{volume (V)} = abc$$

Ring of Rectangular Cross Section

$$\text{volume (V)} = \frac{\pi c}{4}(D^2 - d^2)$$
$$= \left(\frac{D + d}{2}\right)\pi bc$$

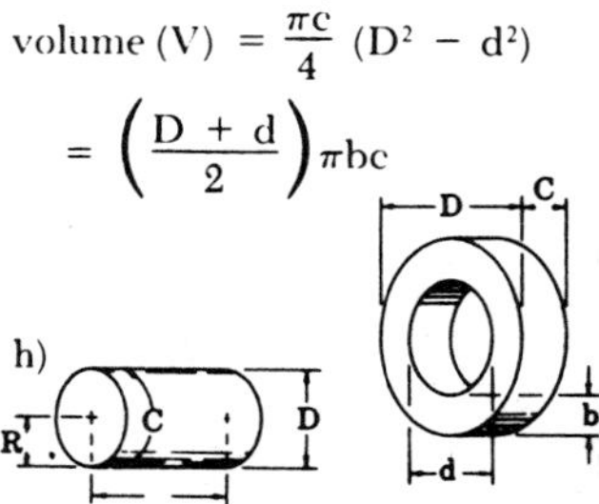

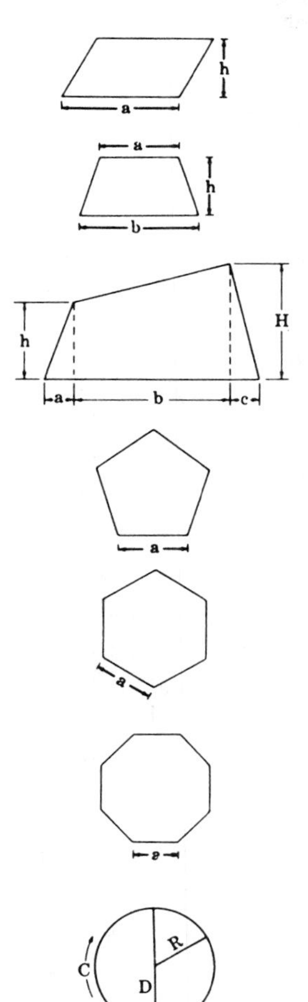

Parallelogram

area (A) $= ah$

Trapezoid

area (A) $= \frac{h}{2}(a + b)$

Trapezium

area (A) $= \frac{1}{2}\,[b(H + h) + ah + cH]$

Regular Pentagon

area (A) $= 1.720\ a^2$

Regular Hexagon

area (A) $= 2.598\ a^2$

Regular Octagon

area (A) $= 4.828\ a^2$

Circle

circumference (C) $= 2\pi R$

$= \pi D$

area (A) $= \pi R^2$

Functions of Numbers

No.	Square	Cube	Square Root	Cubic Root	Logarithm	1000 × Reciprocal	No. = Diameter	
							Circum.	Area
1	1	1	1.0000	1.0000	0.00000	1000.000	3.142	0.7854
2	4	8	1.4142	1.2599	0.30103	500.000	6.283	3.1416
3	9	27	1.7321	1.4422	0.47712	333.333	9.425	7.0686
4	16	64	2.0000	1.5874	0.60206	250.000	12.566	12.5664
5	25	125	2.2361	1.7100	0.69897	200.000	15.708	19.6350
6	36	216	2.4495	1.8171	0.77815	166.667	18.850	28.2743
7	49	343	2.6458	1.9129	0.84510	142.857	21.991	38.4845
8	64	512	2.8284	2.0000	0.90309	125.000	25.133	50.2655
9	81	729	3.0000	2.0801	0.95424	111.111	28.274	63.6173
10	100	1000	3.1623	2.1544	1.00000	100.000	31.416	78.5398
11	121	1331	3.3166	2.2240	1.04139	90.9091	34.558	95.0332
12	144	1728	3.4641	2.2894	1.07918	83.3333	37.699	113.097
13	169	2197	3.6056	2.3513	1.11394	76.9231	40.841	132.732
14	196	2744	3.7417	2.4101	1.14613	71.4286	43.982	153.938
15	225	3375	3.8730	2.4662	1.17609	66.6667	47.124	176.715
16	256	4096	4.0000	2.5198	1.20412	62.5000	50.265	201.062
17	289	4913	4.1231	2.5713	1.23045	58.8235	53.407	226.980
18	324	5832	4.2426	2.6207	1.25527	55.5556	56.549	254.469
19	361	6859	4.3589	2.6684	1.27875	52.6316	59.690	283.529
20	400	8000	4.4721	2.7144	1.30103	50.0000	62.832	314.159
21	441	9261	4.5826	2.7589	1.3222	47.6190	65.973	346.361
22	484	10648	4.6904	2.8020	1.34242	45.4545	69.115	380.133
23	529	12167	4.7958	2.8439	1.36173	43.4783	72.257	415.476
24	576	13824	4.8990	2.8845	1.38021	41.6667	75.398	452.389
25	625	15625	5.0000	2.9240	1.39794	40.0000	78.540	490.874
26	676	17576	5.0990	2.9625	1.41497	38.4615	81.681	530.929
27	729	19683	5.1962	3.0000	1.43136	37.0370	84.823	572.555
28	784	21952	5.2915	3.0366	1.44716	35.7143	87.965	615.752
29	841	24389	5.3852	3.0723	1.46240	34.4828	91.106	660.520
30	900	27000	5.4772	3.1072	1.47712	33.3333	94.248	706.858
31	961	29791	5.5678	3.1414	1.49136	32.2581	97.389	754.768
32	1024	32768	5.6569	3.1748	1.50515	31.2500	100.531	804.248
33	1089	35937	5.7446	3.2075	1.51851	30.3030	103.673	855.299
34	1156	39304	5.8310	3.2396	1.53148	29.4118	106.814	907.920
35	1225	42875	5.9161	3.2711	1.54407	28.5714	109.956	962.113
36	1296	46656	6.0000	3.3019	1.55630	27.7778	113.097	1017.88
37	1369	50653	6.0828	3.3322	1.56820	27.0270	116.239	1075.21
38	1444	54872	6.1644	3.3620	1.57978	26.3158	119.381	1134.11
39	1521	59319	6.2450	3.3912	1.59106	25.6410	122.522	1194.59
40	1600	64000	6.3246	3.4200	1.60206	25.0000	125.66	1256.64
41	1681	68921	6.4031	3.4482	1.61278	24.3902	128.81	1320.25
42	1764	74088	6.4807	3.4760	1.62325	23.8095	131.95	1385.44
43	1849	79507	6.5574	3.5034	1.63347	23.2558	135.09	1452.20
44	1936	85184	6.6332	3.5303	1.64345	22.7273	138.23	1520.53
45	2025	91125	6.7082	3.5569	1.65321	22.2222	141.37	1590.43
46	2113	97336	6.7823	3.5830	1.66276	21.7391	144.51	1661.90
47	2209	103823	6.8557	3.6088	1.67210	21.2766	147.65	1734.94

Functions of Numbers (Cont'd)

No.	Square	Cube	Square Root	Cubic Root	Logarithm	1000 × Reciprocal	No. = Diameter	
							Circum.	Area
48	2304	110592	6.9282	3.6342	1.68124	20.8333	150.80	1809.56
49	2401	117649	7.0000	3.6593	1.69020	20.4082	153.94	1885.74
50	2500	125000	7.0711	3.6840	1.69897	20.0000	157.08	1963.50
51	2601	132651	7.1414	3.7084	1.70757	19.6078	160.22	2042.82
52	2704	140608	7.2111	3.7325	1.71600	19.2308	163.36	2123.72
53	2809	148877	7.2801	3.7563	1.72428	18.8679	166.50	2206.18
54	2916	157464	7.3485	3.7798	1.73239	18.5185	169.65	2290.22
55	3025	166375	7.4162	3.8030	1.74036	18.1818	172.79	2375.83
56	3136	175616	7.4833	3.8259	1.74819	17.8571	175.93	2463.01
57	3249	185193	7.5498	3.8485	1.75587	17.5439	179.07	2551.76
58	3364	195112	7.6158	3.8709	1.76343	17.2414	182.21	2642.08
59	3481	205379	7.6811	3.8930	1.77085	16.9492	185.35	2733.97
60	3600	216000	7.7460	3.9149	1.77815	16.6667	188.50	2827.43
61	3721	226981	7.8102	3.9365	1.78533	16.3934	191.64	2922.47
62	3844	238328	7.8740	3.9579	1.79239	16.1290	194.78	3019.07
63	3969	250047	7.9373	3.9791	1.79934	15.8730	197.92	3117.25
64	4096	262144	8.0000	4.0000	1.80618	15.6250	201.06	3216.99
65	4225	274625	8.0623	4.0207	1.81291	15.3846	204.20	3318.31
66	4356	287496	8.1240	4.0412	1.81954	15.1515	207.35	3421.19
67	4489	300763	8.1854	4.0615	1.82607	14.9254	210.49	3525.65
68	4624	314432	8.2462	4.0817	1.83251	14.7059	213.63	3631.68
69	4761	328509	8.3066	4.1016	1.83885	14.4928	216.77	3739.28
70	4900	343000	8.3666	4.1213	1.84510	14.2857	219.91	3848.45
71	5041	357911	8.4261	4.1408	1.85126	14.0845	223.05	3959.19
72	5184	373248	8.4853	4.1602	1.85733	13.8889	226.19	4071.50
73	5329	389017	8.5440	4.1793	1.86332	13.6986	229.34	4185.39
74	5476	405224	8.6023	4.1983	1.86923	13.5135	232.48	4300.84
75	5625	421875	8.6603	4.2172	1.87506	13.3333	235.62	4417.86
76	5776	438976	8.7178	4.2358	1.88081	13.1579	238.76	4536.46
77	5925	456533	8.7750	4.2543	1.88649	12.9870	241.90	4656.63
78	6084	474552	8.8318	4.2727	1.89209	12.8205	245.04	4778.36
79	6241	493039	8.8882	4.2908	1.89763	12.6582	248.19	4901.67
80	6400	512000	8.9443	4.3089	1.90309	12.5000	251.33	5026.55
81	6561	531441	9.0000	4.3267	1.90849	12.3457	254.47	5153.00
82	6724	551368	9.0554	4.3445	1.91381	12.1951	257.61	5281.02
83	6889	571787	9.1104	4.3621	1.91908	12.0482	260.75	5410.61
84	7056	592704	9.1652	4.3795	1.92428	11.9048	263.89	5541.77
85	7225	614125	9.2195	4.3968	1.92942	11.7647	267.04	5674.50
86	7396	636056	9.2736	4.4140	1.93450	11.6279	270.18	5808.80
87	7569	658503	9.3274	4.4310	1.93952	11.4943	273.32	5944.68
88	7744	681472	9.3808	4.4480	1.94448	11.3636	276.46	6082.12
89	7921	704969	9.4340	4.4647	1.94939	11.2360	279.60	6221.14
90	8100	729000	9.4868	4.4814	1.95424	11.1111	282.74	6361.73
91	8281	753571	9.5394	4.4979	1.95904	10.9890	285.88	6503.88
92	8464	778688	9.5917	4.5144	1.96379	10.8696	289.03	6647.61
93	8649	804357	9.6437	4.5307	1.96848	10.7527	292.17	6792.91

Functions of Numbers (Cont'd)

No.	Square	Cube	Square Root	Cubic Root	Logarithm	1000 × Reciprocal	No. = Diameter	
							Circum.	Area
94	8836	830584	9.6954	4.5468	1.97313	10.6383	295.31	6939.78
95	9025	857375	9.7468	4.5629	1.97772	10.5263	298.45	7088.22
96	9216	884736	9.7980	4.5789	1.98227	10.4167	301.59	7238.23
97	9409	912673	9.8489	4.5947	1.98677	10.3093	304.73	7389.81
98	9604	941192	9.8995	4.6104	1.99123	10.2041	307.88	7542.96
99	9801	970299	9.9499	4.6261	1.99564	10.1010	311.02	7697.69

Metric and English Equivalent Measures

MEASURES OF LENGTH

Metric		English
1 meter	=	39.37 inches, or 3.28083 feet, or 1.09361 yards
.3048 meter	=	1 foot
1 centimeter	=	.3937 inch
2.54 centimeters	=	1 inch
1 millimeter	=	.03937 inch
25.4 millimeters	=	1 inch
1 kilometer	=	1093.61 yards, or 0.62137 mile

MEASURES OF WEIGHT

Metric		English
1 gram	=	15.432 grains
.0648 gram	=	1 grain
28.35 grams	=	1 ounce avoirdupois
1 kilogram	=	2.2046 pounds
.4536 kilogram	=	1 pound
1 metric ton 1000 kilograms	=	.9842 ton of 2240 pounds 9.68 cwt. 2204.6 pounds
1.016 metric tons 1016 kilograms	=	1 ton of 2240 pounds

MEASURES OF CAPACITY

Metric		English
1 liter (= 1 cubic decimeter)	=	61.023 cubic inches .03531 cubic foot .2642 gal. (American) 2.202 lbs. of water at 62° F.
28.317 liters	=	1 cubic foot
3.785 liters	=	1 gallon (American)
4.543 liters	=	1 gallon (Imperial)

English Conversion Table

Length

Inches	×	.0833	= feet
Inches	×	.02778	= yards
Inches	×	.00001578	= miles
Feet	×	.3333	= yards
Feet	×	.0001894	= miles
Yards	×	36.00	= inches
Yards	×	3.00	= feet
Yards	×	.0005681	= miles
Miles	×	63360.00	= inches
Miles	×	5280.00	= feet
Miles	×	1760.00	= yards
Circumference of circle	×	.3188	= diameter
Diameter of circle	×	3.1416	= circumference

Area

Square inches	×	.00694	= square feet
Square inches	×	.0007716	= square yards
Square feet	×	144.00	= square inches
Square feet	×	.11111	= square yards
Square yards	×	1296.00	= square inches
Square yards	×	9.00	= square feet
Dia. of circle squared	×	.7854	= area
Dia. of sphere squared	×	3.1416	= surface

Volume

Cubic inches	×	.0005787	= cubic feet
Cubic inches	×	.00002143	= cubic yards
Cubic inches	×	.004329	= U. S. gallons
Cubic feet	×	1728.00	= cubic inches
Cubic feet	×	.03704	= cubic yards
Cubic feet	×	7.4805	= U. S. gallons
Cubic yards	×	46656.00	= cubic inches
Cubic yards	×	27.00	= cubic feet
Dia. of sphere cubed	×	.5236	= volume

Weight

Grains (avoirdupois)	×	.002286	= ounces
Ounces (avoirdupois)	×	.0625	= pounds
Ounces (avoirdupois)	×	.00003125	= tons
Pounds (avoirdupois)	×	16.00	= ounces
Pounds (avoirdupois)	×	.01	= hundredweight
Pounds (avoirdupois)	×	.0005	= tons
Tons (avoirdupois)	×	32000.00	= ounces
Tons (avoirdupois)	×	2000.00	= pounds

Energy

Horsepower	×	33000.	= ft.-lbs. per min.
B. t. u.	×	778.26	= ft.-lbs.
Ton of refrigeration	×	200.	= B. t. u. per min.

Pressure

Lbs. per sq. in.	×	2.31	= ft. of water (60°F.)
Ft. of water (60°F.)	×	.433	= lbs. per sq. in.
Ins. of water (60°F)	×	.0361	= lbs. per sq. in.
Lbs. per sq. in.	×	27.70	= ins. of water (60°F)
Ins. of Hg (60°F)	×	.490	= lbs. per sq. in.

Power

Horsepower	×	746.	= watts
Watts	×	.001341	= horsepower
Horsepower	×	42.4	= B. t. u. per min.

Water Factors (at point of greatest density—39.2°F.)

Miners inch (of water)	×	8.976	= U. S. gals. per min.
Cubic inches (of water)	×	.57798	= ounces
Cubic inches (of water)	×	.036124	= pounds
Cubic inches (of water)	×	.004329	= U. S. gallons
Cubic inches (of water)	×	.003607	= English gallons
Cubic feet (of water)	×	62.425	= pounds
Cubic feet (of water)	×	.03121	= tons
Cubic feet (of water)	×	7.4805	= U. S. gallons
Cubic inches (of water)	×	6.232	= English gallons
Cubic foot of ice	×	57.2	= pounds
Ounces (of water)	×	1.73	= cubic inches
Pounds (of water)	×	26.68	= cubic inches
Pounds (of water)	×	.01602	= cubic feet
Pounds (of water)	×	.1198	= U. S. gallons
Pounds (of water)	×	.0998	= English gallons
Tons (of water)	×	32.04	= cubic feet
Tons (of water)	×	239.6	= U. S. gallons
Tons (of water)	×	199.6	= English gallons
U. S. gallons	×	231.00	= cubic inches
U. S. gallons	×	.13368	= cubic feet
U. S. gallons	×	8.345	= pounds
U. S. gallons	×	.8327	= English gallons
U. S. gallons	×	3.785	= liters
English gallons (Imperial)	×	227.41	= cubic inches
English gallons (Imperial)	×	.1605	= cubic feet
English gallons (Imperial)	×	10.02	= pounds
English gallons (Imperial)	×	1.201	= U.S. gallons
English gallons (Imperial)	×	4.546	= liters

Metric Conversion Table

Length

Millimeters	×	.03937	= inches
Millimeters	÷	25.4	= inches
Centimeters	×	.3937	= inches
Centimeters	÷	2.54	= inches
Meters	×	39.37	= inches (Act. Cong.)
Meters	×	3.281	= feet

Meters	×	1.0936	= yards
Kilometers	×	.6214	= miles
Kilometers	÷	1.6093	= miles
Kilometers	×	3280.8	= feet

Area

Sq. Millimeters	×	.00155	= sq. in.
Sq. Millimeters	÷	645.2	= sq. in.
Sq. Centimeters	×	.155	= sq. in.
Sq. Centimeters	÷	6.452	= sq. in.
Sq. Meters	×	10.764	= sq. ft.
Sq. Kilometers	×	247.1	= acres
Hectares	×	2.471	= acres

Volume

Cu. Centimeters	÷	16.387	= cu. in.
Cu. Centimeters	÷	3.69	= fl. drs. (U.S.P.)
Cu. Centimeters	÷	29.57	= fl. oz. (U.S.P.)
Cu. Meters	×	35.314	= cu. ft.
Cu. Meters	×	1.308	= cu. yards
Cu. Meters	×	264.2	= gals. (231 cu. in.)
Liters	×	61.023	= cu. in. (Act. Cong.)
Liters	×	33.82	= fl. oz. (U.S.J.)
Liters	×	.2642	= gals. (231 cu. in.)
Liters	÷	3.785	= gals. (231 cu. in.)
Liters	÷	28.317	= cu. ft.
Hectoliters	×	3.531	= cu. ft.
Hectoliters	×	2.838	= bu. (2150.42 cu. in.)
Hectoliters	×	.1308	= cu. yds.
Hectoliters	×	26.42	= gals. (231 cu. in.)

Weight

Grams	×	15.432	= grains (Act. Cong.)
Grams	÷	981.	= dynes
Grams (water)	÷	29.57	= fl. oz.
Grams	÷	28.35	= oz. avoirdupois
Kilograms	×	2.2046	= lbs.

Weight

Kilograms	×	35.27	= oz. avoirdupois

Kilograms	×	.0011023	= tons (2000 lbs.)
Tonneau (Metric ton)	×	1.1023	= tons (2000 lbs.)
Tonneau (Metric ton)	×	2204.6	= lbs.

Unit Weight

Grams per cu. cent.	÷	27.68	= lbs. per cu. in.
Kilo per meter	×	.672	= lbs. per ft.
Kilo per cu. meter	×	.06243	= lbs. per cu. ft.
Kilo per Cheval	×	2.235	= lbs. per h. p.
Grams per liter	×	.06243	= lbs. per cu. ft.

Pressure

Kilograms per sq. cm.	×	14.223	= lbs. per sq. in.
Kilograms per sq. cm.	×	32.843	= ft. of water (60°F.)
Atmospheres (International)	×	14.696	= lbs. per sq. in.

Energy

Joule	×	.7376	= ft. lbs.
Kilogram meters	×	7.233	= ft. lbs.

Power

Cheval vapeur	×	.9863	= h. p.
Kilowatts	×	1.341	= h. p.
Watts	÷	746.	= h. p.
Watts	×	.7373	= ft. lbs. per sec

Standard Tables of Metric Measure

Linear Measure

Unit	Value in meters	Symbol or Abbrev.
Micron	0.000001	μ
Millimeter	0.001	mm.
Centimeter	0.01	cm.
Decimeter	0.1	dm.
Meter (unit)	1.0	m.
Dekameter	10.0	dkm.
Hectometer	100.0	hm.
Kilometer	1,000.0	km.
Myriameter	10,000.0	Mm.
Megameter	1,000,000.0	

Volume

Unit	Value in liters	Symbol or Abbrev.
Milliliter	0.001	ml.
Centiliter	0.01	cl.
Deciliter	0.1	dl.
Liter (unit)	1.0	l.
Dekaliter	10.0	dkl.
Hectoliter	100.0	hl.
Kiloliter	1,000.0	kl.

Surface Measure

Unit	Value in square meters	Symbol or Abbrev.
Square millimeter	0.000001	mm.^{2}
Square centimeter	0.0001	cm.^{2}
Square decimeter	0.01	dm.^{2}
Square meter (centiare)	1.0	m.^{2}
Square dekameter (are)	100.0	a.^{2}
Hectare	10,000.0	ha.^{2}
Square kilometer	1,000,000.0	km.^{2}

Mass

Unit	Value in grams	Symbol or Abbrev.
Microgram	0.000001	μg.
Milligram	0.001	mg.
Centigram	0.01	cg.
Decigram	0.1	dg.
Gram (unit)	1.0	g.
Dekagram	10.0	dkg.
Hectogram	100.0	hg.
Kilogram	1,000.0	kg.
Myriagram	10,000.0	Mg.
Quintal	100,000.0	q.
Ton	1,000,000.0	

Cubic Measure

Unit	Value in cubic meters	Symbol or Abbrev.
Cubic micron	10^{-10}	μ^3
Cubic millimeter	10^{-9}	$mm.^3$
Cubic centimeter	10^{-6}	$cm.^3$
Cubic decimeter	10^{-3}	$dm.^3$
Cubic meter	1	$m.^3$
Cubic dekameter	10^3	$dkm.^3$
Cubic hectometer	10^6	$hm.^3$
Cubic kilometer	10^9	$km.^3$

Metric Ball Bearing Dimensions

Extra Light "100" Series

Basic Bearing Number	Bore		O.D.		Width	
	mm	Inches	mm	Inches	mm	Inches
100	10	0.3937	26	1.0236	8	0.3150
101	12	0.4724	28	1.1024	8	0.3150
102	15	0.5906	32	1.2598	9	0.3543
103	17	0.6693	35	1.3780	10	0.3937
104	20	0.7874	42	1.6535	12	0.4724
105	25	0.9843	47	1.8504	12	0.4724
106	30	1.1811	55	2.1654	13	0.5118
107	35	1.3780	62	2.4409	14	0.5512
108	40	1.5748	68	2.6772	15	0.5906
109	45	1.7717	75	2.9528	16	0.6299
110	50	1.9685	80	3.1496	16	0.6299
111	55	2.1654	90	3.5433	18	0.7087
112	60	2.3622	95	3.7402	18	0.7087
113	65	2.5591	100	3.9370	18	0.7087
114	70	2.7559	110	4.3307	20	0.7874
115	75	2.9528	115	4.5276	20	0.7874
116	80	3.1496	125	4.9213	22	0.8661
117	85	3.3465	130	5.1181	22	0.8661
118	90	3.5433	140	5.5118	24	0.9449
119	95	3.7402	145	5.7087	24	0.9449
120	100	3.9370	150	5.9055	24	0.9449
121	105	4.1339	160	6.2992	26	1.0236

Metric Ball Bearing Dimensions

Light "200" Series

Basic Bearing Number	Bore		O.D.		Width	
	mm	Inches	mm	Inches	mm	Inches
200	10	0.3937	30	1.1811	9	0.3543
201	12	0.4724	32	1.2598	10	0.3937
202	15	0.5906	35	1.3780	11	0.4331
203	17	0 6693	40	1.5748	12	0.4724
204	20	0.7874	47	1.8504	14	0.5512
205	25	0.9843	52	2.0472	15	0.5906
206	30	1.1811	62	2.4409	16	0.6299
207	35	1.3780	72	2.8346	17	0.6653
208	40	1.5748	80	3.1496	18	0.7087
209	45	1.7717	85	3.3465	19	0.7480
210	50	1.9685	90	3.5433	20	0.7874
211	55	2.1654	100	3.9370	21	0.8268
212	60	2.3633	110	4.3307	22	0.8661
213	65	2.5591	120	4.7244	23	0.9055
214	70	2.7559	125	4.9213	24	0.9449
215	75	2.9528	130	5.1181	25	0.9843
216	80	3.1496	140	5.5118	26	1.0236
217	85	3.3465	150	5.9055	28	1.1024
218	90	3.5433	160	6.2992	30	1.1811
219	95	3.7402	170	6.6929	32	1.2598
220	100	3.9370	180	7.0866	34	1.3386
221	105	4.1339	190	7.4803	36	1.4137
222	110	4.3307	200	7.8740	38	1.4961

Metric Ball Bearing Dimensions

Medium "300" Series

Basic Bearing Number	Bore		O.D.		Width	
	mm	Inches	mm	Inches	mm	Inches
300	10	0.0397	35	1.3780	11	0.4331
301	12	0.4724	37	1.4567	12	0.4724
302	15	0.5906	42	1.6535	13	0.5118
303	17	0.6693	47	1.8504	14	0.5512
304	20	0.7874	52	2.0472	15	0.5906

Metric Ball Bearing Dimensions

Medium "300" Series (Cont'd)

Basic Bearing Number	Bore		O.D.		Width	
	mm	Inches	mm	Inches	mm	Inches
305	25	0.9843	62	2.4409	17	0.6693
306	30	1.1811	72	2.8346	19	0.7480
307	35	1.3780	80	3.1496	21	0.8268
308	40	1.5748	90	3.5433	23	0.9055
309	45	1.7717	100	3.9370	25	0.9843
310	50	1.9685	110	4.3307	27	1.0630
311	55	2.1654	120	4.7244	29	1.1417
312	60	2.3622	130	5.1181	31	1.2205
313	65	2.5591	140	5.5118	33	1.2992
314	70	2.7559	150	5.9055	35	1.3780
315	75	2.9528	160	6.2992	37	1.4567
316	80	3.1469	170	6.6929	39	1.5354
317	85	3.3465	180	7.0866	41	1.6142
318	90	3.5433	190	7.4803	43	1.6929
319	95	3.7402	200	7.8740	45	1.7717
320	100	3.9370	215	8.4646	47	1.8504
321	105	4.1339	225	8.8583	49	1.9291
322	110	4.3307	240	9.4480	50	1.9685
324	120	4.7244	260	10.2362	55	2.1654
326	130	5.1181	280	11.0236	58	2.2835
328	140	5.5118	300	11.8110	62	2.4409
330	150	5.9055	320	12.5984	65	2.5591
332	160	6.2992	340	13.3858	68	2.6772
334	170	6.6929	360	14.1732	72	2.8346
336	180	7.0866	380	14.9606	75	2.9528
338	190	7.4803	400	15.7480	78	3.0709
340	200	7.8740	420	16.5354	80	3.1496
342	210	8.2677	440	17.3228	84	3.3071
344	220	8.6614	460	18.1002	88	3.4646
348	240	9.4488	500	19.6850	95	3.7402
352	260	10.2362	540	21.2598	102	4.0157
356	280	11.0236	580	22.8346	108	4.2520

Metric Ball Bearing Dimensions

Heavy "400" Series

Basic Bearing Number	Bore		O.D.		Width	
	mm	Inches	mm	Inches	mm	Inches
403	17	0.6693	62	2.4409	17	0.6693
404	20	0.7874	72	2.8345	19	0.7480
405	25	0.9843	80	3.1496	21	0.8268
406	30	1.1811	90	3.5433	23	0.9055
407	35	1.3780	100	3.9370	25	0.9843
408	40	1.5748	110	4.3307	27	1.0630
409	45	1.7717	120	4.7244	29	1.1417
410	50	1.9685	130	5.1181	31	1.2205
411	55	2.1654	140	5.5118	33	1.2992
412	60	2.3622	150	5.9055	35	1.3780
413	65	2.5591	160	6.2992	37	1.4567
414	70	2.7559	180	7.0866	42	1.6535
415	75	2.9528	190	7.4803	45	1.7717
416	80	3.1496	200	7.8740	48	1.8898
417	85	3.3465	210	8.2677	52	2.0472
418	90	3.5433	225	8.8533	54	2.1260
419	95	3.7402	250	9.8425	55	2.1654
420	100	3.9370	265	10.4331	60	2.3622
421	105	4.1339	290	11.4173	65	2.5591
422	110	4.3307	320	12.5984	70	2.5759

Tap Drill Sizes for American Standard Threads

Diam. of Thread	Threads per Inch	Drill*	Decimal Equiv.
No. 0-.060	80 NF	3/64	.0469
1-.073	64 NC	1.5 MM	.0591
	72 NF	53	.0595
2-.086	56 NC	50	.0700
	64 NF	50	.0700
3-.099	48 NC	5/64	.0781
	56 NF	45	.0820
4-.112	40 NC	43	.0890
	48 NF	42	.0935
5-.125	40 NC	38	.1015
	44 NF	37	.1040
6-.138	32 NC	36	.1065
	40 NF	33	.1130
8-.164	32 NC	29	.1360
	36 NF	29	.1360
10-.190	24 NC	25	.1495
	32 NF	21	.1590
12-.216	24 NC	16	.1770
	28 NF	14	.1820
1/4	20 NC	7	.2010
	28 NF	3	.2130
	32 NEF	7/32	.2188
5/16	18 NC	F	.2570
	24 NF	1	.2720
	32 NEF	9/32	.2812
3/8	16 NC	5/16	.3125
	24 NF	Q	.3320
	32 NEF	11/32	.3438

*To produce approximately 75% full thread.

Tap Drill Sizes for American Standard Threads (Cont'd)

Diam. of Thread	Threads per Inch	Drill*	Decimal Equiv.
7/16	14 NC	U	.3680
	20 NF	25/64	.3906
	28 NEF	Y	.4040
1/2	12 N	27/64	.4219
	13 NC	27/64	.4219
	20 NF	29/64	.4531
	28 NEF	15/32	.4687
9/16	12 NC	31/64	.4844
	18 NF	33/64	.5156
	24 NEF	33/64	.5156
5/8	11 NC	17/32	.5312
	12 N	25/64	.5469
	18 NF	14.5 MM	.5709
	24 NEF	37/64	.5781
11/16	12 N	39/64	.6094
	24 NEF	16.5 MM	.6496
	10 NC	16.5 MM	.6496
3/4	12 N	17 MM	.6693
	16 NF	17.5 MM	.6890
	20 NEF	45/64	.7031
13/16	12 N	18.5 MM	.7283
	16 N	3/4	.7500
	20 NEF	49/64	.7656
7/8	9 NC	49/64	.7656
	12 N	20 MM	.7874
	14 NF	25.5 MM	.8071
	16 N	13/16	.8125
	20 NEF	21 MM	.8268

*To produce approximately 75% full thread.

Tap Drill Sizes for American Standard Threads (Cont'd)

Diam. of Thread	Threads per Inch	Drill*	Decimal Equiv.
15/16	12 N	55/64	.8594
	16 N	7/8	.8750
	20 NEF	22.5 MM	.8858
1	8 NC	7/8	.8750
	12 N	59/64	.9219
	14 NF	23.5 MM	.9252
	16 N	15/16	.9375
	20 NEF	61/64	.9531
1 1/2	6 NC	1 21/64	1.3281
	8 N	1 3/8	1.3750
	12 NF	36 MM	1.4173
	16 N	1 7/16	1.4375
	18 NEF	1 29/64	1.4531
2	4 1/2 NC	1 25/32	1.7812
	8 N	1 7/8	1.8750
	12 N	1 59/64	1.9219
	16 NEF	1 15/16	1.9375
2 1/2	4 NC	2 1/4	2.2500
	8 N	2 3/8	2.3750
	12 N	61.5 MM	2.4213
	16 N	2 7/16	2.4375
3	4 NC	2 3/4	2.7500
	8 N	2 7/8	2.8750
	12 N	74 MM	2.9134
	16 N	2 15/16	2.9375

*To produce approximately 75% full thread.

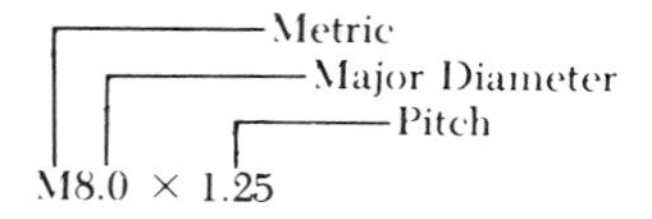

International Standard—Metric Thread Dimensions and Tap Drill Sizes

Major Diameter m/m	Pitch m/m	Minor Diameter m/m	Pitch Diameter m/m	Tap Drill for 75% Thread m/m	Tap Drill for 75% Thread No. of Inches	Clearance Drill Size
2.0	.40	1.48	1.740	1.6	1/16	41
2.3	.40	1.78	2.040	1.9	48	36
2.6	.45	2.02	2.308	2.1	45	31
3.0	.50	2.35	2.675	2.5	40	29
3.5	.60	2.72	3.110	2.9	33	23
4.0	.70	3.09	3.545	3.3	30	16
4.5	.75	3.53	4.013	3.75	26	10
5.0	.80	3.96	4.480	4.2	19	3
5.5	.90	4.33	4.915	4.6	14	15/64"
6.0	1.00	4.70	5.350	5.0	9	1/4"
7.0	1.00	5.70	6.350	6.0	15/64"	19/64"
8.0	1.25	6.38	7.188	6.8	H	11/32"
9.0	1.25	7.38	8.188	7.8	5/16"	3/8"
10.0	1.50	8.05	9.026	8.6	R	27/64"
11.0	1.50	9.05	10.026	9.6	V	29/64"
12.0	1.75	9.73	10.863	10.5	Z	1/2"
14.0*	1.25	12.38	13.188	13.0	33/64"	9/16"
14.0	2.00	11.40	12.701	12.0	15/32"	9/16"
16.0	2.00	13.40	14.701	14.0	35/64"	21/32"
18.0*	1.50	16.05	17.026	16.5	41/64"	47/64"
18.0	2.50	14.75	16.376	15.5	39/64"	47/64"
20.0	2.50	16.75	18.376	17.5	11/16"	13/16"
22.0	2.50	18.75	20.376	19.5	49/64"	57/64"
24.0	3.00	20.10	22.051	21.0	53/64"	31/32"
27.0	3.00	23.10	25.051	24.0	15/16"	1 3/32"
30.0	3.50	25.45	27.727	26.5	1 3/64"	1 13/64"
33.0	3.50	28.45	30.727	29.5	1 11/64"	1 21/64"
36.0	4.00	30.80	33.402	32.0	1 17/64"	1 7/16"
39.0	4.00	33.80	36.402	35.0	1 3/8"	1 9/16"
42.0	4.50	36.15	39.077	37.0	1 29/64"	1 43/64"
45.0	4.50	39.15	42.077	40.0	1 37/64"	1 13/16"
48.0	5.00	41.50	44.752	43.0	1 11/16"	1 29/32"

*Special spark-plug sizes.

Fractional Number and Letter Drill Sizes in Decimals

Drill Size	Decimal	Drill Size	Decimal	Drill Size	Decimal	Drill Size	Decimal
80	.0135	42	.0935	13/64	.2031	X	.3970
79	.0145	3/32	.0938	6	.2040	Y	.4040
1/64	.0156	41	.0960	5	.2055	13/32	.4062
78	.0160	40	.0980	4	.2090	Z	.4130
77	.0180	39	.0995	3	.2130	27/64	.4219
76	.0200	38	.1015	7/32	.2188	7/16	.4375
75	.0210	37	.1040	2	.2210	29/64	.4531
74	.0225	36	.1065	1	.2280	15/32	.4688
73	.0240	7/64	.1094	A	.2340	31/64	.4844
72	.0250	35	.1100	15/64	.2344	1/2	.5000
71	.0260	34	.1110	B	.2380	33/64	.5156
70	.0280	33	.1130	C	.2420	17/32	.5312
69	.0292	32	.1160	D	.2460	35/64	.5469
68	.0310	31	.1200	1/4	.2500	9/16	.5625
1/32	.0312	1/8	.1250	E	.2500	37/64	.5781
67	.0320	30	.1285	F	.2570	19/32	.5938
66	.0330	29	.1360	G	.2610	39/64	.6094
65	.0350	28	.1405	17/64	.2656	5/8	.6250
64	.0360	9/64	.1406				
63	.0370	27	.1440	H	.2660	41/64	.6406
62	.0380	26	.1470	I	.2720	21/32	.6562
61	.0390	25	.1495	J	.2770	43/64	.6719
60	.0400	24	.1520	K	.2810	11/16	.6875
59	.0410	23	.1540	9/32	.2812	45/64	.7031
58	.0420	5/32	.1562	L	.2900	23/32	.7188
57	.0430	22	.1570	M	.2950	47/64	.7344
56	.0465	21	.1590	19/64	.2969	3/4	.7500
3/64	.0469	20	.1610	N	.3020	49/64	.7656
55	.0520	19	.1660	5/16	.3125	25/32	.7812
54	.0550	18	.1695	O	.3160	51/64	.7969
53	.0595	11/64	.1719	P	.3230	13/16	.8125
1/16	.0625	17	.1730	21/64	.3281	53/64	.8281
52	.0635	16	.1770	Q	.3320	27/32	.8438
51	.0670	15	.1800	R	.3390	55/64	.8594
50	.0700	14	.1820	11/32	.3438	7/8	.8750
49	.0730	13	.1850	S	.3480	57/64	.8906
48	.0760	3/16	.1875	T	.3580	29/32	.9062
5/64	.0781	12	.1890	23/64	.3594	59/64	.9219
47	.0785	11	.1910	U	.3680	15/16	.9375
46	.0810	10	.1935	3/8	.3750	61/64	.9531
45	.0820	9	.1960	V	.3770	31/32	.9688
44	.0860	8	.1990	X	.3970	63/64	.9844
43	.0890	7	.2010	25/64	.3906	1	1.0000

Keyway Data

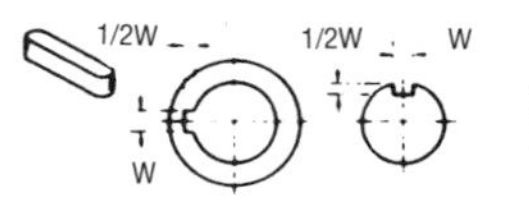

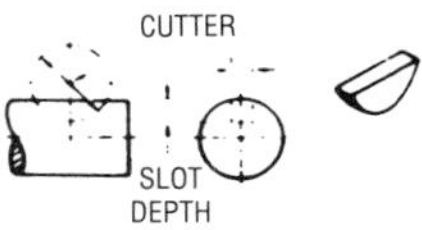

Shaft dia.	Square keyways	Woodruff keyways*			
		Key no.	Thickness	Cutter dia.	Slot depth
0.500	1/8 × 1/16	404	0.1250	0.500	0.1405
0.562	1/8 × 1/16	404	0.1250	0.500	0.1405
0.625	5/32 × 5/64	505	0.1562	0.625	0.1669
0.688	3/16 × 3/32	606	0.1875	0.750	0.2193
0.750	3/16 × 3/32	606	0.1875	0.750	0.2193
0.812	3/16 × 3/32	606	0.1875	0.750	0.2193
0.875	7/32 × 7/64	607	0.1875	0.875	0.2763
0.938	1/4 × 1/8	807	0.2500	0.875	0.2500
1.000	1/4 × 1/8	808	0.2500	1.000	0.3130
1.125	5/16 × 5/32	1009	0.3125	1.125	0.3228
1.250	5/16 × 5/32	1010	0.3125	1.250	0.3858
1.375	3/8 × 3/16	1210	0.3750	1.250	0.3595
1.500	3/8 × 3/16	1212	0.3750	1.500	0.4535
1.625	3/8 × 3/16	1212	0.3750	1.500	0.4535
1.750	7/16 × 7/32				
1.875	1/2 × 1/4				
2.000	1/2 × 1/4				
2.250	5/8 × 5/16				
2.500	5/8 × 5/16				
2.750	3/4 × 3/8				
3.000	3/4 × 3/8				
3.250	3/4 × 3/8				
3.500	7/8 × 7/16				
4.000	1 × 1/2				

*The depth of a Woodruff Keyway is measured from the edge of the slot.

Dimensions of Standard Gib-head Keys, Square and Flat

Approved by ANSI

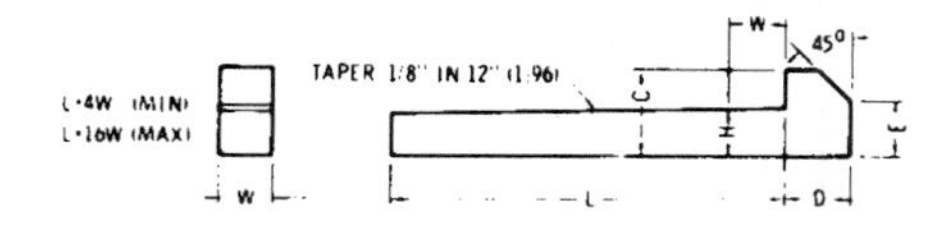

Diameters	Square Type					Flat Type				
	Key		Gib head			Key		Gib head		
of Shafts	W	H	C	D	E	W	H	C	D	E
1/2–9/16	1/8	1/8	1/4	7/32	5/32	1/8	3/32	3/16	1/8	1/8
5/8–7/8	3/16	3/16	5/16	9/32	7/32	3/16	1/8	1/4	3/16	5/32
15/16–1 1/4	1/4	1/4	7/16	11/32	11/32	1/4	3/16	5/16	1/4	3/16
1 5/16–1 3/8	5/16	5/16	9/16	13/32	13/32	5/16	1/4	3/8	5/16	1/4
1 7/16–1 3/4	3/8	3/8	11/16	15/32	15/32	3/8	1/4	7/16	3/8	5/16
1 13/16–2 1/4	1/2	1/2	7/8	19/32	5/8	1/2	3/8	5/8	1/2	7/16
2 5/16–2 3/4	5/8	5/8	1 1/16	23/32	3/4	5/8	7/16	3/4	5/8	1/2
2 7/8–3 1/4	3/4	3/4	1 1/4	7/8	7/8	3/4	1/2	7/8	3/4	5/8
3 3/8–3 3/4	7/8	7/8	1 1/2	1	1	7/8	5/8	1 1/16	7/8	3/4
3 7/8–4 1/2	1	1	1 3/4	1 3/16	1 3/16	1	3/4	1 1/4	1	13/16
4 3/4–5 1/2	1 1/4	1 1/4	2	1 7/16	1 7/16	1 1/4	7/8	1 1/2	1 1/4	1
5 3/4–6	1 1/2	1 1/2	2 1/2	1 3/4	1 3/4	1 1/2	1	1 3/4	1 1/2	1 1/4

*ANSI B17.1—1934. Dimensions in inches.

Taper Pins

All sizes have a taper of 0.250 per foot

TAPER PINS

Size no. of pin	Length of pin	Large end of pin	Small end of reamer	Drill size for reamer
0	1	0.156	0.135	28
1	1 1/4	0.172	0.146	25
2	1 1/2	0.193	0.162	19
3	1 3/4	0.219	0.183	12
4	2	0.250	0.208	3
5	2 1/4	0.289	0.242	1/4
6	3 1/4	0.341	0.279	9/32
7	3 3/4	0.409	0.331	11/32
8	4 1/2	0.492	0.398	13/32
9	5 1/4	0.591	0.482	31/64
10	6	0.706	0.581	19/32
11	7 1/4	0.857	0.706	23/32
12	8 3/4	1.013	0.842	55/64

NEMA Motor Frame Dimensions

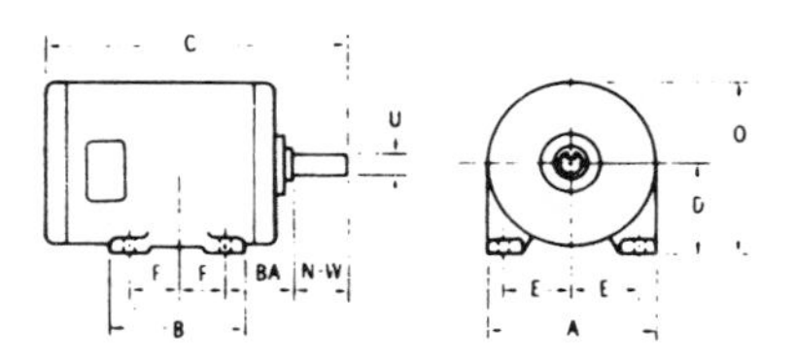

U Nema Motor Frame Dimensions

Horsepower Rating			U Frame No.	U	Shaft Keyseat		Key Length	N-W	A Max.	B Max.	C	D	E	F	BA	O
3600	1800	1200			Width	Depth										
1 1/2	1	3/4	182	7/8	3/16	3/32	1 3/8	2 1/4	8 7/8	6 1/2	12 1/8	4 1/2	3 3/4	2 1/4	2 3/4	8 15/16
2 & 3	1 1/2 & 2	1 & 1 1/2	184	7/8	3/16	3/32	1 3/8	2 1/4	8 7/8	7 1/2	13 1/8	4 1/2	3 3/4	2 3/4	2 3/4	8 15/16
5	3	2	213	1 1/8	1/4	1/8	2	3	10 3/8	7 1/2	15 5/16	5 1/4	4 1/4	2 3/4	3 1/2	10 7/16
7 1/2	5	3	215	1 1/8	1/4	1/8	2	3	10 3/8	9	16 13/16	5 1/4	4 1/4	3 1/2	3 1/2	10 7/16
10	7 1/2	5	254U	1 3/8	5/16	5/32	2 3/4	3 3/4	12 7/16	10 3/4	20 1/4	6 1/4	5	4 1/8	4 1/4	12 1/2
15	10	7 1/2	256U	1 3/8	5/16	5/32	2 3/4	3 3/4	12 7/16	12 1/2	22	6 1/4	5	5	4 1/4	12 1/2
20	15	10	284U	1 5/8	3/8	3/16	3 3/4	4 7/8	13 7/8	12 1/2	23 11/16	7	5 1/2	4 3/4	4 3/4	13 15/16
25	20	....	286U	1 5/8	3/8	3/16	3 3/4	4 7/8	13 7/8	12 1/2	25 3/16	7	5 1/2	5 1/2	4 3/4	13 15/16
....	25	15	324U	1 7/8	1/2	1/4	4 1/4	5 5/8	15 7/8	14	26 3/8	8	6 1/4	5 1/4	5 1/4	15 15/16
30	....	....	324S	1 5/8	3/8	3/16	1 7/8	3 1/4	15 7/8	14	24	8	6 1/4	5 1/4	5 1/4	15 15/16
....	30	20	326U	1 7/8	1/2	1/4	4 1/4	5 5/8	15 7/8	14	27 7/8	8	6 1/4	6	5 1/4	15 15/16
40	....	....	326S	1 5/8	3/8	3/16	1 7/8	3 1/4	15 7/8	14	25 1/2	8	6 1/4	6	5 1/4	15 15/16
....	40	25	364U	2 1/8	1/2	1/4	5	6 3/8	17 5/8	15 1/4	29 3/16	9	7	5 5/8	5 7/8	17 13/16
50	....	....	364US	1 7/8	1/2	1/4	2	3 3/4	17 5/8	15 1/4	26 9/16	9	7	5 5/8	5 7/8	17 13/16
....	50	30	365U	2 1/8	1/2	1/4	5	6 3/8	17 5/8	16 1/4	30 3/16	9	7	6 1/8	5 7/8	17 13/16
60	....	....	365US	1 7/8	1/2	1/4	2	3 3/4	19 3/4	16 1/4	27 9/16	9	7	6 1/8	5 7/8	17 13/16
....	60	40	404U	2 3/8	5/8	5/16	5 1/2	7 1/8	19 3/4	16 1/4	32 7/16	10	8	6 1/8	6 5/8	19 7/8
75	....	....	404US	2 1/8	1/2	1/4	2 3/4	4 1/4	19 3/4	16 1/4	39 9/16	10	8	6 1/8	6 5/8	19 7/8
....	75	50	405U	2 3/8	5/8	5/16	5 1/2	7 1/8	19 3/4	17 1/4	33 15/16	10	8	6 7/8	6 5/8	19 7/8
100	....	....	405US	2 1/8	1/2	1/4	2 3/4	4 1/4	19 3/4	17 1/4	31 1/16	10	8	6 7/8	6 5/8	19 7/8

T Nema Motor Frame Dimensions

Horsepower Rating			T Frame No.	U	Shaft Keyseat		Key Length	N-W	A Max.	B Max.	C	D	E	F	BA	O
3600	1800	1200			Width	Depth										
1 1/2	1	3/4	143T	7/8	3/16	3/32	1 3/8	2 1/4	7	6	12 5/8	3 1/2	2 3/4	2	2 1/4	7
2 & 3	1 1/2 & 2	1	145T	7/8	3/16	3/32	1 3/8	2 1/4	7	6	12 5/8	3 1/2	2 3/4	2 1/2	2 1/4	7
5	3	1 1/2	182T	1 1/8	1/4	1/8	1 3/4	2 3/4	9	6 1/2	12 3/4	4 1/2	3 3/4	2 1/4	2 3/4	9
7 1/2	5	2	184T	1 1/8	1/4	1/8	1 3/4	2 3/4	9	7 1/2	13 3/4	4 1/2	3 3/4	2 3/4	2 3/4	9
10	7 1/2	3	213T	1 3/8	5/16	5/32	2 3/8	3 3/8	10 1/2	7 1/2	15 13/16	5 1/4	4 1/4	2 3/4	3 1/2	10 1/2
15	10	5	215T	1 3/8	5/16	5/32	2 3/8	3 3/8	10 1/2	9	17 5/16	5 1/4	4 1/4	3 1/2	3 1/2	10 1/2
20	15	7 1/2	254T	1 5/8	3/8	3/16	2 7/8	4	12 1/2	10 3/4	20 1/2	6 1/4	5	4 1/8	4 1/4	12 1/2
25	20	10	256T	1 5/8	3/8	3/16	2 7/8	4	12 1/2	12 1/2	22 1/4	6 1/4	5	5	4 1/4	12 1/2
....	25	15	284T	1 7/8	1/2	1/4	3 1/4	4 5/8	14	12 1/2	23 5/16	7	5 1/2	4 3/4	4 3/4	14
30	25	15	284TS	1 5/8	3/8	3/16	1 7/8	3 1/4	14	12 1/2	22	7	5 1/2	4 3/4	4 3/4	14
....	30	20	286T	1 7/8	1/2	1/4	3 1/4	4 5/8	14	14	24 7/8	7	5 1/2	5 1/2	4 3/4	14
40	30	20	286TS	1 5/8	3/8	3/16	1 7/8	3 1/4	14	14	23 1/2	7	5 1/2	5 1/2	4 3/4	14
....	40	25	324T	2 1/8	1/2	1/4	3 7/8	5 1/4	16	14	26 1/2	8	6 1/4	5 1/4	5 1/4	16
50	40	25	324TS	1 7/8	1/2	1/4	2	3 3/4	16	14	24 5/8	8	6 1/4	5 1/4	5 1/4	16
....	50	30	326T	2 1/8	1/2	1/4	3 7/8	5 1/4	16	15 1/2	27 3/4	8	6 1/4	6	5 1/4	16
60	50	30	326TS	1 7/8	1/2	1/4	2	3 3/4	16	15 1/2	26 1/8	8	6 1/4	6	5 1/4	16
....	60	40	364T	2 3/8	5/8	5/16	4 1/4	5 7/8	18	15 1/4	28 3/4	9	7	5 5/8	5 7/8	18
75	60	....	364TS	1 7/8	1/2	1/4	2	3 3/4	18	15 1/4	26 9/16	9	7	5 5/8	5 7/8	18
....	75	50	365T	2 3/8	5/8	5/16	4 1/4	5 7/8	18	16 1/4	29 3/4	9	7	6 1/8	5 7/8	18
100	75	....	365TS	1 7/8	1/2	1/4	2	3 3/4	18	16 1/4	27 9/16	9	7	6 1/8	5 7/8	18

Wire Gauge Standards

Decimal parts of an inch							
Wire gauge no.	American or Brown & Sharpe	Birmingham or Stubs wire	Washburn & Moen on steel wire gauge	American S. & W. Co.'s music wire	Imperial wire gauge	Stubs steel wire	U.S. standard for plate
00000	0.516549	0.500	0.4305	0.005	4.432	.....	0.43775
0000	0.460	0.454	0.3938	0.006	0.400	.....	0.40625
000	0.40964	0.425	0.3625	0.007	0.372	.....	0.375
00	0.3648	0.380	0.3310	0.008	0.348	.....	0.34375
0	0.32486	0.340	0.3065	0.009	0.324	.....	0.3125
1	0.2893	0.300	0.2830	0.010	0.300	0.227	0.28125
2	0.25763	0.284	0.2625	0.011	0.276	0.219	0.265625
3	0.22942	0.259	0.2437	0.012	0.252	0.212	0.250
4	0.20431	0.238	0.2253	0.013	0.232	0.207	0.234375
5	0.18194	0.220	0.2070	0.014	0.212	0.204	0.21875
6	0.16202	0.203	0.1920	0.016	0.192	0.201	0.203125
7	0.14428	0.180	0.1770	0.018	0.176	0.199	0.1875
8	0.12849	0.165	0.1620	0.020	0.160	0.197	0.171875
9	0.11443	0.148	0.1483	0.022	0.144	0.194	0.15625
10	0.10189	0.134	0.1350	0.024	0.128	0.191	0.140625
11	0.090742	0.120	0.1205	0.026	0.116	0.188	0.125
12	0.080808	0.109	0.1055	0.029	0.104	0.185	0.109375
13	0.071961	0.095	0.0915	0.031	0.092	0.182	0.09375
14	0.064084	0.083	0.0800	0.033	0.080	0.180	0.078125
15	0.057068	0.072	0.0720	0.035	0.072	0.178	0.0703125
16	0.05082	0.065	0.0625	0.037	0.064	0.175	0.0625
17	0.045257	0.058	0.0540	0.039	0.056	0.172	0.05625
18	0.040303	0.049	0.0475	0.041	0.048	0.168	0.050
19	0.03589	0.042	0.0410	0.043	0.040	0.164	0.04375
20	0.031961	0.035	0.0348	0.045	0.036	0.161	0.0375
21	0.028462	0.032	0.0317	0.047	0.032	0.157	0.034375
22	0.025347	0.028	0.0286	0.049	0.028	0.155	0.03125
23	0.022571	0.025	0.0258	0.051	0.024	0.153	0.028125
24	0.0201	0.022	0.0230	0.055	0.022	0.151	0.025
25	0.0179	0.020	0.0204	0.059	0.020	0.148	0.021875
26	0.01594	0.018	0.0181	0.063	0.018	0.146	0.01875
27	0.014195	0.016	0.0173	0.067	0.0164	0.143	0.0171875
28	0.012641	0.014	0.0162	0.071	0.0149	0.139	0.015625
29	0.011257	0.013	0.0150	0.075	0.0136	0.134	0.0140625
30	0.010025	0.012	0.0140	0.080	0.0124	0.127	0.0125

Wire Gauge Standards (Cont'd)

Decimal parts of an inch							
Wire gauge no.	American or Brown & Sharpe	Birmingham or Stubs wire	Washburn & Moen on steel wire gauge	American S. & W. Co.'s music wire	Imperial wire gauge	Stubs steel wire	U.S. standard for plate
31	0.008928	0.010	0.0132	0.085	0.0116	0.120	0.0109375
32	0.00795	0.009	0.0128	0.090	0.0108	0.115	0.01015625
33	0.00708	0.008	0.0118	0.095	0.0100	0.112	0.009375
34	0.006304	0.007	0.0104	....	0.0092	0.110	0.00859375
35	0.005614	0.005	0.0095	....	0.0084	0.108	0.0078125
36	0.005	0.004	0.0090	....	0.0076	0.106	0.00703125
37	0.004453	....	0.0085	....	0.0068	0.103	0.006640625
38	0.003965	....	0.0080	....	0.0060	0.101	0.00625
39	0.003531	....	0.0075	....	0.0052	0.099	
40	0.003144	....	0.0070	....	0.0048	0.097	

Metal Weights

Material	Chemical Symbol	Weight, in Pounds Per Cubic Inch	Weight, in Pounds Per Cubic Foot
Aluminum	Al	0.093	160
Antimony	Sb	0.2422	418
Brass	—	0.303	524
Bronze	—	0.320	552
Chromium	Cr	0.2348	406
Copper	Cu	0.323	450
Gold	Au	0.6975	1205
Iron (cast)	Fe	0.260	450
Iron (wrought)	Fe	0.2834	490
Lead	Pb	0.4105	710
Manganese	Mn	0.2679	463
Mercury	Hg	0.491	849
Molybdenum	Mo	0.309	534
Monel	—	0.318	550
Platinum	Pt	0.818	1413
Steel (mild)	Fe	0.2816	490
Steel (stainless)	—	0.277	484
Tin	Sn	0.265	459
Titanium	Ti	0.1278	221
Zinc	Zn	0.258	446

Commercial Pipe Sizes and Wall Thicknesses

The following table lists the pipe sizes and wall thicknesses currently established as standard, or specifically:

1. The traditional standard weight, extra strong, and double extra strong pipe.
2. The pipe wall thickness schedules listed in ANSI B36.10, which are applicable to carbon steel and alloys other than stainless steels.
3. The pipe wall thickness schedules listed in ANSI B36.19, which are applicable only to stainless steels.

Nominal Pipe Size	Outside Diam.	Nominal Wall Thickness For													
		Sched. 5	Sched. 10	Sched. 20	Sched. 30	Standard	Sched. 40	Sched. 60	Extra Strong	Sched. 80	Sched. 100	Sched. 120	Sched. 140	Sched. 160	XX Strong
1/8	0.405	—	0.049	—	—	0.068	0.068	—	0.095	0.095	—	—	—	—	—
1/4	0.540	—	0.065	—	—	0.088	0.086	—	0.119	0.119	—	—	—	—	—
3/8	0.675	—	0.065	—	—	0.091	0.091	—	0.126	0.126	—	—	—	—	—
1/2	0.840	—	0.083	—	—	0.109	0.109	—	0.147	0.147	—	—	—	0.187	0.294
3/4	1.050	0.065	0.083	—	—	0.113	0.113	—	0.154	0.154	—	—	—	0.218	0.308
1	1.315	0.065	0.109	—	—	0.133	0.133	—	0.179	0.179	—	—	—	0.250	0.358
1 1/4	1.660	0.065	0.109	—	—	0.140	0.140	—	0.191	0.191	—	—	—	0.250	0.382
1 1/2	1.900	0.065	0.109	—	—	0.145	0.145	—	0.200	0.200	—	—	—	0.281	0.400
2	2.375	0.065	0.109	—	—	0.154	0.154	—	0.218	0.218	—	—	—	0.343	0.436
2 1/2	2.875	0.083	0.120	—	—	0.203	0.203	—	0.276	0.276	—	—	—	0.375	0.552
3	3.5	0.083	0.120	—	—	0.216	0.216	—	0.300	0.300	—	—	—	0.438	0.600
3 1/2	4.0	0.083	0.120	—	—	0.226	0.226	—	0.318	0.318	—	—	—	—	—

Commercial Pipe Sizes and Wall Thicknesses (Cont'd)

Nominal Pipe Size	Out-side Diam.	Nominal Wall Thickness For													
		Sched. 5	Sched. 10	Sched. 20	Sched. 30	Stand-ard	Sched. 40	Sched. 60	Extra Strong	Sched. 80	Sched. 100	Sched. 120	Sched. 140	Sched. 160	XX Strong
4	4.5	0.083	0.120	—	—	0.237	0.237	—	0.337	0.337	—	0.438	—	0.531	0.674
5	5.563	0.109	0.134	—	—	0.258	0.258	—	0.375	0.375	—	0.500	—	0.625	0.750
6	6.625	0.109	0.134	—	—	0.280	0.280	—	0.432	0.432	—	0.562	—	0.718	0.864
8	8.625	0.109	0.148	0.250	0.277	0.322	0.322	0.406	0.500	0.500	0.593	0.718	0.812	0.906	0.875
10	10.75	0.134	0.165	0.250	0.307	0.365	0.365	0.500	0.500	0.593	0.713	0.843	1.000	1.125	—
12	12.75	0.156	0.180	0.250	0.330	0.375	0.406	0.562	0.500	0.687	0.843	1.000	1.125	1.312	—
14 O.D.	14.0	—	0.250	0.312	0.375	0.375	0.438	0.593	0.500	0.750	0.937	1.093	1.250	1.406	—
16 O.D.	16.0	—	0.250	0.312	0.375	0.375	0.500	0.656	0.500	0.843	1.031	1.218	1.438	1.593	—
18 O.D.	18.0	—	0.250	0.312	0.438	0.375	0.562	0.750	0.500	0.937	1.156	1.375	1.562	1.781	—
20 O.D.	20.0	—	0.250	0.375	0.500	0.375	0.593	0.812	0.500	1.031	1.281	1.500	1.750	1.968	—
22 O.D.	22.0	—	0.250	—	—	0.375	—	—	0.500	—	—	—	—	—	—
24 O.D.	24.0	—	0.250	0.375	0.562	0.375	0.687	0.968	0.500	1.218	1.531	1.812	2.062	2.343	—
26 O.D.	26.0	—	—	—	—	0.375	—	—	0.500	—	—	—	—	—	—
30 O.D.	30.0	—	0.312	0.500	0.625	0.375	—	—	0.500	—	—	—	—	—	—
34 O.D.	34.0	—	—	—	—	0.375	—	—	0.500	—	—	—	—	—	—
36 O.D.	36.0	—	—	—	—	0.375	—	—	0.500	—	—	—	—	—	—
42 O.D.	42.0	—	—	—	—	0.375	—	—	0.500	—	—	—	—	—	—

Pipe Hangers, Clips, Clamps, etc.

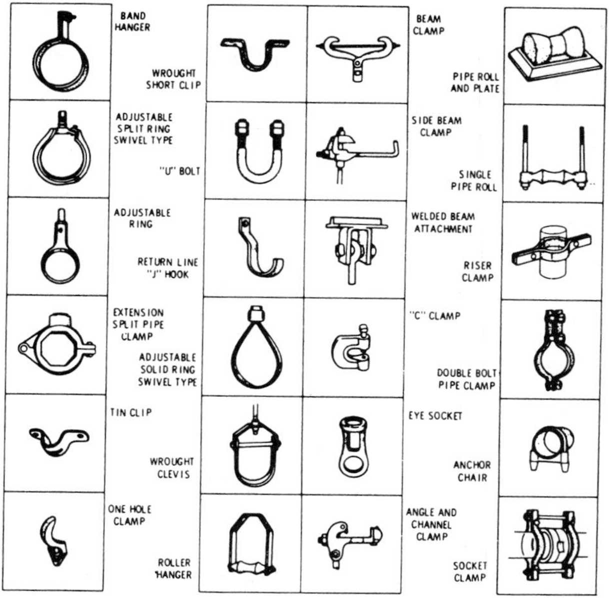

BAND HANGER

WROUGHT SHORT CLIP

BEAM CLAMP

PIPE ROLL AND PLATE

ADJUSTABLE SPLIT RING SWIVEL TYPE

"U" BOLT

SIDE BEAM CLAMP

SINGLE PIPE ROLL

ADJUSTABLE RING

RETURN LINE "J" HOOK

WELDED BEAM ATTACHMENT

RISER CLAMP

EXTENSION SPLIT PIPE CLAMP

ADJUSTABLE SOLID RING SWIVEL TYPE

"C" CLAMP

DOUBLE BOLT PIPE CLAMP

TIN CLIP

WROUGHT CLEVIS

EYE SOCKET

ANCHOR CHAIR

ONE HOLE CLAMP

ROLLER HANGER

ANGLE AND CHANNEL CLAMP

SOCKET CLAMP

Welding Symbols

TYPE OF WELD								FIELD WELD	WELD ALL AROUND	FLUSH
BEAD	FILLET	GROOVE					PLUG AND SLOT			
		SQUARE	V	BEVEL	U	J				

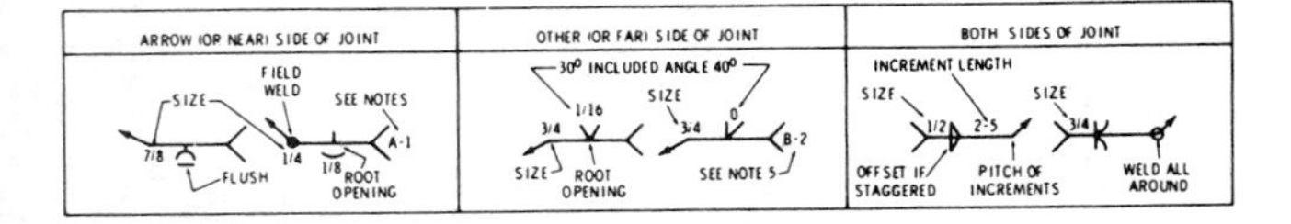

Bolting Dimensions for 150 lb. Flanges

Nom. Pipe Size	150 LB. STEEL FLANGES: Diam. of Bolt Circle	Diam. of Bolts	No. of Bolts	Length of Studs 1/4" Raised Face	Bolt Length for 125 Lb. Cast Iron Flanges
1/2	2 3/8	1/2	4	2 1/4	
3/4	2 3/4	1/2	4	2 1/4	
1	3 1/8	1/2	4	2 1/2	1 3/4
1 1/4	3 1/2	1/2	4	2 1/2	2
1 1/2	3 7/8	1/2	4	2 3/4	2
2	4 3/4	5/8	4	3	2 1/4
2 1/2	5 1/2	5/8	4	3 1/4	2 1/2
3	6	5/8	4	3 1/2	2 1/2
3 1/2	7	5/8	8	3 1/2	2 3/4
4	7 1/2	5/8	8	3 1/2	3
5	8 1/2	3/4	8	3 3/4	3
6	9 1/2	3/4	8	3 3/4	3 1/4
8	11 3/4	3/4	8	4	3 1/2
10	14 1/4	7/8	12	4 1/2	3 3/4
12	17	7/8	12	4 1/2	3 3/4

Blind Flanges

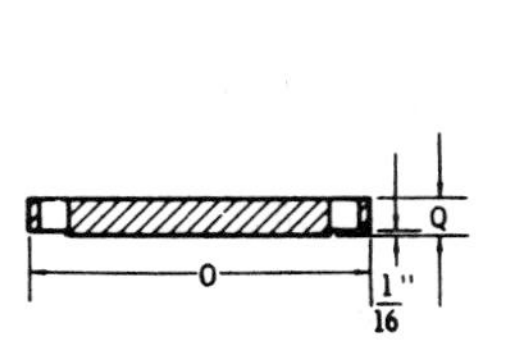

Nom. Pipe Size	150 Lb. Outside Diam. of Flange O	150 Lb. Thickness Q	300 Lb. Outside Diam. of Flange O	300 Lb. Thickness Q
1/2	3 1/2	7/16	3 3/4	9/16
3/4	3 7/8	1/2	4 5/8	5/8
1	4 1/4	9/16	4 7/8	11/16
1 1/4	4 5/8	5/8	5 1/4	3/4
1 1/2	5	11/16	6 1/8	13/16
2	6	3/4	6 1/2	7/8
2 1/2	7	7/8	7 1/2	1
3	7 1/2	15/16	8 1/4	1 1/8
3 1/2	8 1/2	15/16	9	1 3/16
4	9	15/16	10	1 1/4
5	10	15/16	11	1 3/8
6	11	1	12 1/2	1 7/16
8	13 1/2	1 1/8	15	1 5/8
10	16	1 3/16	17 1/2	1 7/8
12	19	1 1/4	20 1/2	2

Welding Neck Flanges

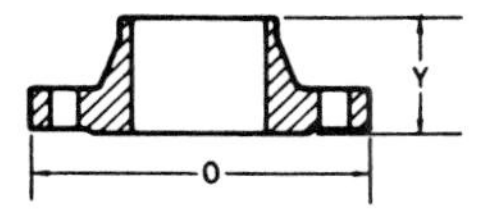

Nom. Pipe Size	150 Lb. Outside Diam. of Flange O	150 Lb. Length Thru Hub Y	300 Lb. Outside Diam. of Flange O	300 Lb. Length Thru Hub Y
1/2	3 1/2	1 7/8	3 3/4	2 1/16
3/4	3 7/8	2 1/16	4 5/8	2 1/4
1	4 1/4	2 3/16	4 7/8	2 7/16
1 1/4	4 5/8	2 1/4	5 1/4	2 9/16
1 1/2	5	2 7/16	6 1/8	2 11/16
2	6	2 1/2	6 1/2	2 3/4
2 1/2	7	2 3/4	7 1/2	3
3	7 1/2	2 3/4	8 1/4	3 1/8
3 1/2	8 1/2	2 13/16	9	3 3/16
4	9	3	10	3 3/8
5	10	3 1/2	11	3 7/8
6	11	3 1/2	12 1/2	3 7/8
8	13 1/2	4	15	4 3/8
10	16	4	17 1/2	4 5/8
12	19	4 1/2	20 1/2	5 1/8
14	21	5	23	5 5/8
16	23 1/2	5	25 1/2	5 3/4
18	25	5 1/2	28	6 1/4

Standard Cast Iron Companion Flanges and Bolts

(For working pressures up to 125 psi steam, 175 psi WOG)

Size in Inches	Diam. of Flange, in Inches	Bolt Circle, in Inches	No. of Bolts	Size of Bolts, in Inches	Length of Bolts, in Inches
¾	3½	2½	4	⅜	1⅜
1	4¼	3⅛	4	½	1½
1¼	4⅝	3½	4	½	1½
1½	5	3⅞	4	½	1¾
2	6	4¾	4	⅝	2
2½	7	5½	4	⅝	2½
3	7½	6	4	⅝	2½
3½	8½	7	8	⅝	2½
4	9	7½	8	⅝	2¾
5	10	8½	8	¾	3
6	11	9½	8	¾	3
8	13½	11¾	8	¾	3¼
10	16	14¼	12	⅞	3½
12	19	17	12	⅞	3¾
14	21	18¾	12	1	4¼
16	23½	21¼	16	1	4¼

Extra Heavy Cast Iron Companion Flanges and Bolts

(For working pressures up to 250 psi steam, 400 psi WOG)

Pipe Size Inches	Diam. of Flanges	Diam. of Bolt Circle	No. of Bolts	Diam. of Bolts	Length of Bolts
1	4⅞	3½	4	⅝	2¼
1¼	5¼	3⅞	4	⅝	2½
1½	6⅛	4½	4	¾	2½
2	6½	5	8	⅝	2½
2½	7½	5⅞	8	¾	3
3	8¼	6⅝	8	¾	3¼
3½	9	7¼	8	¾	3¼
4	10	7⅞	8	¾	3½
5	11	9¼	8	¾	3¾
6	12½	10⅝	12	¾	3¾
8	15	13	12	⅞	4¼
10	17½	15¼	16	1	5
11	20½	17¾	16	1⅛	5½
14 O.D.	23	20¼	20	1⅛	5¾
16 O.D.	25½	22½	20	1¼	6

Feet Head of Water to PSI

Feet Head	Pounds Per Square Inch	Feet Head	Pounds Per Square Inch
1	.43	100	43.31
2	.87	110	47.64
3	1.30	120	51.97
4	1.73	130	56.30
5	2.17	140	60.63
6	2.60	150	64.96
7	3.03	160	69.29
8	3.46	170	76.63
9	3.90	180	77.96
10	4.33	200	86.62
15	6.50	250	108.27
20	8.66	300	129.93
25	10.83	350	151.58
30	12.99	400	173.24
40	17.32	500	216.55
50	21.65	600	259.85
60	25.99	700	303.16
70	30.32	800	346.47
80	34.65	900	389.78
90	38.98	1000	433.00

Note: One foot of water at 62° Fahrenheit equals .433 pound pressure per square inch. To find the pressure per square inch for any feet head not given in the table above, multiply the feet head by .433.

Water Pressure to Feet Head

Pounds Per Square Inch	Feet Head	Pounds Per Square Inch	Feet Head
1	2.31	100	230.90
2	4.62	110	253.98
3	6.93	120	277.07
4	9.24	130	300.16
5	11.54	140	323.25
6	13.85	150	346.34
7	16.16	160	369.43
8	18.47	170	392.52
9	20.78	180	415.61
10	23.09	200	461.78
15	34.63	250	577.24
20	46.18	300	692.69
25	57.72	350	808.13
30	69.27	400	922.58
40	92.36	500	1154.48
50	115.45	600	1385.39
60	138.54	700	1616.30
70	161.63	800	1847.20
80	184.72	900	2078.10
90	207.81	1000	2309.00

Note: One pound of pressure per square inch of water equals 2.309 feet of water at 62° Fahrenheit. Therefore, to find the feet head of water for any pressure not given in the table above, multiply the pressure pounds per square inch by 2.309.

Boiling Points of Water at Various Pressures

Vacuum, in Inches of Mercury	Boiling Point	Vacuum, in Inches of Mercury	Boiling Point
29	76.62	7	198.87
28	99.93	6	200.96
27	114.22	5	202.25
26	124.77	4	204.85
25	133.22	3	206.70
24	140.31	2	208.50
23	146.45	1	210.25
22	151.87	Gauge Lbs.	
21	156.75	0	212.
20	161.19	1	215.6
19	165.24	2	218.5
18	169.00	4	224.4
17	172.51	6	229.8
16	175.80	8	234.8
15	178.91	10	239.4
14	181.82	15	249.8
13	184.61	25	266.8
12	187.21	50	297.1
11	189.75	75	320.1
10	192.19	100	337.9
9	194.50	125	352.9
8	196.73	200	387.9

Total Thermal Expansion of Piping Material in Inches per 100 ft. above 32°F.

Temperature °F	Carbon and Carbon Moly Steel	Cast Iron	Copper	Brass and Bronze	Wrought Iron
32	0	0	0	0	0
100	0.5	0.5	0.8	0.8	0.5
150	0.8	0.8	1.4	1.4	0.9
200	1.2	1.2	2.0	2.0	1.3
250	1.7	1.5	2.7	2.6	1.7
300	2.0	1.9	3.3	3.2	2.2
350	2.5	2.3	4.0	3.9	2.6
400	2.9	2.7	4.7	4.6	3.1
450	3.4	3.1	5.3	5.2	3.6
500	3.8	3.5	6.0	5.9	4.1
550	4.3	3.9	6.7	6.5	4.6
600	4.8	4.4	7.4	7.2	5.2
650	5.3	4.8	8.2	7.3	5.6
700	5.9	5.3	9.0	8.5	6.1
750	6.4	5.8	...	...	6.7
800	7.0	6.3	...	...	7.2
850	7.4	...	...	...	...
900	8.0	...	...	...	...
950	8.5	...	...	...	...
1000	9.0	...	...	...	...

Specific Gravity of Gases

(At 60°F and 29.92" Hg)

Gas	Formula	Specific Gravity
Dry air (1 cu. ft. at 60°F. and 29.92" Hg. weighs .07638 pound)		1.000
Acetylene	C_2H_2	0.91
Ethane	C_2H_6	1.05
Methane	CH_4	0.554
Ammonia	NH_2	0.596
Carbon dioxide	CO_2	1.53
Carbon monoxide	CO	0.967
Butane	C_4H_{10}	2.067
Butene	C_4H_8	1.93
Chlorine	Cl_2	2.486
Helium	He	0.138
Hydrogen	H_2	0.0696
Nitrogen	N_2	0.9718
Oxygen	O_2	1.1053

Specific Gravity of Liquids

Liquid	Temp. °F	Specific Gravity
Water (1 cu. ft. weighs 62.41 lb.)	50	1.00
Brine (Sodium Chloride 25%)	32	1.20
Pennsylvania Crude Oil	80	0.85
Fuel Oil No. 1 and 2	85	0.95
Gasoline	80	0.74
Kerosene	85	0.82
Lubricating Oil SAE 10-20-30	115	0.94

Tempering and Heat Colors

	Color	Degrees	
		Fahrenheit	Centigrade
Temper Colors	Faint straw	400	205
	Staw	440	225
	Deep straw	475	245
	Bronze	520	270
	Peacock blue	540	280
	Full blue	590	310
	Light blue	640	340
Heat Colors	Faint red	930	500
	Blood red	1075	580
	Dark cherry	1175	635
	Medium cherry	1275	690
	Cherry	1375	745
	Bright cherry	1450	790
	Salmon	1550	840
	Dark orange	1680	890
	Orange	1725	940
	Lemon	1830	1000
	Light yellow	1975	1080
	White	2200	1200

TYPICAL BTU VALUES OF FUELS

ASTM Rank Solids	BTU Values Per Pound
Anthracite Class I	11,230
Bituminous Class II Group 1	14,100
Bituminous Class II Group 3	13,080
Sub-Bituminous Class III Group 1	10,810
Sub-Bituminous Class III Group 2	9,670

Liquids	BTU Values Per Gal.
Fuel Oil No. 1	138,870
Fuel Oil No. 2	143,390
Fuel Oil No. 4	144,130
Fuel Oil No. 5	142,720
Fuel Oil No. 6	137,275

Gases	BTU Values Per Cu. Ft.
Natural Gas	935 to 1132
Producers Gas	163
Illuminating Gas	534
Mixed (Coke oven and water gas)	545

INDEX

A

B

C

D

E

N

O

P

T

U

V

W